Praise for Larrie D. Ferreiro's *Measure of the Earth*

"Delightful and perceptive. . . . Scientific history has rarely been more entertaining, or seemed more relevant."—*Physics World*, Best Books of 2011

"Ferreiro . . . marvelously details an almost doomed 18th-century geodesic expedition to South America to determine Earth's shape. Ferreiro's skill as storyteller and scholar is displayed in full vigor. Easy to read and fast moving, the book is often dramatic. . . . Rarely does a history of science volume discuss such events, and rarely does its author present them so well. Ferreiro also masterfully blends political and scientific history, going to lengths to place the expedition's people and events in context."
—*Library Journal*, starred review

"Bringing the first half of the 18th century to life is a tall order, but historian/naval architect Ferreiro has succeeded. . . . Highly recommended."
—*Choice*

"It's impossible not to be impressed with the operation's technical achievement while reading the author's fascinating and clearly written account. . . . Impressive . . . Ferreiro recounts with straightforward enthusiasm."—*Publishers Weekly*

"A sophisticated work tracing the arduous mid-18th-century international expedition to the Latin American equator to determine the 'figure of the earth.' . . . A fascinating, absorbing journey."—*Kirkus Reviews*

"In *Measure of the Earth*, author Larrie Ferreiro transports readers to an intriguing world of colonial politics and scientific competition, blundering incompetence, dedication, and hardship."—*Prism*

"The fractious, disaster-ridden, yet ultimately successful mission unfolds in scrupulous detail. Ferreiro enlivens his account with the fruits of years spent translating letters and memoirs produced by the mission in French and Spanish."—*Literary Review*

"Ferreiro dutifully records the scientific advance . . . but rightly stresses the adventure."—*Maclean's*

“Ferreiro gives life to three French scientists and their 1736 geodesic mission to the equator.”—*Science News*

“Who would have thought that a book about an obscure eighteenth-century surveying expedition could be a good read? The author is a proficient storyteller and an assiduous researcher, and this book was clearly a labor of love. Ferreiro, I think, will persuade readers of the lasting stature of some very resourceful, stalwart men.”—***Earth Island Journal***

“A great story of an important scientific achievement and the huge amount of effort it took for the simple purpose of satisfying scientific curiosity.”—***Commercial Dispatch***

“Ferreiro succeeds in blending exciting narrative along with erudite science, including one of the most concise and clear descriptions of triangulation surveying I have encountered, leaving the reader both educated and entertained.”—***eHistory***

“Science and history holdings alike should consider this a ‘must.’”
—***Bookwatch***

“Larrie D. Ferreiro tells us that Voltaire could make difficult subjects accessible to everyone. In *Measure of the Earth* Ferreiro shows that he can do the same, with his Voltairean gifts of mastery of material and fluency in prose.”—**Felipe Fernández-Armesto**

“The greatest achievement of Larrie D. Ferreiro’s wonderful book is to walk us with perfect ease through remote locales and arcane subjects. Mr. Ferreiro seems no less at home in Guayaquil than in Paris or London and no less lucid in explaining the debates over the shape of the earth between Newtonians and Cartesians than in describing the intrigues surrounding the French Academy or the excruciating logistics of a scientific mission unfolding in colonial South America.”
—**Andrés Reséndez, author of *A Land So Strange***

“Doing science in the eighteenth century demanded almost unbearable sacrifices for distant rewards and only the most dedicated could handle the challenges. Larrie Ferreiro’s deep research has produced a highly readable account of one of the great scientific expeditions of the age of

the Enlightenment, a venture all the more riveting since it unfolded amidst imperial contests and devastating tragedy and tested the psychological and physical limits of those keen to expand knowledge of the shape of the Earth."
—Peter C. Mancall, author of
Fatal Journey: The Final Expedition of Henry Hudson

"The story of the race to determine the shape of the Earth is one of history's most engaging yet least-known stories. In *Measure of the Earth*, Larrie Ferreiro takes us inside the scientific expedition that set off from France to South America in the 18th century to discover the answer. Ferreiro not only brings to life the band of characters that embarked on this journey, with all of their intrigues and rivalries, but he also details the huge stakes involved. Whichever country discovered the Earth's correct shape would take a giant leap forwards in enhancing their military and economic power. A fascinating account of scientific inquiry thoroughly enmeshed in the race for power and empire."
—Kim MacQuarrie, author of *The Last Days of the Incas*

"Ferreiro's *Measure of the Earth* nicely captures the scientific complexity and physical difficulty of this extraordinary expedition. At the same time, the author provides richly textured portraits of all the principal protagonists, whose personal foibles and rivalries sometimes undercut their professional skills. This is a compelling tale of international politics, Enlightened science, and human drama, played out on both sides of the Atlantic."—Carla Rahn Phillips, University of Minnesota, Twin Cities

"In *Measure of the Earth*, Larrie Ferreiro tells the dramatic story of the first international scientific expedition to South America to establish the precise dimensions of the globe. The French scientists who led the expedition to the Andes overcame incredible adversities traversing the jungles and highlands of equatorial Peru, surviving near mutiny, attacks by local inhabitants, war, siege, and the skepticism of fellow academicians in their homeland to complete their mission and achieve lasting fame. Beautifully written, Ferreiro's book provides an authoritative and gripping account of one of the greatest scientific breakthroughs of the Enlightenment."
—James Horn, author of *A Kingdom Strange*
and *A Land as God Made It*

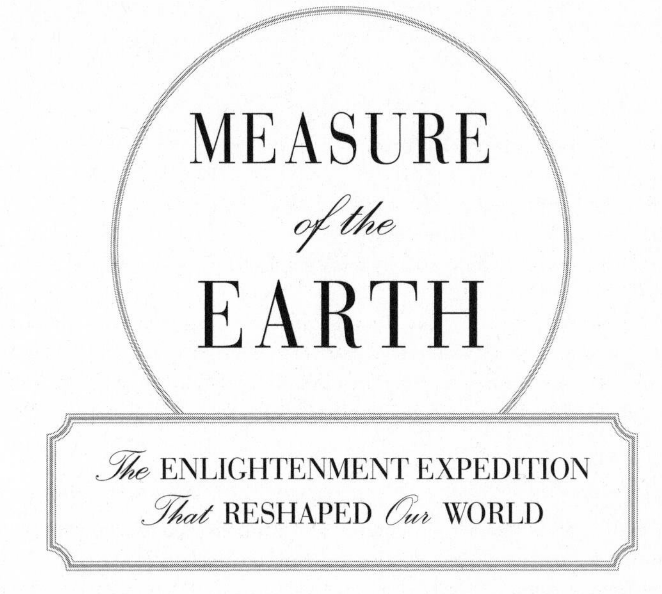

LARRIE D. FERREIRO

BASIC BOOKS
A Member of the Perseus Books Group
New York

Hardcover first published in 2011 by Basic Books,
A Member of the Perseus Books Group
Paperback first published in 2013 by Basic Books

Typeset in ITC Caslon 224 Book

The Library of Congress has cataloged the hardcover as follows:

Ferreiro, Larrie D.
Measure of the Earth : the enlightenment expedition that reshaped our world / Larrie D. Ferreiro.
p. cm.
Includes bibliographical references and index.
ISBN 978-0-465-01723-2 (hardback)—ISBN 978-0-465-02345-5 (e-book)
1. Geodesy—Europe—History—18th century. 2. Scientific expeditions—Europe—History—18th century. I. Title.
QB296.E9F47 2011
526'.609033—dc22

2011007173

ISBN 978-0-465-06381-9 (paperback)

10 9 8 7 6 5 4 3 2 1

To my wife, Mirna,
and our sons, Marcel and Gabriel:
This book is about the past,
but they are the present and the future.

CONTENTS

PRINCIPAL CHARACTERS

Members of the Geodesic Mission

- Louis Godin (1704–1760): astronomer and original leader of the mission
- Pierre Bouguer (1698–1758): astronomer, mathematician, and hydrographer; de facto leader after Godin's fall from authority
- Charles-Marie de La Condamine (1701–1774): scientist and adventurer
- Jorge Juan y Santacilia (1713–1773): Spanish naval officer and astronomer
- Antonio de Ulloa y de la Torre-Guiral (1716–1795): Spanish naval officer and astronomer
- Joseph de Jussieu (1704–1779): doctor and botanist
- Jean Seniergues (1704–1739): surgeon
- Jean-Joseph Verguin (1701–1777): engineer and cartographer
- Jean-Louis de Morainville (1707–circa 1765): draftsman and artist
- Théodore Hugo (died circa 1781): instrument maker
- Jean-Baptiste Godin des Odonais (1713–1792): assistant
- Jacques Couplet-Viguier (circa 1718–1736): assistant

Political Figures

- Jean-Frédéric Philippe Phélypeaux, Comte de Maurepas (1701–1781): French minister of the navy and sponsor of the Geodesic Mission
- José Antonio de Mendoza Caamaño y Sotomayor, Marqués de Villagarcía de Arousa (1667–1746): Spanish viceroy of Peru during the mission
- Dionisio de Alsedo y Herrera (1690–1777): president of the *Audiencia* of Quito at the arrival of the mission

- José de Araujo y Río (died 1754): succeeded Alsedo as president of Quito during the mission

Others

- Voltaire, or François-Marie Arouet (1694–1778): author, friend of La Condamine
- Pierre-Louis Moreau de Maupertuis (1698–1759): astronomer, adversary of Bouguer
- Pedro Vicente Maldonado (1704–1748): politician and geographer, accompanied La Condamine down the Amazon
- Isabel Godin des Odonais (1728–1792): wife of Jean Godin des Odonais, made harrowing journey down the Amazon

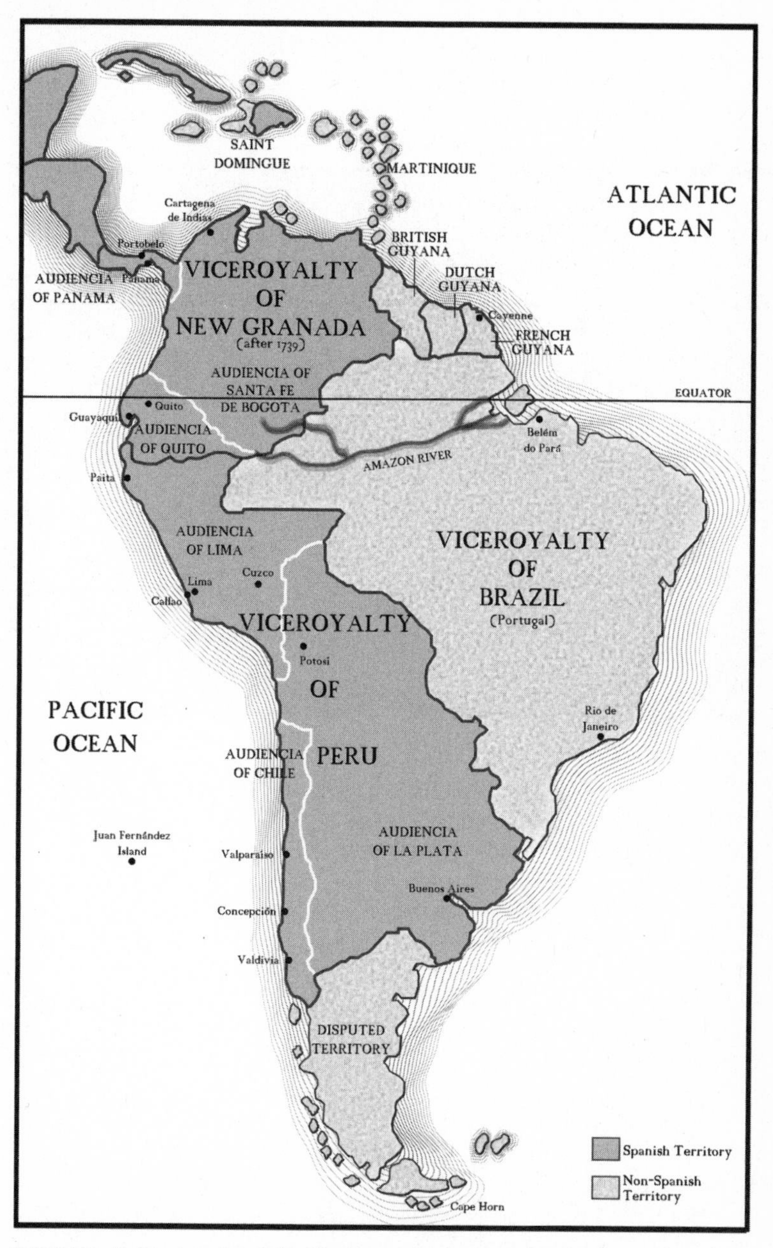

Colonial South America, circa 1740. Illustration: Eliecer Vilchez Ortega.

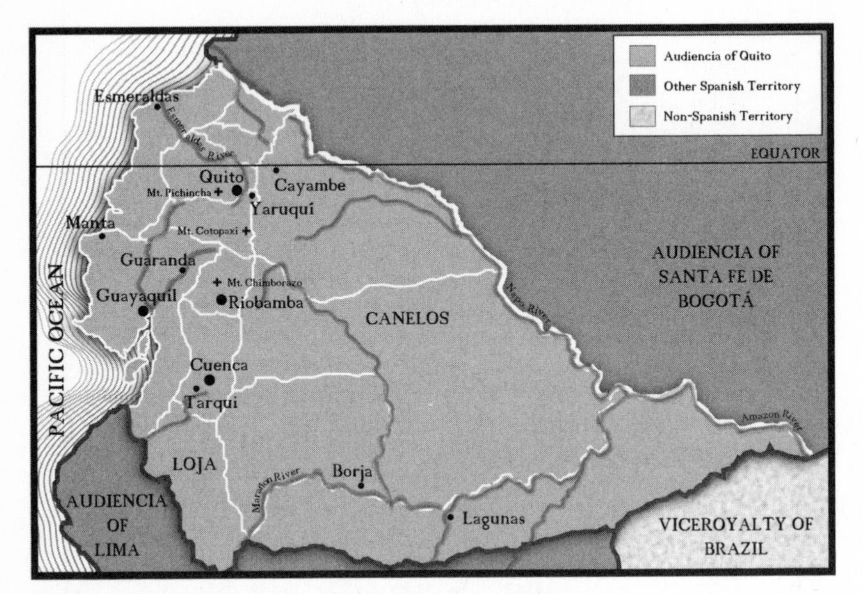

Audiencia of Quito, circa 1740. Illustration: Eliecer Vilchez Ortega.

Introduction

THE BASELINE AT YARUQUÍ

Each day at first light, long before the sun peered over the eastern ridge of the Andes, the two scientists were at work in the open fields, adjusting the wedges and planks under their measuring poles to keep them level. The three wooden poles, each twenty feet long and painted a different color, with paddle-like copper plates capping both ends, were arrayed end to end along the ground. They followed the survey baseline that had been scraped out of the landscape several weeks before: a dusty brown ribbon of bare earth, arrow-straight and barely as wide as a forearm, running from one horizon to the other. A thin cotton cord, pulled taut between two stakes and checked for level by three assistants, guided the men as they picked up each pole and moved it forward. They carefully placed the forward pole so that it barely touched the one behind it, ensuring that its exact position would not be disturbed by the movement. More wedges were placed to level the poles, and then each man meticulously penned the measurement in his little notebook. They continually made corrections for the length of the poles as they expanded in the rising equatorial heat, comparing them with a precisely calibrated six-foot iron rod called a *toise*, which they kept cool in the shade to prevent it, too, from expanding. During the month of October 1736, they repeated these steps thousands of times, working with some haste to survey the baseline from north to south before the rains began.[1]

Pierre Bouguer, the senior of the two scientists and, at age thirty-eight, the oldest member of the group, labored under the unaccustomed physical effort. At almost two miles of altitude, his energy was quickly sapped by the thin atmosphere, which offered little oxygen and even less protection against the sun. It would hardly have seemed fair; only a few miles to the south was a seemingly perpetual cloud cover, which obscured the mountains and cooled the hilly green lands underneath.

The scientists had chosen the plateau at Yaruquí, about twelve miles east of Quito, the provincial capital of northern Peru, to lay out the baseline for their measurements because it was relatively flat and had clear views to the peaks around them. But the plateau was lower than the surrounding hills, and they soon found that it had its own microclimate, further hampering their task; though cool at night, it became very hot during the day and was exposed to high winds that sometimes generated towering whirlwinds of sand and dust, one of which had recently killed a local Indian.

This dichotomy between expectations and reality was already becoming a hallmark of their mission. Plans that seemed to be ideal on first inspection were plagued with overwhelming problems when actually executed. These problems went far beyond the normal, expected setbacks of any scientific expedition. It was almost as if the Earth itself was refusing to reveal its true measure.

Yet it was the measure of the Earth that they had journeyed thousands of miles to seek. Bouguer was one of three members of the French Academy of Sciences who, with the agreement and protection of the king of Spain, had been sent to the Spanish Viceroyalty of Peru in order to carry out precise measurements of the Earth. Traveling with several aides and two doctors, and accompanied by two young Spanish navy officers, the scientists had arrived in Quito in May 1736 after two years of planning and a year's voyage from Europe. Their orders were to determine the length of a degree of latitude at the equator, a measurement that could be compared with a degree of latitude that had already been established in France. By this comparison, the shape and measure of the Earth would be known with certainty for the first time.

The true shape and exact measure of the globe—generally referred to as the "figure of the Earth"—had been the subject of an ongoing debate for several years now. It had recently become obvious that the Earth was not a perfect sphere. The century-old system of the French philosopher René Descartes, some scientists now argued, implied that the Earth was elongated at the poles like an egg. They were supported in this by survey results that appeared to show a distinct lengthening of the planet along its axis. In contrast, the more recent theories developed by the British mathematician Isaac Newton suggested that the Earth's spin caused it to bulge at the equator and flatten at the poles, like a *boule* of bread that some giant hand had pushed down. This theory explained why careful experiments showed that the tug of gravity appeared to diminish near the equator.

The French Academy of Sciences stood at the center of the debate over the Earth's shape. Scientists such as Johann Bernoulli on the European continent supported Descartes' theory and an elongated Earth. Across the English Channel, off Fleet Street in London, the members of the Royal Society vigorously defended Newton's flattened Earth. The French Academy, split roughly equally between Cartesians and Newtonians, was a haven for discourse between the two polarized communities. Its members carried on lively discussions inside the Academy (housed within the palace of the Louvre) and outside at the cafés and salons of Paris. If the question were to be resolved, it seemed certain that it would happen here.

The debate over the Earth's shape would have remained an arcane scientific discourse but for the growing interest of Jean-Frédéric Philippe Phélypeaux, Comte de Maurepas, a young but powerful minister in the court of Louis XV. Maurepas had been president and vice president of the Academy of Sciences and remained its principal supporter, but his interest in the debate was pragmatic. As minister of the navy and minister of colonies he knew that without an accurate understanding of the size and shape of the earth, navigation on the high seas would continue to be a hit-or-miss affair. Maurepas fully grasped the political and military consequences of this geodesic knowledge: The nation that could accurately locate its ships at sea could control an empire.

For France and Britain, imperial expansion was as much about security as financial gain. Although the two nations were now at a time of relative peace, this surely would not last; the century-old conflict over empire was merely ticking over. Even as the Academy's scientists debated the finer points of geodesy, Maurepas was preparing for future battles with the British. Scientists in Britain, too, were attempting to resolve questions of navigation as a necessary component of global domination; for the past twenty years Britain had sponsored the famous £20,000 Longitude Prize, as yet unclaimed. The war for knowledge would be fought in the halls of the Louvre and in the meeting rooms off Fleet Street, as well as across the oceans.

Mindful of his opponents on the other side of the Channel, Maurepas had enthusiastically embraced proposals put before the Academy in 1734 to send a scientific mission to measure a degree of latitude at the equator. But the equatorial African coast was still hostile, and the tropical Asian islands were too far. The only accessible place on the equator was Peru, the principal source of wealth for the Spanish empire, and Spain was closely linked with France through the Bourbon family alliance. Besides promising to reveal the Earth's true shape, this mission would serve two of Maurepas' other strategic goals: as minister of the navy, he was keen to strengthen the military alliance with Spain as a counterweight to Britain, and as minister of colonies, he was equally anxious for a close look at the famous riches of South America, perhaps even to open trade between the Spanish colonies there and France.

The Geodesic Mission to the Equator, as it became known, was unprecedented: It was the world's first international scientific expedition, involving official cooperation between two nations as well as participation by members of both countries. Maurepas had thrown himself into the planning of the mission, arranging for transport and provisions from the French navy and money from the treasury, obtaining passports from the king of Spain, and hand-selecting the French members of the mission—including Bouguer, now laboring under the equatorial sun and perhaps wondering why Maurepas had chosen him and why he had accepted.

The scientist working alongside Bouguer stood in complete contrast to him. Charles-Marie de La Condamine was a relative novice in the scientific world, one of the newest members of the Academy of Sciences. He had made a name for himself not as a scholar, but as an adventurer. A former soldier who had fought against Spain, and more recently a corsair and explorer in the Mediterranean, La Condamine was the polar opposite of the studious Bouguer, a child prodigy who became a full professor of navigation at age sixteen and who had spent his entire life toiling at science and mathematics in a solitary corner of France.

Starting from the southern end of the baseline, a second party of scientists was measuring the same line as Bouguer and La Condamine but in the opposite direction. It was headed by Louis Godin, the academician who had originally proposed this mission and was nominally leader of the overall expedition. Having the longest tenure with the Academy of Sciences, Godin was assumed to have seniority, even though it was painfully obvious he had neither experience nor skill in leading men. By contrast, La Condamine was a military man, and Bouguer had long been in charge of students often twice his age; both understood that leadership was more than simply issuing orders and were chafing under Godin's inept authority.

Despite their differences, the three men would need to work together if they stood any chance of completing their task. It was simple, in theory, to measure the length of a degree of latitude, which denotes an angular position on the sphere of the Earth and is described in degrees north or south of the equator (Paris, for example, is at about 49° north latitude; the North Pole is 90° north). It was, however, enormously complex in practice. It was accomplished using long-distance surveying by triangulation, the principles of which had been worked out a century earlier, and which had been used to create the first accurate maps of France. The premise was as old as Euclid: Given the length of only one side of a triangle (the baseline), and the measure of two angles, the entire triangle can be constructed using trigonometric formulas.

This meant that a survey over a long distance could be carried out by constructing a "chain" of triangles dozens or even hundreds of miles in length. By plotting these triangles across the terrain, and observing with an accurate instrument called a quadrant the angles between large, visible markers at each of the vertices, surveyors could compute the overall linear distance (in miles) of the chain of triangles. They would then take star sightings to determine the latitude at each end of the chain. The difference between the two latitudes gave the angular distance (in degrees) of the chain. Dividing the linear distance by the number of degrees would then give them the length of a single degree of latitude.

The plan conceived at the French Academy of Sciences was to use the summits of the Andes, the double chain of mountainous volcanoes in Peru, as the principal vertices of the triangles, with at least one baseline to be laid out on a flat plain. Once they had arrived in the country, the expedition members had selected the Yaruquí plateau as the best site for the seven-mile-long baseline, and they were now laboriously measuring it in order to begin plotting the enormous chain of triangles that would ultimately stretch over two hundred miles south along the Andes. A precise measure of that baseline was critical, for with this triangulation, the scientists were attempting a level of accuracy that had never been achieved.

Godin, Bouguer, and La Condamine had prepared for their journey as scientists, not as explorers—and this would nearly be their undoing. They could anticipate the problems with their instruments, the exacting computations, the physical exertions of the survey; these were knowable, calculable, and solvable by reason and the application of the scientific method. They were completely unprepared for the random, often perverse catastrophes that dogged their every step.

The expedition members arrived in Peru without fully understanding that, even after two centuries of Spanish rule, it was a foreign and potentially hostile territory. Outsiders were greeted with as much trepidation as fascination; the local population, long accustomed to staving off smugglers and pirates, brushed aside French protests that they were on a purely scientific mission and—believing the foreigners were really on a search for treasure—made even the simplest act a bureaucratic nightmare. This hostility could quickly shift from mere obstruction to outright brutality: The scientists came armed against attacks by beasts and bandits but would find themselves using those arms against a mob intent on killing them. The land itself continually threw unforeseen obstacles in their path; imagining steaming jungles, they would find themselves freezing on mountaintops.

Many of the roadblocks the voyagers encountered they set themselves, through weak leadership, greed, vengeance, and a barely disguised contempt for the locals and their customs. The Europeans saw no need to adapt to the land or its culture; they thought they would be gone at most three years—six months out, two years for the survey, and six months on the voyage home. They could not foresee that the Geodesic Mission would keep them in Peru almost a decade, that some of them would not return for almost forty years, and that some would return not at all.

By October 1736, as the scientists were measuring the baseline, they had already been gone eighteen months; they should have been halfway down the avenue of volcanoes stretched out before them. Looking south as the clouds dispersed, they would have beheld those ancient volcanoes with ancient names: Pichincha, Pambamarca, Guamaní, and, majestically, Cotopaxi, the perfect snow-capped cone rising thirty miles distant. But they and their assistants still would need to climb each of those mountains, and return from each peak with perfect measurements, in order to complete the great chain of triangles that would reveal the true figure of the Earth. Cotopaxi would have seemed impossibly far away, yet they would have known that they had eight times farther to go before they were through; and even that distance paled in comparison with the long journey home. But they also knew they would first have to complete the measurement of the baseline at Yaruquí, all seven miles of it, twenty feet at a time. For now, they remained in their little world of measuring poles, moving the next one into its place.

I

The Problem of the Earth's Shape

The Geodesic Mission to the Equator was the culmination of two thousand years of effort to determine the precise measure of the Earth. Since the earliest days of Greece and Rome, the twin sciences of astronomy (the measure of the sun, stars, and sky) and geodesy (the measure of the Earth) directly served the geography of empire. As conquests stretched their power into far-flung realms, rulers relied on precise physical knowledge of their territories in order to exploit their holdings and on accurate long-range navigation to dispatch military forces where needed, ensuring a steady flow of commerce.

The earliest measurements of the Earth were used for imperial cartography and navigation. By 240 BCE, the Greek mathematician Eratosthenes had already estimated the size of the Earth and used his findings to construct a detailed atlas of the Greek Empire built by Alexander the Great and his successors. Early Greek philosophers had known that the Earth was spherical (for one thing, it always cast a circular shadow on the moon during a lunar eclipse), but their estimates of its size were pure guesses. Eratosthenes' methods for calculating the Earth's size were rudimentary, but his logic was keen. He knew that on the June solstice the sun shone vertically down a well in modern-day Aswan, Egypt. On the same day, he noted that, in his home of Alexandria, a vertical stick cast a shadow measuring about 7° (one-fiftieth of a circle). Knowing that the two cities were about 500 miles apart, he calculated the Earth's circumference to be about 25,000 miles—remarkably close to the actual figure. In the second century CE, the Roman geographer Claudius Ptolemy expanded upon Eratosthenes' and his successors' calculations to create an updated atlas, called *Geography*, taking the added step of incorporating lines of latitude and longitude to specifically locate Roman navigation routes and trading entrepôts as far east as China.[1]

When the Roman Empire ended in 476 CE, so too did European advancement in geography and navigation. The European interest in long-distance navigation was only reawakened in the 1400s with the slow choking of the overland Silk Routes that had for centuries connected Europe to the riches of Asia. With travel becoming perilous as a result of the disintegration of the Mongolian Empire and the order it had imposed along the trade routes, merchants needed a way to reach Asia by sea. Explorers flying the Portuguese flag laboriously clawed their way south and east around Africa to reach the Indian Ocean. Christopher Columbus, sailing under the Spanish flag (and under a delusion about the true size of the Earth—it was much bigger than he guessed), took the novel approach of sailing west to reach Asia. He landed in the Americas instead.

The long-distance European ocean voyages of the fifteenth century were aided and abetted by an intellectual renaissance in science. Ptolemy's *Geography*, recopied through the ages, became the touchstone for early modern mapmakers, who slowly filled in their charts with reports from explorers in order to create a more accurate portrait of the known world. Even while maps grew more sophisticated, astronomical techniques for navigation improved. Columbus, for example, famously brought with him a newfangled marine quadrant for taking sun or star sightings. The quadrant, a precursor to the modern sextant, allowed a navigator to accurately establish latitude by measuring the angle between Polaris and the horizon; since Polaris is almost directly above the north pole, that angle is effectively the same as one's latitude north of the equator. But it was difficult to make precise measurements while standing on a rolling, pitching ship. Columbus himself relied on more traditional navigational methods, reading off a table that listed latitudes by the hours of daylight throughout the year. Nevertheless, having the quadrant aboard gave confidence to both his crew and his royal sponsors that he would arrive at his destination and return safely.[2]

Exploration quickly gave way to empire. After Columbus returned from scouting the area in 1493, Spain immediately began settling the Caribbean basin. In 1494, the Treaty of Tordesillas separated the globe into Spanish and Portuguese halves, effectively "giving" Spain most of the Americas, where it continued to colonize huge swathes of territory. In 1503 Spain established the powerful House of Trade (Casa de Contratación) to regulate all sailing and commerce with its colonies. Its new territories at first yielded only modest production of cotton and tobacco, but they would soon bring unheard-of wealth to the kingdom. The discovery of enormous deposits of silver in Mexico and Peru in the mid-1500s quickly turned Spain's colonial empire into a vast moneymaking enterprise, at its height furnishing a quarter of Spain's total revenue.

Spain's success in the New World soon attracted the envy of other European powers, many of which were just establishing their own maritime colonial empires. The early American colonies of England, France, and the Netherlands yielded no gold or silver, but by the 1600s, all three nations found profit in the Caribbean, establishing sugar-cane plantations that replenished national coffers on the backs of millions of enslaved Africans. These European competitors also preyed on the Spanish treasure galleon fleets that regularly sailed from Mexico and Panama, bringing bullion to Seville and Cadiz in protected convoys.[3]

Besides exposing the riches of the American continents, the overseas expansion of European powers transformed naval warfare. Large, ocean-going warships replaced coastwise craft, and permanent navies were established to protect sea lanes and colonial territories as well as to escort cargo ships carrying goods and bullion to and from the colonies. The control of distant trade routes became of primary importance for nations whose incomes were increasingly dependent on colonial imports and exports, and battles over these routes were often won or lost at sea. France, recognizing that the Atlantic Ocean had become the principal field of battle, established the naval fortress of Louisbourg in Nova Scotia to protect fisheries and provide a staging area to attack English interests, while England secured Jamaica as a similar base of operations in the Caribbean.

Naval conflict between France and England became the dominant theme during this period, but the situation was as volatile as the Atlantic, and allies and adversaries could switch from one year to the next. In 1672, for example, Louis XIV's French navy joined the English fleet in the third Anglo-Dutch war—fought, in part, over trade access. In 1689, however, France went to war against an alliance of English and Dutch forces, during which time France tried and failed to mount an invasion of Ireland. From 1701 to 1714, France again fought England and the Netherlands in the War of the Spanish Succession, which ended in France's favor with Felipe V, the grandson of Louis XIV, retaining the Spanish throne. By 1716, France and England were exhausted from fighting each other and therefore created the Anglo-French Alliance as a hedge against the newly upstart Spain, which was making claims for territory it had lost during the War of the Spanish Succession. This alliance between the two superpowers would bring relative peace for almost two decades.

In both wartime and peacetime, the French navy was usually outnumbered. France was primarily a land power, with adversaries on all sides that required it to maintain a large and costly army at the expense of its naval forces. Meanwhile, England (later Britain, after the 1707 merger with Scotland), surrounded by water, poured most of its budget into its "wooden walls" of warships for protection. To counter this numerical advantage, the French navy turned to science as a military force multiplier, a way to augment the fighting power of each ship. Once again, the twin sciences of astronomy and geodesy were called on to improve ocean navigation, ensuring that each ship got to its destination faster and with less chance of loss. Science had become, like war, politics carried out by other means.

Science was the touchstone of the Age of Enlightenment, begun in the seventeenth century as a general movement toward reason, not faith, as the gateway to knowledge. One French leader in particular viewed the quest for scientific knowledge as part of the same intractable competition with the English for commerce, territory, and influence around the globe. Jean-Baptiste Colbert was Louis XIV's hyperactive minister of finance as well as minister of the navy, and he eyed with envy the scientific developments across the Channel—especially the Royal Society of London for Improving Natural Knowledge, which had been chartered in 1660. The Society was "Royal" in name only, since it had no direct government support; its generally wealthy members, some of whom could be described as enthusiastic amateurs, paid dues for the rent of the headquarters and for publishing books and a scientific journal titled *Philosophical Transactions*. Colbert quickly decided to go the British one better, and in 1666 he founded—with generous government funding—the Royal Academy of Sciences of Paris, more commonly known as the French Academy of Sciences.

In establishing the Academy, Colbert erected science as an institutional arm of the French state, but his vision went beyond the basic needs of the military and industry. Colbert sought to tie science to the state "for sound political reasons. . . . He knew that the sciences and arts make a reign glorious; that they spread the language of a nation perhaps even more than do conquests; that they give the reign a control over knowledge and industry which is just as prestigious and useful; that they attract to the country a multitude of foreigners who enrich it by their talents, adopt its character and become attached to its interests."[4] For Colbert, the French Academy would be a political force multiplier, a way to ensure that scientific investments would benefit the government on the national as well as international fronts.

Colbert spared no expense filling the ranks of his new Academy. Specifically rejecting the English model (as he saw it) of a dispersed body of enthusiastic amateurs, he wanted—and was willing to pay for—top-flight scientists who could make specific discoveries in astronomy, mathematics, and physics that would improve navigation, the military arts, and commerce. His first priority was to build a state-of-the art observatory on the outskirts of Paris—no wood to fuel fires, no metal to cause magnetic disturbances. He also spent large sums to attract talent from abroad, including the renowned Italian astronomer Giovanni Domenico Cassini and the Dutch astronomer Christiaan Huygens, further burnishing the Academy's stature.[5]

Within a few years of its founding, the Academy had begun to sponsor international scientific missions aimed at improving France's navigational techniques. In 1671, Colbert sent a young astronomer, Jean Richer, to Cayenne in the French colony of Guyana, close to the equator in South America. He was ordered to carry out a whole menu of astronomical observations, including the creation of an accurate map of the southern sky, increasingly important as French warships and merchant vessels extended their voyages across the globe and became ever more reliant on the stars for guidance. Richer was in Cayenne for two years, performing astronomical observations and experiments on atmospheric refraction. To complete these experiments, he was equipped with some of the most accurate instruments available, including two finely tuned pendulum clocks, a fairly recent invention of Christiaan Huygens.[6]

Richer's pendulum clocks were intended to assist in his astronomical observations, but they ended up revolutionizing the debate about the Earth's shape. A pendulum clock keeps accurate time because the pendulum arm is precisely fabricated to swing one beat (a single motion from left to right) during a precise interval, typically one second. The period is also directly related to the length of the pendulum; as any pianist knows, a metronome beats faster when the pendulum is effectively shortened by moving the weight down. Richer's pendulum clocks had been carefully calibrated in Paris by marking the passage of individual stars over many nights. On arriving in Cayenne, Richer set up those clocks as a necessary prelude to his research, but he found to his dismay that they were losing about two minutes and twenty-eight seconds a day, when compared with a locally made clock (and further checked against the stars). To make his "seconds clock" beat correctly, he had to speed up the oscillation by shortening the three-foot-long pendulum about a twelfth of an inch.[7]

When Richer returned to Paris in 1673, none of the French Academy scientists could explain the discrepancy between his pendulum clock's behavior in Guyana and in France. The difference was too large to have been caused by expansion of the pendulum from heat, and while the force of gravity was known to affect pendulum clocks' oscillation—lower gravity caused them to beat more slowly, a fact described by Huygens—it did not seem logical that the gravity in Cayenne should be different from that in Paris.[8]

Across the Channel, an English scientist named Isaac Newton had a sharply different reaction to news of Richer's findings. Newton, a mathematics professor at the University of Cambridge, believed that the slower oscillation of the pendulum was, in fact, caused by the force of gravity being less at the equator. He postulated, moreover, that this diminished gravity was the direct result of the centrifugal force of rotation causing the Earth to bulge out at the equator. A relatively minor scientific anomaly had generated an entirely new conception of the figure of the Earth.

When he first learned of Richer's findings, Newton had been thinking about gravitation for a decade. He had famously deduced that the fall of an apple is due to the same principle of universal attraction that holds the moon in its orbit. Gravity, he believed, is an innate property of a body, which attracts other bodies as a proportion of their mass and diminishes as the square of their distance from each other. The Earth and the apple tug at each other, causing the apple to appear to fall to Earth (in reality, they fall toward each other). By the same mechanism, the attraction of gravity holds the moon in its orbit around the Earth; without this attraction, the moon's inertia would carry it off in a straight line as a result of centrifugal force. The principle of universal attraction was Newton's Theory of Everything, and he seized on Richer's findings as further proof of his ideas.[9]

While Newton may have been satisfied with his own explanation of gravity, others were not so easily convinced. When Newton published his massive three-volume *Principia mathematica* (Principles of Mathematics) in 1687, it was greeted by, in the words of the science historian I. Bernard Cohen, "a wholly unexpectant and unprepared audience who did not, in actual fact, know what to make of it or how to use it for some time."[10] In particular, Newton's principle of universal attraction required a significant leap of faith that many of the greatest physicists of his day were unable to make. According to the prevailing notions of physics, all motion had to come from contact between bodies. Many scientists could not accept the idea of a spooky, unseen force (which even Newton could not define) that attracted far-off bodies with no visible means of communication. The Swiss mathematician Johann Bernoulli, at the time one of Newton's few rivals as a mathematical genius, found the principle of universal attraction "incomprehensible."[11]

In *Principia mathematica*, Newton unveiled his argument about the figure of the Earth. Newton placed Richer's anomalous results front and center in the first volume, presenting them as the clearest demonstration of his theory of universal attraction and arguing that this attraction had pulled the planet into a spherical shape. The sphere, however, was not perfect; Newton meticulously calculated that the Earth's rotation generated a centrifugal force that caused the equator to bulge out slightly, so that the Earth measured about 3,984 miles in diameter at the equator, but only 3,966 miles through the poles; in other words, it was flattened at the poles by 1 part in 230 (1/230). This centrifugal force, acting opposite to the Earth's attraction, would also cause the effective gravity to be measurably smaller at the equator, as Richer had demonstrated.[12]

Newton's colleagues at the Royal Society were immediately receptive to the mathematically based philosophy inherent in the work, even if they were often frustrated in understanding the formulas. British scientists particularly embraced the concept of attraction as a guide to the general study of matter. Working doggedly through the intricacies of Newton's dense mathematics, they sought to find practical applications for the laws of attraction.[13]

By contrast, much of the scientific community on the European continent was highly skeptical of Newton's claims, particularly in France. There scientists found this newfangled concept of attraction, and the corollary of a flattened Earth with variable gravity, completely at odds with the commonsense model espoused nearly half a century before by their own countryman, René Descartes. According to Descartes' monumental work *Principia philosophiae* (Principles of Philosophy, 1644), the Earth and its moon, the planets, and the stars are immersed in a vast, invisible fluid he called "ether," which God had set into circular motion at the creation and whose great vortices continue to swirl. In Descartes'

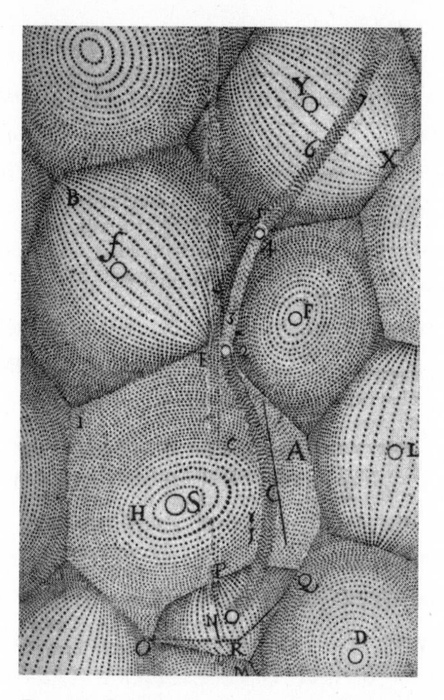

Figure 1.1 Descartes' planetary vortices (S=sun). From René Descartes, *Principia philosophiae* (1685). Credit: Special Collections & Archives, Nimitz Library, U.S. Naval Academy.

system, planets are moved by these cosmic vortices, which also cause gravity by pushing objects toward Earth (see Figure 1.1). He studiously avoided any mathematical details in his reasoning, instead using a series of analogies with water eddies and magnetism to explain his theory.

Two generations before Newton, Descartes had furnished his own Theory of Everything, one that relied on contact between objects and transfer of momentum to explain both the orbit of the moon and the fall of the apple. It made sense both mechanically and theologically; the universe, originally set in motion by the hand of God, continued to turn in a predetermined fashion. Most importantly, Descartes did not violate the church's doctrine of an immutable universe, since his vortices were simply God's original motions carried forward in time.[14] Descartes himself did not propose any shape of the Earth other than spherical, but French and Continental scientists later used his theory of vortices to ascribe to the Earth an elongated shape, as a counterweight to Newton.

Descartes' theory of vortices gained a wide following in the years after the release of *Principia philosophiae*. It was widely promoted by the playwright Bernard de Fontenelle, who would go on to become the secretary of the French Academy of Sciences and play a pivotal role in the Cartesian-versus-Newtonian debate. In 1686, just as Newton's *Principia* was being prepared for publication, Fontenelle came out with his chatty novel *Conversations on the Plurality of Worlds*, in which he described recent scientific discoveries in lay terms. With a playwright's view of astronomy, he described Descartes' vortices as the machines that move the scenery around the universe's stage. Even the stars, said Fontenelle, were proof of Descartes' theory. "The inhabitants of a planet in one of these infinite vortices," he explained, "see on all sides the lighted centers of the vortices surrounding them."[15]

Thanks to advocates like Fontenelle, Descartes' theory had pervaded the public consciousness on both sides of the Channel, making it quite difficult for Newton's ideas to get traction. As one of the first popularizations of science ever written, *Conversations* was an overnight success and was quickly translated into every major European language. Vortices were on the program of every afternoon salon and evening lecture from Paris to Amsterdam. Descartes' lack of mathematical rigor was a positive blessing, as neither aristocrats nor wealthy merchants wanted their physics weighed down with equations and numbers. Therefore, when Newton's mathematically based (and conspicuously secular) system of physics—in particular, his theory of a bulging Earth flattened at the poles—came up for discussion in these very salons, it was greeted with widespread skepticism, religious indignation, and even outright hostility.

Intellectual and spiritual dogmatism was not the only reason that Newton's ideas received such a cool reception in France and elsewhere on the European continent. At precisely the same time that Newton was presenting his theory in *Principia*, the French Academy of Sciences was leading a new line of research that would soon pose a further challenge to Newton. The Academy, in addition to supporting astronomy to improve ocean navigation, had carried out geodesic surveys of France to develop better maps for the army and for tax assessors. These surveys would indicate that the Earth was elongated at the poles, in direct contrast to Newton's flattened Earth. The first such surveys, carried out in 1670 by the astronomer Jean Picard, were too small in scope to show this. A subsequent series of wars and famines bled the nation dry and prevented the Academy, starved of funds and in disarray, from carrying out more extensive work for more than twenty years.

By the mid-1690s, the political and economic situation in France had improved, and the Academy's fortunes had brightened along with it. The powerful Phélypeaux family took control of the Academy, reversing its long decline. One of the Phélypeaux family members, Jean-Paul Bignon, became president, while Bernard de Fontenelle was appointed secretary for his ability to clearly explain science to an increasingly literate and interested public—including, most importantly, to the members of the royal court. Where scientists and mathematicians wrote long, dense papers for each other in the Academy's annual *Memoirs*, Fontenelle wrote breezy, easy-to-understand summaries in the *Memoirs*' History section. He also wrote the eulogies of recently deceased Academicians, which became something of a national treasure for their warm and often witty portrayals of their subjects.

Now under a strong and popular leadership, the French Academy also enjoyed a change of scenery as the eighteenth century approached. In 1699 the Academy moved from the small house that headquartered it into the Louvre palace, which was no longer a royal residence—Louis XIV had decamped to Versailles some twenty years earlier—but rather a vast workshop, filled with marble dust, wood chips, and the smell of paints from the Royal Academy of Painting, the Academy of Architecture, and the Royal Manufactures housed there. The Academy of Sciences was placed in a small bedroom antechamber—nowadays it is Room 33 of the Sully Wing—and sessions were often crowded affairs.[16]

With the Academy back on firm footing, it now undertook to extend Picard's survey of France using the same geometrical techniques, albeit with more modern instruments. The method for carrying out long-distance surveys had been around for several centuries and used a basic Euclidean premise: Given two angles of a triangle and the length of one side, the remaining sides and angle can be computed. This principle could be employed to measure over long distances by establishing a geodesic chain of triangles between two fixed points (see Figure 1.2). A team of surveyors, beginning at one end of the chain, would lay out and measure a baseline (AB) of several miles using long measuring rods. Because the baseline had to be perfectly straight, it was laid out on fairly flat, open ground. Next, they would use a survey quadrant (larger and more accurate than a marine quadrant) to measure the angles from each end of the baseline to the apex of the triangle (C), which was a visible marker many miles away, such as a large boulder, tree, or church spire. They then used trigonometry to calculate the length of the legs of the triangle.

After using angle measurements to calculate the length of each side of the triangle, the team of surveyors would repeat the process until they had a chain of triangles extending for the distance they wished to measure. Working off the first triangle, they would select another visible marker for the apex (D) of a second triangle. The surveyors would then measure the angles from points B and C to D, then from points C and D to apex E of the next triangle, and so on, further extending the chain of triangles. When the surveyors finally arrived at the endpoint of the chain, they sometimes measured the baseline of the final triangle to verify accuracy. Then, using long, iterative trigonometric calculations, they could measure the distance between the extreme ends of the chain of triangles (C and L).

Besides measuring the distance between two points, geodesic triangulation would also be used for determining the latitude of different sites, vital to correctly charting positions on a map. Early surveyors would use their large quadrants, turned vertically, to measure the angle between Polaris and the horizon. This was easier and more accurate than with a small marine quadrant, since the terrain does not pitch and roll. This procedure also allowed the surveyors to determine the length of a single degree of latitude at a particular site, by taking the angular difference in latitude at each end and dividing by the total length of the chain.

The famed Italian astronomer Giovanni Domenico Cassini, whom Colbert had brought to the Academy upon its founding but who was now almost seventy years old, led the new survey. He leaned heavily on his son Jacques Cassini as well as on Claude-Antoine Couplet, an engineer and a founding member of the Academy of Sciences. The three men tri-

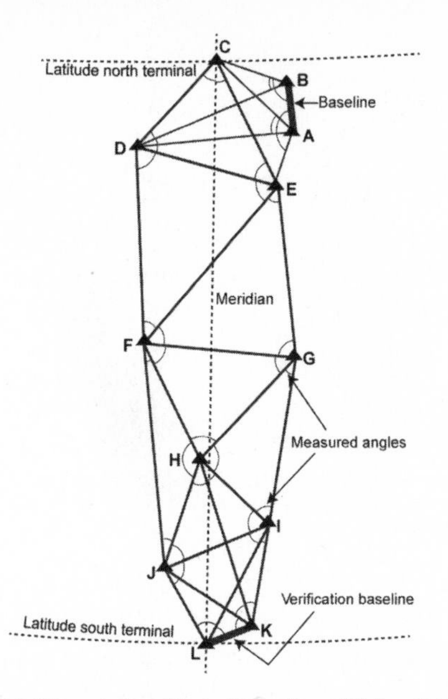

Figure 1.2 Triangulation. Illustration: Eliecer Vilchez Ortega.

angulated along the Paris meridian, starting at the capital and working south to the Pyrenees for some four hundred miles. They reported the results to the Academy in 1701. As they finished the surveys and began their calculations, they saw that the length of a degree of latitude they measured in the south of France appeared noticeably longer than the one Picard had measured in the north in 1670. Cassini reported this fact with little fanfare.[17]

Bernard de Fontenelle, long a Cartesian supporter, seized on Cassini's longer degree of latitude as a riposte to Newton's flattened Earth, claiming that Cassini had clearly showed the Earth to be elongated toward the poles. Fontenelle was resting his argument on the well-understood principle that, in order to determine if the Earth were flattened or elongated, one could compare the lengths of a degree of latitude at two widely separated points on the globe, one near the equator, the other closer to the poles (see Figure 1.3). If the Earth were flattened (oblate), the length of a degree of latitude would be greatest at the poles; if it were elongated (prolate), the length of a degree of latitude would be greatest at the equator and diminish toward the poles. Since Cassini had showed that the northernmost

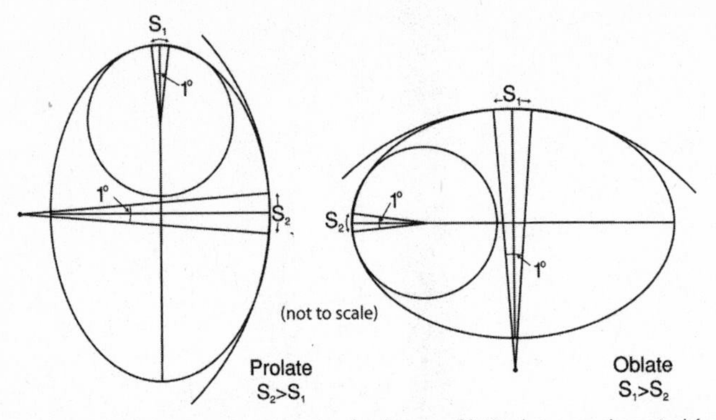

Figure 1.3 Comparative arc measurements of a degree of latitude on an elongated (prolate) Earth and a flattened (oblate) Earth, after James R. Smith, *From Plane to Spheroid* (1986). Illustration: Eliecer Vilchez Ortega.

measurement of latitude was smaller than the southern one, argued Fontenelle, the Earth must be prolate, making Newton wrong.[18]

Jacques Cassini, who took over the Paris observatory after his father's death in 1712, reinforced Fontenelle's argument with a strongly worded historical examination of the figure of the Earth. He noted that various measurements of latitude taken since ancient times showed a distinct pattern of decreasing toward the north, clearly proving that the Earth was prolate. In 1718, Cassini was commissioned by the Academy to extend the original chain of triangles northward to the coast at Dunkirk, thus completing the full geodesic survey of France. When all the measurements were in, Jacques Cassini announced that a degree of latitude at Dunkirk in the north of France was about 900 feet shorter than in the Pyrenees, eight hundred miles to the south, thus firmly (and finally, in his view) concluding that the Earth was noticeably prolate.[19]

During this time, sentiment in Britain was shifting from Cartesianism to Newtonianism. Many of Descartes' original ideas did not hold up to close scrutiny. His vortex theory, as Newton had noted, predicted that planets farther away from the sun would revolve in their orbits much faster than they actually do. Even the Cassinis' elongated Earth was held in doubt by the Royal Society, which pointed out that their survey instruments were not nearly accurate enough to account for a difference of nine hundred feet in a degree of latitude, over the surveyed distance of eight hundred *miles*—equivalent to the width of a human hair in the quadrant's eyepiece.[20]

By 1720, a fissure had developed between British and French science. On one side of the Channel, where Newton's theory of universal attraction held sway, the Earth was flattened at the poles. On the other side, where Descartes' vortices had long swirled, the French Academy had apparently proven their countryman right; based on the hard physical evidence of surveys, they had determined that the Earth indeed seemed to be elongated. While French scientists at the Academy continued to debate the issue, their colleagues' research—and the vociferous arguments of such Cartesian advocates as Jacques Cassini and Fontenelle—appeared more convincing than the strange theories emanating from a lone professor in Cambridge.

Even as they gravitated toward opposing notions of the Earth's shape, the British and French camps had yet to decisively prove that either system was correct. The problem facing British science was the lack of any observations that could prove Newton's theory. The problem for French scientists was that nothing in Descartes' theory suggested an elongated Earth; in fact, Christiaan Huygens, using purely Cartesian mechanics, had independently predicted a flattened Earth to explain Richer's pendulum clock results.[21] Lacking evidence that would lay the issue to rest once and for all, each side glowered warily at the other across the Channel.

Just when the debate over the earth's shape seemed to be reaching a deadlock, a political sea change offered new hope to scientists in both countries. The Anglo-French Alliance, which had begun in 1716 as a military and political treaty designed to avoid the continuation of costly fighting between the two nations as well as to check the resurgent ambition of Spain, was blossoming into a wider cultural détente between the two superpowers. This relaxation allowed an unprecedented exchange of people and ideas, fostering renewed scientific cooperation between erstwhile foes. Travelers crisscrossed the Channel for long stays on either side, and the exchange of letters between the Royal Society and the French Academy of Sciences grew warmer and more frequent.[22]

While the détente between Britain and France allowed scientists in the two nations to compare notes on their findings about the figure of the Earth, it did not mean that they were in any way agreed on the debate. During the first years of the Anglo-French Alliance, scientists on both sides of the Channel exchanged volleys on the discrepancies between Cassini's arguments for an elongated Earth and Richer's pendulum clock experiments, the latter of which both Newton and Huygens claimed as proof for a flattened Earth. The first broadside, a memoir by Jean-Jacques de Mairan delivered before the French Academy of Sciences, argued both sides of the debate by suggesting that the Earth was prolate, as Cassini said, but that the original stationary Earth must have been even *more* elongated than Newton claimed, somehow squeezed into an egg shape as a result of the pressure of the celestial vortex. The spinning of the Earth, Mairan said, caused the centrifugal force that explained Richer's results. The Royal Society returned fire a few years later, with a paper that vigorously disputed this logic. John Desaguliers, a French émigré living in London, argued that a stationary Earth would be spherical like a drop of water, and trotted out a machine composed of spinning hoops to show that the Earth's rotation would cause it to bulge out at the equator and flatten it at the poles.[23]

Up until the mid-1720s, supporters of Descartes seemed to have the upper hand in the culture war that was swirling around the scientific debate. Bernard de Fontenelle, now the French Academy's chief Cartesian apologist, went beyond merely contrasting the two sets of physical observations. He called into doubt Newton's theory on both physical and philosophical grounds, questioning the whole idea of attraction. "If Gravity is an attraction," he asked, "if at the center of the Earth there is some virtue [physical property] that attracts bodies . . . then what is that virtue? What is attraction?" The fact that the theory of universal attraction led to the notion of "variable gravity" was, for him, beyond "intelligible."[24] Fontenelle continued to flourish his rhetorical pen in the pages of the Academy's annual memoirs, defending his beloved Descartes while smoothly erasing any serious consideration of "attraction" in his brisk summaries of scientific papers. Fontenelle's wit and style seemed to carry the day.

In May 1726, an unexpected ally appeared on the Newtonian front. Like Fontenelle, he would deploy his well-honed literary skills in defense of a bewildering scientific theory. François-Marie Arouet, a French poet and playwright who made a highly successful career out of poking the aristocracy in the eye, had brashly insulted one too many members of the nobility. Rather than serving yet another jail sentence, Arouet decamped to London. Word of his new smash-hit play *Oedipus* had already reached the British capital, so when he arrived, Arouet was welcomed under his nom de plume—Voltaire.[25]

During his almost two-year exile in England, Voltaire imbibed English, becoming fluent not only in the language but also in its way of thinking. He was endlessly fascinated by the nation's freedom of religion and of commerce, which were closely controlled in France. He was struck by the openness of scientific debate in the Royal Society, which, unlike the French Academy, was not an official government institution. Here in London, scientific arguments and disagreements were quite openly discussed in public forums. Back in Paris, Fontenelle, the former playwright, carefully stage-managed the dispute, allowing debate within the chambers but keeping a tight lid on public spectacles and editing papers to remove any appearance of discord. Voltaire was amazed by the adulation given to Isaac Newton, all the more so since Newton did not frequent cafés and salons the way well-known French scientists had to in order to stay in the public eye. Indeed, although they never met, Voltaire was the first to report the story of the falling apple that had inspired Newton's thoughts on gravity. When Newton passed away in March 1727, Voltaire was still ensconced in London's cultural milieu and witnessed firsthand the adulation given to the noted scientist that would normally have been reserved for kings.

Voltaire would soon become the Newtonians' public mouthpiece, as Fontenelle was for the Cartesians. Voltaire found that, like Fontenelle, he could translate technical jargon into prose that was accessible to everyone. He became increasingly interested in doing so and therefore put his skills to use explaining Newton's science and mathematics. When he returned to France in November 1728, the year following Newton's death, Voltaire was brimming not just with new ideas for plays and poems but also with the notion that the universe could be precisely, even mathematically, described. This was a great leap from Descartes' hand-waving philosophy and was difficult to explain to the more conservative French public, so when, five years later, Voltaire finally published his description of these new British ideas, he first did so not in his native French, but in English. In 1733 Voltaire published in London a small but widely read book titled *Letters Concerning the English Nation*. In this series of essays describing his impressions of the nation, its people, and its ideas, Voltaire was openly admiring of the British philosophies he found so different from French ones. His work naturally garnered accolades in London.

When *Letters Concerning the English Nation* was published in French the following year, in 1734, it had the opposite effect. Voltaire had again raised the ire of the Paris elite that had driven him from the country some eight years before. Yet for many of those French intellectuals, it was their first thoughtful exposure to the new physics of Newton. "A Frenchman in London finds everything different, in philosophy as in all else," noted Voltaire. "He left behind a world that was full; here it is empty. In Paris one sees a universe composed of vortices of subtle matter; in London one sees none of that. . . . For our Cartesians, everything is done by impulsions that they barely understand; for Mr. Newton, it is by an attraction whose cause is not any better known. In Paris you figure the Earth as shaped like a melon; in London it is flattened on both sides." Voltaire, by publicly favoring Newtonianism, was once again sticking a finger in the eye of the Parisian elite.[26]

Voltaire knew a good bet, and Newton was clearly the man to follow. If he could explain the arcane philosophies of this Englishman to the French, Voltaire would be placed to take the mantle of Fontenelle as the leading popularizer of science in the cafés and salons of Paris. Fontenelle, after all, was now seventy-five years old, and Descartes' ideas were even older; a change was needed, not just of the messenger but of the message. But Voltaire knew he could not do this alone—he needed to expand his circle of friends and acquaintances beyond the literary and into the scientific.

In order to find someone who could strengthen his advocacy of Newton's ideas, Voltaire began cultivating the friendship of Pierre-Louis Moreau de Maupertuis. Maupertuis, an up-and-coming mathematician, had been an accomplished musician before turning to mathematics and becoming a member of the Academy of Sciences in 1723. As part of this growing interest in the physical sciences, he had gone to London for several months while Voltaire was in exile there, although they apparently never met during that time. Like Voltaire, Maupertuis had been on an intellectual adventure in Britain, wanting to see how things were done in a different way; also like Voltaire, he had stumbled on Newton's system and found its attraction palpable. On his return to the continent, he had traveled to Basel to study with the great Swiss mathematician Johann Bernoulli, who, although he vigorously rejected many of Newton's ideas, gave Maupertuis the mathematical tools to evaluate them for himself.[27]

It had taken Maupertuis several years, but he eventually became convinced that Newton had it right. In 1732, after forewarning Johann Bernoulli, he outlined his views of the effect of gravity and centrifugal force on rotating bodies in a paper to the Royal Society, instead of in a memoir to the French Academy, because, as he acknowledged to his mentor, "it would be better received in England than here." A few months later, knowing he could write reasonably freely about Newton's theories in a book, whereas a French Academy memoir would first have to pass muster with Fontenelle, Maupertuis explained his Newtonian theories to his countrymen in a work published independently in France, *Discourse on the Different Figures of Celestial Bodies*.[28]

His book gave Maupertuis scientific notoriety, which immediately attracted Voltaire's attention. In late 1732, while he was drafting his *Letters Concerning the English Nation*, Voltaire wrote to Maupertuis to ask for clarification on the concept of gravitational attraction, and Maupertuis replied by patiently explaining the relevant parts of Newton's *Principia*. Voltaire, who knew how to sweet-talk a fellow artist, compared Maupertuis to Newton and gushed, "your first letter baptized me in the Newtonian religion; your second gave me my confirmation. I thank you for your sacraments."[29] Their growing friendship was mutually beneficial; Voltaire now had a scientifically literate colleague who could interpret Newton for him, and Maupertuis had a writer who could give popular voice to his ideas. Together with the brilliant Emilie du Châtelet, their mutual friend and lover who would later write the standard translation of the *Principia*, they became the principal force behind the popularization and eventual acceptance of Newtonianism on the Continent.[30]

The combined force of Voltaire and Maupertuis was needed to keep Newton's ideas alive in France, for the political relationship between

France and Britain had changed considerably since the two men were in London in the 1720s. For almost a decade, the Anglo-French Alliance had given rise to an almost unheard-of level of cordiality between the scientific elite. Although Fontenelle had continued to dismiss many of Newton's ideas when he eulogized him in 1727, he nevertheless put the Englishman on the same plane as Descartes, the highest compliment Fontenelle could give. Within a few years, however, a political rift developed between the two nations, as Britain entered into a new alliance with Austria, at the time France's long-time adversary. The Anglo-French Alliance collapsed in 1731, followed by the cultural détente and, within a few years, the decades-long peace itself.[31]

The strains of the political rupture between France and Britain quickly showed in the debate over the figure of the Earth. The correspondence between the scientists of both nations dropped off, and the scientific debate between Newtonian and Cartesian advocates at the French Academy of Sciences took on increasingly strident tones. Fontenelle now shifted his stage directions, from sweeping the dispute under a rug to wrapping it in the white fleur-de-lys flag of the monarchy. Whereas previously the war of words had been directed at opposing conceptual and mathematical systems, it now took on a distinctly political overtone—a change that had the effect of stifling scientific discussion. In 1731, Maupertuis specifically referred to the "dispute between the English and the French" when describing Mairan's earlier paper to Johann Bernoulli. Fontenelle cast suspicion on Maupertuis' patriotism, wondering if he cared to "claim a glory for his fatherland, or justify the English at the expense of the French."[32]

Voltaire frequently commented on the Newtonian-Cartesian debate in letters to his influential friends who helped shape public opinion. He was quick to notice the nationalistic language that now entered the argument. He observed that the very word "attraction" was politically charged; because it arose in London, he said, it was now considered "a ridiculous idea in Paris."[33] Sensing that a great plot line was emerging, he cast the conflict as between the old Cartesians (represented by Fontenelle) and the young Newtonians, led by Maupertuis. He wisely chose to frame the debate in terms of the figure of the Earth, on the grounds that most people would find the mathematics of planetary motion far too obscure. There was one glaring hole in Voltaire's otherwise lucid summation of the controversy: No one had yet been able to clearly articulate why the Newtonian-Cartesian debate had anything to do with the problem of the Earth's shape. While Newton's theories clearly pointed to a flattened Earth, it was still a leap to say that Descartes' system of vortices implied that it was elongated.

It was Maupertuis' mentor, Johann Bernoulli, who finally provided the mathematical "proof" to the already widely held conjecture that Descartes' vortices implied an elongated Earth. In a 1734 memoir that won a French Academy of Sciences prize, Bernoulli likened the figure of the Earth to the elongated shape of a sailing ship, arguing that the Earth, pushed by Descartes' vortices, drifted at an angle to the plane of the solar equator for the same reason that a ship drifts at an angle when pushed by the wind. Bernoulli calculated that the drift of the Earth's orbit confirmed that the planet was elongated at the poles and not at the equator. Europe's greatest living mathematician thus gave his firm support to the "egg-shaped Earth" results of Cassini's geodesic surveys, linking his results directly to Cartesian physics.[34] The fissure between Newtonians and Cartesians had now widened to a chasm.

Perhaps the one man in the entire French Academy who remained unconvinced by either side was Pierre Bouguer. A supremely self-confident and ambitious young mathematician, Bouguer, like Maupertuis, hailed from Brittany (and thus pronounced his name "boo-GAYR" in the Breton manner). He was the son of Jean Bouguer, a former pilot who had lost a leg during the aborted French invasion of Ireland in 1689 and who later became a professor of hydrography, the art of navigation and piloting. Pierre was born in the pleasant coastal town of Le Croisic in Brittany on February 16, 1698, and from an early age was surrounded by his father's students—Jean's "school" was simply one room in their modest home—and their astronomical and nautical instruments, which he played with as toys. His childhood and lifelong friend Paul Desforges-Maillard (who lived just around the corner) would later become a famous poet, publicly duping Voltaire in the process.[35]

From the beginning, Pierre Bouguer was recognized as a child prodigy. Although he attended the Jesuit college in nearby Vannes, his

Figure 1.4 Pierre Bouguer, age fifty-five. Pastel by Jean-Baptiste Perroneau (1753), Louvre. Credit: Art Resource, New York.

father also taught him a great deal about mathematics, hydrography, and astronomy. Pierre would soon make good use of this knowledge. In 1714 Jean Bouguer died, leaving the family without income. Pierre, just sixteen years old, applied for his father's now-vacant post, and after passing a rigorous exam he was made royal professor of hydrography at Le Croisic, teaching hundreds of students, often twice his age, the fundamentals of navigation, astronomy, and piloting.

While still in his early twenties, Bouguer was introduced to the French Academy of Sciences after impressing a professor of mathematics with his self-taught knowledge of the calculus. He was introduced to Academy member Jean-Jacques de Mairan, who had just finished his paper attempting to reconcile Cassini's geodesic measurements with Richer's pendulum experiments. Even while Bouguer was still teaching in Le Croisic, Mairan enthusiastically engaged the young professor to work on naval problems such as the ideal placement of masts and sails, for which Bouguer won an important Academy prize in 1727. His work netted him a new, more prestigious posting at Le Havre, leaving his brother Jan to continue the family school at Le Croisic. Pierre Bouguer continued to win prizes, and in 1731 he was given the mid-level position of associate mathematician at the Academy, replacing Maupertuis, who had been promoted to the highest position, that of pensioner.

Bouguer entered a French Academy divided between the opposing Newtonian and Cartesian theories, and he quickly realized he could not rise in its ranks simply by becoming one of Maupertuis' Newtonian acolytes. At the same time, he saw that the Cartesian camp was running out of steam. Bouguer therefore sought greater recognition—and more upward mobility—by casting himself as a skeptical neutral in the Newtonian-Cartesian debate. He addressed that debate in two papers that appeared in 1734, at the height of the conflict. The first paper, examining the path of the Earth in its orbit, competed for (but lost) the Academy prize that Johann Bernoulli won by linking vortices with an elongated Earth. The next paper modeled the shape of a spinning Earth as a fluid sphere. In both papers Bouguer took new and bold lines of reasoning that, by giving equal considerations to both Newton and Descartes, marked him as an intelligent and objective observer, and the man to watch in the ongoing debate.

Bouguer's auspicious entrance into the Academy won him the admiration of many of its members—but not all. Maupertuis clearly detested the young upstart who had occupied his old position and appeared to threaten the older man's standing in the Academy. Maupertuis attempted to undermine Bouguer by taking his results and reformulating them as his own, without crediting Bouguer. Bouguer returned the sentiment if not the actions.[36] While certainly quick to undercut perceived opponents like Bouguer, Maupertuis was a strong leader of the Academy Newtonians, and actively cultivated his protégés, inviting them to lively dinners at which they strategized on how to stand up to Fontenelle and his old guard Cartesians.[37]

One of Maupertuis' closest friends in his circle of protégés was neither mathematician nor astronomer, but he was easily one of the most interesting and charismatic members of the entire Academy. Charles-Marie de La Condamine was several years younger than Bouguer and came from a considerably more well-to-do family. The product of a happy May-December marriage, La Condamine was born in Paris on January 27, 1701, to a prosperous tax collector and wanted for nothing. Charles-Marie was unusually close to his younger sister Anne-Marie; in later years he would marry her daughter, his niece, which at the time was not seen as particularly aberrant. Raised in a doting family, he was not driven to excellence and did not closely follow any one subject at Louis Le Grand, the elite Jesuit college he attended (the equivalent of a modern-day preparatory school), opting rather to pursue whatever interested him at the moment.

La Condamine's professional career had followed the same capricious path as his early life. At age seventeen he had joined his uncle in the Army of Roussillon in the southwest of France; Britain and France became allied in war against Spain soon after, so in August 1719 La Condamine and his fellow troops marched across the Pyrenees to besiege the Catalan city of Rosas. During the siege he had demonstrated the combination of curiosity, bravery, and sheer idiocy that would mark his entire career. As the combined land and sea assault unfolded, he climbed to a high vantage point to "amuse himself by observing [the enemy] firing rounds from an artillery battery, whose shells fell around him," ignoring the fact that his bright purple cape stood out against the landscape and made him an obvious target. Despite the bombardment, La Condamine only descended when finally ordered to do so.[38]

Although no records remain to indicate when, exactly, La Condamine resigned his military commission, the regimented life of an officer could not keep him interested for long, and he apparently returned to Paris after peace was declared in 1720. He probably contracted smallpox during the city's 1723 outbreak, and the disease scarred and roughened his lively but delicate features. He and his father both became caught up in the Mississippi Company scheme invented by the Scottish-born economist John Law, who had been appointed by the bankrupt French government as controller-general of finances during the period of improved relations between France and newly expanded Britain. As it would turn out, Law's appointment was one of the more unfortunate consequences of the Anglo-French Alliance. He had created a government-backed investment scheme around the apparently limitless resources of French Louisiana, with the predictable result that many French speculators, including the La Condamines, lost their shirts when the bubble burst in 1720.

Figure 1.5 Charles-Marie de La Condamine, age fifty-two. Pastel by Maurice Quentin de la Tour (1753). Credit: Frick Art & Historical Center, Pittsburgh.

Despite the failure of the Mississippi Company scheme, La Condamine could not resist underfunded government-backed speculations, and he more than regained his fortune—and also won a noteworthy new ally—on the next one. In 1728 the city of Paris defaulted on municipal bonds that had been issued to cover the Mississippi Company debts, and many wealthy Parisians who had bought the bonds were now holding piles of worthless paper. The city held yet another lottery for these bondholders with a promise of large payout to the winners, in hopes they would reinvest. At some point in late 1728, the idea struck La Condamine that if he were to buy up all the bonds at their cut-rate prices, he would have a lock on winning the lottery, which promised to pay out far more than the bonds were worth. At a dinner at the house of the Academy chemist Charles du Fay, La Condamine met Voltaire, who had recently returned from Britain. Both were graduates of the Louis Le Grand college, though as Voltaire was six years La Condamine's senior, the two had probably not known each other well at the time. Nevertheless, their shared experience drew them together, and La Condamine discussed his lottery idea with Voltaire. From February 1729 to June 1730, the two bought up all the available bonds and easily won the lottery, with a net gain of around 500,000 livres; each man now had the equivalent of over $2 million in his coffers. La Condamine and Voltaire were now inseparable, and Voltaire would often look to his friend's exploits as inspiration for his plays and writings.[39]

La Condamine's friendships landed him not only in Voltaire's oeuvre but also in the French Academy of Sciences itself. Through his friendship with du Fay, La Condamine was appointed to the Academy as adjunct chemist in 1730. Of course he did absolutely no chemistry, instead convincing the Academy to sponsor his expedition to accompany the famous corsair René Duguay-Trouin, who was escorting French merchant ships around the Mediterranean. From May 1731 to June 1732, La Condamine rode with Duguay-Trouin on what would be the privateer's last voyage; they visited the historical sites of Carthage, Alexandria, Jerusalem, and Constantinople (Voltaire immortalized his friend's voyage by setting his latest play, *Zaïre*, in Constantinople). La Condamine did not see any military action on his journeys but rather took astronomical measurements and made copious notes on the geography, botany, and natural history of the regions, all of which he duly reported back to the Academy.[40] Soon after La Condamine's return to Paris, du Fay and Voltaire introduced him to Maupertuis. La Condamine almost immediately sensed that the debate over the figure of the Earth would become his next great adventure.

Besides La Condamine, Maupertuis also recruited another young Academician into his ever-widening circle of acolytes. Louis Godin was something of a sensation at the Academy of Sciences, only twenty-nine and having risen all the way to the top level of pensioner astronomer in just eight years. He had been born in Paris in 1704 and groomed to become a lawyer in the family tradition, but he had other ideas. In his eulogy many years later, it would be remembered that he had decided to study astronomy at the Royal College "despite the protests of his father . . . sacrificing all other occupations to this favorite study, to which he gave everything without reserve." But the Royal College offered far more than mere instruction; many of its professors were members of the French Academy of Sciences, able to pick and choose among their students for entry into that privileged body. Godin, in spite of his youth, was possessed of "the

Figure 1.6 Louis Godin, about age fifty-two. Oil by Nicolas Henri Jeaurat de Bertry (circa 1756). Credit: Bibliothèque de l'Observatoire de Paris.

fire of imagination . . . married with the justice of reason," and was marked early for admission.[41]

Godin had demonstrated stunning ambition upon entering the Academy in 1725 and had risen quickly through its ranks. From his first position as an adjunct astronomer—the lowest rung on the Academy's ladder—he was soon promoted to editor in chief of the Academy's historical memoirs for 1666–1699, which had been gathering dust for a generation without being published. Despite this administrative workload he carried on his astronomical observations, carefully reporting on various eclipses and meteors, and updating astronomical tables that were used for navigation. Godin showed himself to be a competent practitioner and even invented several novel instruments to take sun and star sightings.

Godin did not live for science alone, for he was very much a ladies' man. Tall and handsome, with well-hewn features and penetrating eyes, he wooed and married the strong-willed Rose Angélique Le Moine in 1728, shortly after his admittance into the Academy, and together they had a son and daughter. The family lived in the well-to-do Left Bank district near the Sorbonne University on the rue de Postes, a fashionable address for many members of the Paris scientific community. Godin's privileged birth and early, comfortable position at the Academy meant that he had never needed to leave the observatory to do any astronomical fieldwork—or, for that matter, much of anything that required teamwork. He was used to setting and carrying out his own agenda, and he did not brook anyone questioning his judgment.

By the time Godin, La Condamine, and Bouguer were beginning their careers at the French Academy, the Newtonians and the Cartesians had reached a dead end in their arguments over the differing theories of gravity. The problem wasn't simply that the opposing sides had worn each other out; even Bouguer, in his even-handed examination of the subject, had not arrived at any definitive conclusions. Maupertuis, however, was determined to win the debate for the Newtonians, and he decided that they could make headway against the Cartesians by changing the terms of the discussion.

Maupertuis now switched his argument to geodesy—the shape of the earth—to prove the validity of the broader Newtonian system. If he could prove that, contrary to the Cassinis' findings, the length of a degree of latitude actually *increased* as one moved northward, Maupertuis could show that the earth was slightly flattened—a shape that could be explained by Newton's theory of gravity. Maupertuis' strategy was to cast doubt on the Cassinis' triangulation from Dunkirk to the Pyrenees, completed back in 1719, which had shown a miniscule shortening of a degree of latitude going south along the north-south meridian. But Maupertuis did not wish to simply remeasure the Paris meridian; the Cassinis' original results had been roundly criticized for being flawed in their fundamental assumption that the flattening of the Earth could be detected over such a short distance.

The Newtonians knew that their best chance of successfully contesting the Cassinis' measurement would be to measure the length of a degree of latitude at some distant part of the globe, to compare with the length of a degree in France. This would ideally be done at the equator, where (in theory) the difference would be most pronounced. Such a far-flung journey appeared impossible; France's sole possession near the equator was Guyana, where Richer had made his measurements, but its jungle was thought to be impenetrable. Besides, it was several degrees north of the line, and the exacting academicians preferred that their measurements be taken right at the equator. Spain and Portugal were thought to be too possessive of their South American colonies (Peru and Brazil) to allow a French expedition there. Africa's equatorial coast was well-known to the French for slave trading, but apart from a few widely scattered fortifications, it was a mysterious and hostile land. The Asian islands of Borneo and Sumatra were almost never visited by Europeans. The equator seemed hopelessly out of reach.

With the ideal surveying route cut off to French scientists, Maupertuis' best option to determine the Earth's true figure would be to take a measurement within France. Since the Cassinis had exhausted the latitudinal measurements, Maupertuis and his brightest disciples, Godin and La Condamine, jointly developed a plan to measure the length of a degree of longitude on an east-west line though Paris. Appearing before the French Academy of Sciences in June 1733, they read three papers in quick succession, describing the practical points of this method. Using the same principles as those commonly held about the correlation between latitude and spherical shapes, they argued that, if the Earth were egg-shaped, the length of a degree of longitude would be shorter than for a perfectly spherical earth; if flattened, a degree would be longer.

Maupertuis, Godin, and La Condamine knew Jacques Cassini had already begun an east-west longitudinal survey of France when they presented their papers, but they believed they would have time to complete their procedure before he did. Unfortunately for them, by November 1733 Cassini had finished his measurements and determined that a degree of longitude through Paris was in fact shorter than for a perfect spheroid—not a surprising conclusion, as it confirmed his previous calculations that the Earth was egg-shaped. This was not at all the answer that Maupertuis and his cohorts were looking for, and it should have dashed their hopes for any further geodesic surveys that could prove Newton correct.[42]

Just as Cassini's results were being discussed in Paris—just as it seemed as if the Cartesians had found even more empirical evidence to strengthen their case—a new and unexpected ally emerged on the Newtonian side of the debate. The new player, Minister of the Navy Comte de Maurepas, would provide the scientists with the critical support for their dream of going to the equator to conduct a geodesic survey unlike any other to date. While the previous measurements in France had seemed to raise more questions about the shape of the Earth than they had answered, this new survey had the potential to resolve the debate once and for all.

II

Preparations for the Mission

With the collapse of the Anglo-French Alliance in 1731, France had been left without a powerful ally to face a newly hostile world. King Louis XV knew his position was untenable. Immediately across the Channel to the north, Britain had entered into an alliance with Austria, with the intentions of checking France's power on the European continent. Spain, France's neighbor to the south, had found its own ambitions checked time and again by these same powers. It was perhaps inevitable that the two nations, sharing a border and united by a common adversary and family ties (King Felipe V of Spain was the uncle of France's Louis XV, the two having been descended from Louis XIV in the royal House of Bourbon), would join forces.

On November 7, 1733, the Bourbon Family Compact between the two nations was signed in the Spanish king's palace of El Escorial outside Madrid. The agreement would be the first of three family compacts that France would sign with Spain, and it would allow the two nations to gain new territories—the Duchy of Lorraine and the kingdoms of Naples and Sicily, respectively. The Treaty of El Escorial (the official title of the first compact) was quite far-reaching, providing, among other things, that France would support Spain's claim on Gibraltar, if war with Britain should occur. This was a particularly sensitive issue for the British and marked France's clear break with its former ally. It also meant that any French ships sailing across the Atlantic would now have to keep a weather eye open in case hostilities with Britain suddenly flared up. But the treaty had far more immediate consequences for the scientists in both nations, who were still embroiled in the debate over the figure of the Earth.

Late in November 1733, the same month that Jacques Cassini's east-west survey had once again shown that the Earth was egg-shaped, news of the Bourbon Compact arrived in Paris. Scientists at the French Academy immediately grasped the implication of the treaty. It had long been clear that, in order to most decisively ascertain the true shape of the Earth, the Academy would need a latitudinal measurement from the equator to compare against the ones taken in France. Spain had prime real estate right on the equator: The Viceroyalty of Peru, long fabled for its riches, sat astride the line. But while France and Britain were allied against Spain, Peru had been inaccessible to the Academy, and none of the other known equatorial sites—in Africa, Brazil, or Asia—were considered feasible, either. Most French scientists, therefore, had dismissed the idea of an expedition to the equator as a pipe dream. Now, with the political landscape dramatically changed, an expedition to the equator had suddenly become a real possibility.

A small group of French scientists immediately began planning a geodesic mission to the legendary kingdom at the middle of the Earth. The Newtonians felt the need for an equatorial expedition most urgently. The old-guard Cartesians appeared to have the upper hand; they had the backing of the Academy officials, the apparently ironclad geodesic results of the Cassinis, and a supporting theory from Johann Bernoulli, Europe's greatest living mathematician. The younger generation of Newtonians, on the other hand, was championing a controversial, foreign-born theory that had no supporting empirical results. They could only hope for a trump card that would save their weak hand.

A geodesic mission to establish the length of a degree of latitude at the equator offered the Newtonians an opportunity to discredit their opponents, but both sides of the debate saw the advantages of such an expedition. Immediately obvious to both camps was the opportunity to give their nation a leg up on the high seas while also promoting *la gloire de la France*. For the old guard, who thought that a mission to the equator would assuredly prove Descartes correct, the undertaking would confirm the primacy of French science. The younger generation also had nationalistic motivations; they intended to steal a march on Britain, for "in carrying out the measure of a degree of the meridian, the French will confer a greater honor upon their own nation, than the English do theirs in glorifying the discoveries of Newton."[1]

Louis Godin, an affirmed Newtonian, was quick to take advantage of the new opening to propose a mission to equatorial Peru. Although Maupertuis and La Condamine had included him in their original plans to conduct an east-west geodesic survey of France, Godin elected to go behind the backs of his mentor and his fellow protégé and exclude them from the planning of his Peru expedition. He turned to his friend and colleague at the Academy, Jean-Paul Grandjean de Fouchy, who had been a fellow astronomy student at the Royal College. The two men were also neighbors on Paris's glamorous rue de Postes, so they naturally talked shop at home.

From Godin and Grandjean de Fouchy's strategizing, a plan quickly emerged to conduct a geodesic survey along the Andes. Peru was regularly visited by Spanish ships, so getting there from Europe would pose no problem. Examining the few maps of Peru available, Godin noted that the double chain of mountains extending south from the equator—the eastern and western cordilleras of the Andes, separated by a wide, flat valley—would make excellent vantage points to carry out a triangulation. Along this valley they could survey a chain of triangles spanning up to three degrees of latitude (about two hundred twenty miles), which should provide the requisite accuracy for calculating the exact length of a degree at the equator. Logistics were always a major problem with these types of expeditions, but supplies and manpower would be readily available at the two cities, Quito and Cuenca, that would mark the endpoints of the chain of triangles. A third city, Riobamba, was sited conveniently midway between them.

As Godin and Grandjean de Fouchy discussed Godin's ideas for an expedition, one of their neighbors began to take an interest in the project. Jean-Nicolas, Chevalier de Pimodan, was neither an academician nor a scientist, but he had studied sciences at the Sorbonne and had other reasons to be interested in the scientific mission. He was an officer of the Swiss bodyguards of Louise Elisabeth d'Orléans, who had briefly been queen of Spain before her husband (the son of Felipe V) died, after which she slunk back to Paris. Guarding a dispossessed, depressed queen who broke wind in public was hardly glamorous work, and Pimodan eagerly latched on to the daring project that his two neighbors were engineering.[2]

Throughout December the three neighbors honed Godin's plans, probing for flaws and finding none. "The [plans] appeared so solid that we did not hesitate for a moment to approve them and offer our assistance in executing them," Grandjean de Fouchy later recalled. On December 23, 1733, the Academy of Sciences assembled in the Louvre for its regular twice-weekly meeting. Thirty-seven members were there, including Godin, Fontenelle, Jacques Cassini, Maupertuis, La Condamine, and Grandjean de Fouchy. Bouguer was most likely away at his teaching post in Le Havre, and Pimodan was evidently excluded because he wasn't a member. Administrative matters were taken care of first, followed by a discourse on comets by Maupertuis. Then Godin stood up and began to read from his prepared text. It must have captivated the entire audience, for the scribe, normally quite detailed in his minutes of the meetings, wrote only a single sentence at the top of an otherwise empty page: "M. Godin began to read a Text on the utility of a Voyage under the Equator."[3]

Godin's proposal had an electric and immediate effect, in spite of the Christmas holiday. On January 2, 1734, a week after Godin's address to the Academy, Maupertuis reported to Bernoulli: "A propos of M. Godin, they are talking in the Academy about sending him with a few others to Peru to measure several degrees of the Equator."[4] Godin quickly began making preparations for the Geodesic Mission to the Equator, as it became known. In January and February he wrote to the Royal Society of London, inquiring about the purchase of several precision astronomical devices. Despite the political differences between the two nations, the French Academy knew it had to reach out to its British counterpart, which could connect the French with London instrument makers, widely acknowledged to be the finest in Europe. The expedition would need astronomical tools of exacting standards if the survey were to succeed.[5]

Immediately after Godin's letter to the Royal Society, the official record of the preparations for the mission goes silent. For the whole of the year 1734, the subject of the Geodesic Mission does not come up once in the minutes (or in the memoirs) of the Academy of Sciences. But behind the scenes, a powerful new player had taken the reins of the planning process and had set the wheels of diplomacy and bureaucracy in motion to prepare for the expedition's departure.

Minister of the Navy Jean-Frédéric Philippe Phélypeaux, Comte de Maurepas, had joined the Academy's preparations at exactly the right moment. It had been obvious from the start that the French Academy did not have anything like the diplomatic or financial resources to carry out a long-distance mission like the one Godin had proposed; rather, the Academy would need the government of Louis XV to make the appropriate entreaties to Spain and to assemble the needed equipment, transportation, and money. The Academy naturally turned to Maurepas for this support, as the project fell under his direct jurisdiction. The scion of the extremely rich and powerful Phélypeaux family, which owned the immense Ponchartrain and Maurepas estates near Versailles, Maurepas had inherited the titles of minister of the navy and of colonies, as well as several other posts (including secretary of state, which controlled the Academy of Sciences), from his father and grandfather. Maurepas had assumed these titles at age fifteen, after the death of Louis XIV and the installation of a caretaker government for Louis XV, who was only two years old at the time. It was not until Louis XV reached his majority that Maurepas, then only twenty-two, was fully instated as minister of the navy, the position once held by Jean-Baptiste Colbert.[6]

Throughout his lifetime, Maurepas enjoyed playing the merry fool and the simpleton, though he was anything but. He kept the court off guard with his light banter and wicked satire, often penned anonymously against his opponents. They in turn railed against his inexperience, accusing him, as navy minister, of "not even knowing what color the ocean was."[7] But Maurepas could read people the way great navigators read the winds and currents, and he would unerringly chart his way through court intrigues, pinpoint the weakness in an adversary's position, and map out his own subterfuges, all while accounting for the changing political tides. He depended on these instincts to decide whose advice he would take, and as honorary vice president of the Academy of Sciences, he had many experts to choose from to ensure that Science served the needs of the State. This mandate was of course made easier by the fact that his cousin, Jean-Paul Bignon, was the president of the Academy.

Knowing the importance of having Maurepas on their side, the Academy quickly got to work apprising him of the advantages of a geodesic mission to the equator. The enhancement of navigation was a priority for Maurepas; like Colbert before him, Maurepas was faced with a numerically superior British navy, and with a limited budget he had to make every vessel count. Ships were routinely lost, waylaid, or dashed against a badly charted shoreline. If Maurepas could reduce that number, it would be the same as building more ships, but without the added expense. Knowing that the navigation issue was the best way to convince the minister, the chemist Charles du Fay wrote a memoir for Maurepas' consumption that cast the enterprise as a means to improve ocean navigation. He explained why Peru was the best possible choice for the geodesic measurements and suggested that the Spanish court be requested to provide protection for this mission.

Maurepas immediately grasped the benefits that a geodesic mission could have for his naval and commercial fleets, but he also understood the delicacy of asking the Spanish court to allow Frenchmen into Peru. Even after two centuries of Spanish occupation and a continuous procession of treasure galleons, Peru's gold and silver mines remained a primary source of Spain's wealth and the principal lure for European pirates and smugglers. Since the beginning of the eighteenth century, merchantmen from various nations were making use of the route around Cape Horn and into the South Pacific, from which they could ply the west coast of South America, on which Peru was located. France alone had sent over a hundred forty ships to Peru and back, bringing almost $2 billion into French coffers.[8]

From the Spanish point of view, this trade was a disaster. Fully two-thirds of all traffic in the South Pacific—French and otherwise—was illicit, meaning that the colonial government could neither control nor tax it properly. Foreign ships smuggled in scarce commodities ranging from rope to flour to munitions in exchange for gold and silver, bringing the goods ashore in remote places with the tacit approval of local authorities, who expected their cut of the proceeds. After several attempts by the Spanish monarchy to stop the contraband traffic, by 1726 most European countries had agreed to Spanish demands and forbade any further voyages, essentially turning the South Pacific into a Spanish lake.[9] The Spanish viceroys, moreover, jealously guarded their coasts and trade routes. Maurepas knew that, in order to have any hope of getting French scientists into Peru, he would have to approach the Madrid court bearing gifts.

On February 27, 1734, Maurepas made his first overture to the Spanish government. He sent to Madrid a packet containing a copy of du Fay's memoir and a covering letter explaining the significance of the recent geodesic findings, noting that—if the scientists were correct and the world was not a perfect sphere—navigators could be off by three hundred miles or more on a voyage between Europe and the Americas. A mission to the equator was vital to resolving these errors, Maurepas said, before arriving at his point: "Several academicians propose to go to Peru to make astronomical observations there, the utility of which would be advantageous to the Spanish as well." Maurepas sweetened the proposal by suggesting that his astronomers determine the exact position of the coast of Peru while in the region, as an aid to Spain's navigators. To allay fears of smuggling, he offered to have the astronomers' baggage opened upon entry to Peru, to show they only contained scientific instruments and not contraband goods.

On March 28, Maurepas' packet reached José Patiño Rosales, his opposite number as minister of the navy and the colonies, and Spain's prime minister in all but name. Just as Maurepas had anticipated, Patiño immediately saw the navigational benefits to his own fleet. As the man behind the Bourbon Family Compact, he also saw the opportunity for a clear demonstration of the newfound trust between France and Spain, and immediately sent the request through the Spanish bureaucracy that connected the government's Royal Palace in Madrid to the king's residence at El Escorial, thirty miles distant. Within a few days Patiño received preliminary approval from the king, and the following week he forwarded the request to the Council of the Indies, the legislative and judicial body that oversaw the Spanish colonies.[10]

Spain's Council of the Indies deliberated over the French proposal for a month and finally returned a tentative yes—along with several additions and restrictions to the plan. Despite the alliance with the French, it seemed that suspicions about their intentions still persisted in Spain. The chief astronomer of the council recommended that "two intelligent Spaniards accompany the said scientists, so that they may be educated under [the Frenchmen's] tutelage," thus ensuring that any scientific knowledge gained from the expedition would find its way into Spanish navigation.[11] A marginal note on his proposal emphasized that these officials would also keep an eye on the French to tamp down smuggling. The council demanded further safeguards against smuggling, such as the right of colonial officials to open the astronomers' baggage at every port and city.

By May 3 a letter was on its way to Maurepas from Madrid, signaling that the council had granted the French request pending some minor administrative details and asking for the names of the travelers who would need passports to Peru.

The roster of the proposed expedition was under debate even before word of Spain's acceptance of the plan reached France. In Paris, speculations were rife over whom Maurepas would select to see the riches of Peru. Even though France had long been barred from the Spanish colony, its very name had become synonymous with wealth: *Il vaut le Pérou* (it's worth all of Peru) meant one had struck pay dirt. It was even rumored that Cuenca, the southernmost point of the triangulation, once had been the mythical city of El Dorado. Even if no deliberations occurred in the French Academy itself during the spring of 1734, scientists were almost certainly talking among themselves about this fabulous voyage. The Café Gradot on the Quai d'École was a favorite spot for academicians to gather and gossip, since it was just a block away from the Louvre near Pont

Neuf. Their conversations were a mix of scientific observations; debate on the current political situation between France, Spain, and Britain; and discussion of the latest plays by Voltaire, who was also a regular at Gradot.

At the café the scientists would have discussed the measurements to be taken at the equator and would have ridiculed a proposal, sent in by a French astronomer living in Russia, to extend them thousands of miles to Tierra del Fuego (the proposal was ultimately rejected).[12] They might have talked about the qualifications (or lack thereof) of the proposed expedition members. The founders of the expedition, Godin and Grandjean de Fouchy, had no astronomical field experience, and Pimodan was not even a member of the Academy of Sciences. One selectee would have appeared to be an obvious choice: Jean de La Grive, the mathematician and geographer who at the moment was conducting a new survey of France with Jacques Cassini. But the selection of La Condamine, another regular at Gradot, might have been greeted with general puzzlement: He wasn't even an astronomer, after all, and had no mathematical background to speak of.

As Maurepas was casting about for men to go to Peru, he would have been weighing many more factors than just mathematical ability. The Spanish kingdom was a little-understood and potentially hazardous destination, requiring a man with a cool head and good judgment. La Condamine, a veteran of battle and a privateering expedition, had shown himself to be intrepid and daring. His detailed reports from Jerusalem and Constantinople marked him as scientifically curious and a keen observer. As a military man he knew how to lead as well as follow orders, a decided contrast and much-needed counterweight to Godin, who had never carried out a large-scale operation in his life. Finally, in reflecting on La Condamine's brush with death at the siege of Rosas, and his remarkable winnings in the Paris lottery, Maurepas may have also decided that he was uniquely endowed with luck, the attribute of a leader that Napoleon would later prize above all else. He would do.

Voltaire was absolutely smitten with his friend La Condamine's mission to Peru. Now happily living with Emilie du Châtelet in the rolling hills of Champagne, he began work on a new play titled *Alzire, ou Les américains*, a religious and moral tragedy loosely based on *Zaïre* but set in Peru at the time of the conquistadors. *Alzire* tells the story of Gusman, a wicked Spanish ruler, who wants to marry the Peruvian princess Alzire. She refuses to conspire with the invaders, instead supporting her Peruvian lover in a failed uprising against them. Gusman orders them killed but then is himself mortally wounded and retracts his order, instead asking Alzire and her lover to rule the country. His sudden repentance convinces the Peruvians to consider converting to Christianity.[13] When the play was finally produced in 1736, it became an instant hit and, as Voltaire intended, kept Paris society focused on their scientists in Peru.

By the end of June 1734 the plans for the Geodesic Mission were well advanced. Maurepas included a botanist in the group to catalog and bring home samples of the bewildering variety of South American flora. Maurepas probably made this addition at the instigation of Charles du Fay, in his second role as intendant (administrator) of the Royal Garden. The Royal Garden was an appendage of the Academy of Sciences in all but name—both fell under Maurepas' control—and was the government's research center for medicinal plants, crop disease, and identifying and growing new plants for commerce and trade.[14] Voyagers to the French colonies as well as explorers of new worlds were routinely instructed by the Royal Garden to "botanize"—to collect and bring back seeds and entire plants for careful study. Jean Richer, for example, had brought plant samples back to France from his 1672 expedition to Cayenne.

The Jussieu family from Lyon was becoming a dynasty within the Royal Garden, and given the family's friendship with Louis Godin, it was logical that they should supply a botanist to the expedition. Antoine-Laurent de Jussieu and Bernard de Jussieu, two of the oldest brothers, had made their names by botanizing up and down Europe, reporting the results in large, best-selling books. Their younger brother Joseph was just beginning his career as a botanist, and a voyage to Peru would unquestionably bring him the fame and comfortable future that his elder brothers already enjoyed. But thus far in his meandering career, Joseph de Jussieu had shown no interest in either fame or fortune. He had originally followed in his brothers' footsteps and intended to practice medicine (botanists at the time were initially trained as doctors since most medicines were herbal, and many botanists practiced both professions), but then he changed his mind and studied physics and engineering. He ultimately switched back to medicine, graduating with a doctorate from the University of Reims in 1729. He joined the faculty of medicine at the University of Paris in 1732, while also assisting his brothers with their botanical studies. Now just twenty-nine years old, shy, sensitive, and often melancholy, Jussieu seemed to be searching for his path in life when the position on the Geodesic Mission came open. Whether nudged by his brothers or of his own volition, Joseph volunteered and was accepted for the position of botanist and doctor.[15]

With the addition of Joseph de Jussieu, the French membership of the Geodesic Mission had been finalized. In July, Maurepas sent the list to the Spanish court, naming each member along with his title: Godin, Grandjean de Fouchy, and La Condamine, astronomers; La Grive, Pimodan, and Jussieu, mathematicians and botanist; plus servants.

Despite some last-minute misgivings by the Council of the Indies, which was still suspicious of foreigners on Spanish soil, the Spanish court issued two Royal Decrees (*Reales Cédulas*) on August 14 and 20, 1734, authorizing the passage of the French scientists to Peru to conduct astronomical and physical observations. The decrees accepted the French scientists' offer to determine the exact position of the coast of Peru and also set out the conditions of their passage. The Spanish government ordered a close inspection of all the Frenchmen's baggage and equipment, but also bade Spanish colonies and banks to open their coffers to the scientists for any outstanding expenses, with the understanding that they would be repaid by the French government.

Shortly before the Royal Decrees were issued, a marginal note was added to them, stating that one or two Spanish officials would accompany the scientists to help search for medicinal plants. This last-minute addition, although firmly grounded in the Council of the Indies' original recommendation, came as a surprise to the French, who nevertheless accepted it with good grace.[16] The selection of these two officials would have profound and lasting repercussions for both the mission and the Spanish court itself.

Upon receiving word of the decree and the open credit line from Spain, Maurepas began negotiating the financial aspects of the mission, a project that would occupy much of his efforts through the autumn of

1734. Maurepas left no detail to chance, doing everything he could to smooth the passage of the scientists into Peru. He sent instructions to the French consul at Cadiz, the financial center of Spain's colonial enterprise and home of Spain's powerful House of Trade, which was surrounded by a multitude of banks, shipping companies, and merchant navies, over 40 percent foreign-owned (mostly French), all engaged in buying and selling cargo carried in Spanish hulls to and from the Americas. As a precautionary measure against any last-minute problems with the Spanish authorities, Maurepas instructed the French consul to arrange for 4,000 pesos credit (roughly 20,000 livres, about $180,000 today) with the French trading firm of Casaubon, Béhic et Compagnie, via their correspondents in the Spanish colonies. Maurepas also sent dispatches to the governors and intendants of the French Caribbean colonies where the scientists would first land after their transatlantic voyage, instructing the colonial administrators to pay for any unforeseen expedition expenses as well as the scientists' passage to Panama, from which they would have to trek overland in order to find passage to Peru.[17]

Although Maurepas was making extensive allowances for the expedition's finances, La Condamine was not about to leave his fortunes in the hands of the French government—having been bankrupted once for doing just that. His friend Philibert Orry, the French minister of finance who had spent his youth in Spain, had confirmed that La Condamine's wariness was justified, advising him that the expedition's costs could far outrun the original budget, since, in the colonies, "the most necessary items would come at an excessive price." Concerned that the expedition would run out of money, La Condamine hedged against this by arranging for personal letters of credit worth 100,000 livres (almost a million dollars) that he could exchange in several South American and Caribbean cities through the powerful Castanier Bank of Paris, which in partnership with the financier Samuel Bernard had dealings with financial and trading companies all over the world.[18] La Condamine would not regret this foresight.

As the diplomatic and financial arrangements for the Geodesic Mission were being finalized, the first setback occurred: Several of the team's original members unexpectedly withdrew. The losses left the expedition shorthanded, although they were for the most part unavoidable. The cartographer La Grive was still mapping France with Jacques Cassini and could not be detached from the task; Maurepas suggested that Jacques' son César-François Cassini join the expedition instead, but he, too, could not be detached from the survey. Pimodan, meanwhile, had been promoted to commander of the Swiss Guards, who were protecting the increasingly despondent Louise Elisabeth d'Orléans; despite any personal hesitation, this was an honor he could not turn away from, so he (likely with some regret) bowed out of the expedition. Grandjean de Fouchy was increasingly suffering from kidney or bladder stones, and his doctors forced him to abandon the mission that he helped propose.[19]

The expedition was now down to just three men, and substitutions had to be found quickly. To replace La Grive, Maurepas appointed two men: a draftsman and a naval engineer. Of the draftsman Jean-Louis de Morainville little is known except that he was born in 1707, he was a talented artist, and he was married; his birthplace is unknown, although his surname was common in Normandy. Somewhat more can be discovered about Jean-Joseph Verguin, the engineer and principal surveyor for the mission. Born in Toulon in 1701 and married with young children, he was trained in astronomy and cartography, which he had already put to use in the New World. In 1720 he had accompanied a fleet to survey the Caribbean and French Louisiana, where he drew plans for the Spanish colonial city Cartagena de Indias (in modern-day Colombia) as well as the Mississippi Delta. On his return, he had worked as an architect, draftsman, and civil engineer in the Toulon dockyard. In 1731 he had served as cartographer with the Chevalier de Caylus on a pirate-hunting voyage along the Barbary Coast (modern-day Tunisia), at exactly the same time La Condamine was on his Mediterranean cruise. Verguin's strategic maps of the enemy harbors and moorings, made under wartime conditions, greatly impressed Maurepas: "I found the maps quite precise, and I am grateful to Mr. Verguin for the detail he has brought to the task," he wrote Caylus. Maurepas would ensure that Verguin's skills were put to use in mapping the expedition's survey through the Andes.[20]

The expedition gained its two youngest members after Pimodan bowed out. Intended to take his place, the two were selected by Louis

Godin as favors to his family and friends and were perhaps even less qualified for their posts than Pimodan had been. Jean-Baptiste Godin des Odonais was Louis' first cousin, like him the son of a lawyer, but more interested in science than law. Unlike his hard-working, careerist cousin, Odonais was something of a dreamer; born in 1713 in Saint-Amand-Montrond in the heart of France, he had spent his twenty-one years wandering the fields of the Odonais family property along the river Cher, thinking of far-off places. His role in the expedition reflected his rudderless youth; with no experience in astronomy, surveying, or botany, he would be expected to quickly learn the details of those professions as he assisted in whatever tasks were given him. For his own part, he hoped to study the indigenous languages of South America (principally Quechua) and write a grammar upon his return.[21]

The second replacement for Pimodan had an illustrious pedigree, but in reality he was no more prepared for his role than Odonais. Jacques Couplet-Viguier was the grandson of Claude-Antoine Couplet, who had assisted in the Cassini surveys of France, and a nephew of the treasurer of the Academy of Sciences, Nicolas (Pierre) Couplet de Tartreaux, a veteran explorer and astronomer who had gone to Brazil in the seventeenth century to make astronomical and pendulum observations in much the same manner as Jean Richer. Nicolas Couplet was very close to Louis Godin and his family; Godin's wife, Rose Angelique, would later be the main beneficiary in his last will and testament.[22] It is possible that Nicolas saw himself in his nephew Jacques and persuaded the father to send his son, probably no more than seventeen or eighteen years old, on this voyage to Peru.

Maurepas decided to add two more members to the expedition, probably at the behest of the Jussieu family, who appeared to have known them well: the instrument maker Théodore Hugo (sometimes spelled Hugot) and the surgeon Jean Seniergues. We know little of Hugo, not even his age, but his job was critical. He had to keep the various astronomical and survey instruments in good repair and build new ones when needed. He was not particularly qualified for the post, however, previously having been a clock-maker who had never built a telescope or quadrant in his life. Seniergues was a native of a hamlet called Bonneval in the province of Quercy in southwest France. He was the same age as his friend Joseph de Jussieu, but as a surgeon he did not have Jussieu's training or education, being considered more of a manual laborer (surgeons had only recently become a separate trade from barbers) rather than part of the elite profession of medical doctor. Both Hugo and Seniergues were unmarried, and both were lured by the riches that Peru promised. Hugo's aspirations were modest, perhaps a pension or a reward for his service. Seniergues' avarice was apparent to all; he would later brag that as highly sought-after men of medicine, he and Jussieu "make money wherever [they] go." He looked forward to finding his first gold mine and advised his less-greedy friend that "it's always better to prefer a *doblón* to a seashell."[23] Despite his mercenary streak, Seniergues acted as something of a protector to the still-immature Jussieu, making him a welcome companion on the mission.

The last man to be chosen for the Geodesic Mission was intended to replace Grandjean de Fouchy as astronomer yet would end up playing an outsized role in the expedition. Selected by Maurepas himself, he would become the mission's de facto leader and the man responsible for the incredible accuracy of its observations. Pierre Bouguer, the professor and astronomer who advocated neither vortices nor attractions, was the logical choice to counterbalance the pro-Newtonian scientists Godin and La Condamine. He was also, at first, the man most reluctant to join the mission.

Bouguer had become one of Maurepas' most trusted advisors and enjoyed his behind-the-scenes backing at the Academy of Science, despite the astronomer's youth and the opposition he encountered from the arch-Newtonian Maupertuis. Bouguer had come to Maurepas' attention early in his career as someone who could bring practical science to bear on the navy's most pressing problems. In addition to his Academy work on mathematics and astronomy, which he pursued while still teaching navigation at the Le Havre school of hydrography, Bouguer was single-handedly creating a new science of ship design that relied not on rule-of-thumb principles but on calculating displacement, stability, and speed even before the ship was built.[24]

Bouguer was clearly on the rise, yet he remained a mystery to many of his colleagues. Like a number of the Academy members, he shunned marriage and even sexual relations in favor of intellectual passions.[25] He was gentle with his friends, but—as his stern portrait suggests—he was fiercely vain of his intellect, often arrogant and condescending, and did not suffer fools gladly. By this point in his career, Bouguer had shed his immature role of child prodigy but was not yet crowned with the mature title of genius. He could not attain that title merely on paper; in the Paris crowd, one had to shine in person, and for Bouguer this was almost impossible. He was hamstrung by not being in the thick of things, since he lived an expensive three-day coach ride from the capital, and money for him was always in short supply.

Even while in Paris, Bouguer was immersed in work, instead of idling away the hours at the Café Gradot or in one of the fashionable salons run by immensely powerful women such as Françoise de Graffigny (a close friend of La Condamine and Voltaire). Although Bouguer was politically astute, he never fully grasped the intricate social rituals of the Academy, believing that scientific reasoning should stand on its own merits. In vying for recognition, he also had to contend with his adversary Maupertuis, who had attempted to steal the spotlight on more than one occasion.

Maurepas knew that he would have a hard time convincing Bouguer to join the Geodesic Mission. Bouguer, who had taught hundreds of mariners how to navigate the high seas, was reluctant to travel outside his beloved France; as he later put it, he had "never thought to take part in this enterprise" because of the "repugnance that I always had for ocean voyages on account of my weak constitution." But Maurepas knew of Bouguer's hunger for recognition as a top-flight scientist, and he cleverly played on Bouguer's aspirations. He began enticing Bouguer by ordering the Academy of Sciences to send the penurious astronomer four crates of expensive astronomical instruments that he had been unable to afford for himself. Maurepas then asked his cousin Jean-Paul Bignon to dangle before Bouguer the promise of promotion to pensioner astronomer if he would join the mission, knowing that Bouguer had been competing for the title against two more experienced astronomers. Maurepas sweetened the deal by guaranteeing that all of Bouguer's expenses would be paid while he still retained his professor's salary.

In other circumstances, the cautious Bouguer might have refused to be drawn away from his steady rise within the Academy. But the Geodesic Mission, he realized, would cement his reputation and vault him into the upper ranks of the scientific elite, and this opportunity changed his outlook completely. Just before Christmas, Maurepas invited Bouguer to come to Paris to discuss the mission with Bignon. On December 29, 1734, Bignon reported to Maurepas, "Mr. Bouguer arrived here yesterday, and I saw in him the greatest resolve to accomplish the voyage with the most marvelous zeal. It would be difficult to find a more determined voyager." With Bouguer finally on board, the French side of the roster was now complete.[26]

In Spain, the other side was being filled at the same time as the French were fine-tuning their selections. The Council of the Indies had insisted on having a pair of Spanish officials accompany the French to Peru but had left the selection to Minister of the Navy Patiño. At the suggestion of the Lima-born aristocrat José Augustín Pardo de Figueroa, Patiño ordered the Academy of Navy Guards in Cadiz to appoint two cadets to the Geodesic Mission.[27]

By insisting that the Spanish expedition members came from the Academy of Navy Guards, Patiño was ensuring that the French would be accompanied by the best the military had to offer. He had created the Navy Guards as an elite body of naval officers, whose cadet training included not only shipboard cruises but also a year at the Academy receiving intense instruction in mathematics, navigation, engineering, shipbuilding, and the sciences. They were cadets rather than commissioned officers, so that they would be junior enough to take orders from their senior French counterparts, while also having the requisite knowledge to assist the French scientists in their astronomical observations. They were also young and healthy, thus able to endure the hardships of the expedition—or, "in case of need, to take the place and act in the stead of any Academician who goes absent or dies," as Patiño presciently described in his orders.[28]

Patiño left the selection of the cadets to the battle-hardened commander of the Navy Guards and the cerebral Academy director, and in seeking out both military prowess and scholarly distinction, they chose far better than they knew. The first to be selected was Jorge Juan y Santacilia, then just twenty-one years old but already known to his classmates as "Euclid" for his brilliance in mathematics. Born in the Mediterranean city of Novelda in 1713, he had lost his father at age three and was raised by his paternal uncles, who also covered his expenses in the form of loans. He was sent to Malta at age twelve to become a knight of the Order of Malta, which required of its members lifelong celibacy. In 1729, at age sixteen, he entered the Academy of Navy Guards, but his studies were interrupted by several years of sea duty against corsairs, where he undoubtedly honed his renowned skills as a swordsman. In 1732 he participated in the invasion of Oran, a city in modern-day Algeria held for a time by the Ottoman Empire until the Spanish landing force captured it, under the leadership of the fearsome admiral Blas de Lezo y Olavarrieta. Though wounded in action, Jorge Juan (as he was usually called) soon returned to the Academy to complete his studies. He was strong of jaw but not otherwise remarkable, with a medium build and an easygoing temperament, often pausing for long periods during a conversation to reflect on the issues at hand. This mild exterior belied a sharp intellect, absolute fearlessness, and supreme self-confidence, all of which had caught the attention of his superior officers.

Jorge Juan's companion on this expedition would be Antonio de Ulloa y de la Torre-Guiral, born into the Seville aristocracy in 1716. Antonio's father, Bernard de Ulloa, was a noted economist who advised the king on revitalizing manufacture and commerce, instilling within the young Antonio a passion for reforming outdated institutions. His father had used his royal connections to send Ulloa to sea at age thirteen as an "Adventurer" (as he later referred to himself) with the treasure galleon fleet, making several voyages around the Caribbean and to Cartagena de Indias. Ulloa had entered the Academy of Navy Guards in 1733 and received his own baptism of fire in combat against Austrian forces in early 1734, shortly before being selected for the Geodesic Mission. His portrait depicts a rather schoolmasterly naval officer, quick to spot any errors; indeed Ulloa's keen eye as well as his ability to assess a situation in short order neatly balanced Jorge Juan's more analytical and detail-oriented character, and the two Spanish officers would become close friends during the mission.

Figure 2.1 Jorge Juan y Santacilia. Oil by Rafael Tejeo (1828), modeled after his death mask. Credit: Museo Naval, Madrid.

Figure 2.2 Antonio de Ulloa y de la Torre-Guiral, age sixty-nine. Oil by José Roldán y Martínez (circa 1840). Credit: Museo Naval, Madrid.

In January 1735, Ulloa and Jorge Juan were vaulted four grades to the rank of lieutenant (*teniente de navío*) and received a commensurate rise in pay to 60 escudos per month (about $54,000 per year). Their orders for the mission were to assist the French scientists in their astronomical observations, to make maps of ports and cities as required, and to fix the position of coastal towns to aid in navigation. The expectation was that the young officers, after having worked alongside the astronomers, would bring their hard-won knowledge and experience into the service of Spain.[29]

The Spanish cadets soon learned that, by great good fortune, they were to sail to their rendezvous with the French scientists in the naval convoy that would carry the new viceroy of Peru to his kingdom. At age sixty-eight, José Antonio de Mendoza Caamaño y Sotomayor, Marqués de Villagarcía de Arousa, was well prepared for the position, having lived through many upheavals in Spanish politics and having served as viceroy of Catalonia and ambassador to Venice. His reign in Peru would turn out to precisely overlap the Geodesic Mission. The Spanish lieutenants knew that they would have to immediately develop a good rapport with Villagarcía to ensure that the French scientists could carry out their mission unhindered. Even though the local Peruvian officials had received orders from the king to protect the Frenchmen and provide all possible assistance, those officials often took to heart the colonial maxim "*Obedezco pero no complo*" (I obey, but I do not comply). The naval officers would need to pave the way at almost every step.

Jorge Juan, being the older—and hence higher-ranking—officer, would accompany Villagarcía aboard the sixty-four-gun ship *Nuevo Conquistador*; Ulloa would sail on the smaller fifty-gun *Incendio*. After the fleet arrived in Cartagena de Indias (on the Caribbean coast of present-day Colombia), Jorge Juan and Ulloa would wait for the French scientists, then travel with them across Panama, down the Pacific to the coast of Peru, and overland to Quito, where they would begin their survey.

As the Spanish were selecting their team members that winter, the French scientists were working to outfit the mission and hone its methods. In December 1734 Louis Godin wrote to the Royal Society of London, apprising it of his upcoming visit in early January 1735 to purchase the astronomical instruments that were critical to the expedition. These included several telescopes, an octant (a precursor to the sextant) built by the instrument-maker John Hadley, and—most importantly—a pendulum clock and precision twelve-foot zenith sector for measuring star angles created by another renowned artisan, George Graham.[30]

While in London, Godin also met with Edmond Halley, the astronomer royal at Greenwich. Although no record of their meeting survives, it is likely that they discussed the technicalities of carrying out astronomical and physical observations in distant lands. Halley, after all, was the veteran of several scientific expeditions, including a famous voyage around the South Atlantic in 1698 to measure magnetic variations, where he commanded the navy ship *Paramour*. Their discussions were in French, since all educated men—especially officers of the Royal Society like Halley—spoke the language at least comprehensibly; Halley himself had lived in Paris for half a year and was quite fluent. Despite the growing political tensions between France and Britain, the Royal Society elected Godin as a fellow.[31]

Back in Paris, Bouguer and La Condamine hurriedly prepared the instruments and procedures for the mission. They undoubtedly consulted extensively with Jacques Cassini on the practicalities of undertaking such a long survey, since he was now the veteran of six separate survey expeditions around France. How long should the baseline for triangulation be? (Cassini's were between six and nine miles long.) And also: How should the baseline be measured? (Cassini used pairs of twenty-four-foot-long poles placed end to end.) Meanwhile, Cassini wrote long memoirs for the Academy proceedings on the additional astronomical observations they should undertake while in Peru.[32]

Even with Godin's successful purchases in London, a variety of survey and astronomical instruments still had to be bought or made. The workhorses for the survey would be the portable quadrants, the large quarter-circles typically between two and three feet in radius that were used for measuring angles. La Condamine bought a refurbished one from the estate of an astronomer, and three others were ordered from the instrument maker Claude Langlois, at a cost of around 1,500 livres ($14,000) apiece. Langlois also carefully forged and ground an iron bar called a *toise* (about 6.4 feet, the conventional unit of length in France in the eighteenth century), which was to be used to calibrate all measurements during the mission.[33] As these instruments were delivered, they were carefully crated and sealed alongside the growing mountain of supplies needed for the long voyage to the New World.

The Geodesic Mission was to be the first voyage by Frenchmen to South America in over a decade and the first extended trip into Peru's interior by trained observers of any nationality. The French had been frequent visitors in South Pacific waters from the turn of the century—often to the great dismay of the Spanish government, which saw them as usurpers. But not all of the voyages by the French had been devoted to trade and smuggling. Two expeditions in particular were well known to the Academicians for their scientific and geographic intelligence on Peru.

The first foray into South America by a French scientist had been made from 1707 to 1711 by the priest and astronomer Louis Éconches Feuillée, a disciple of Giovanni Cassini. Departing from the south of France, Feuillée rounded Cape Horn and explored the coasts of Chile and Peru. He made several trips to Lima, calculating the city's longitude and latitude, conducting various astronomical observations, and drawing maps of the city and its environs. He collected plants and made notes on the wildlife, including the spectacular condor. On his return to France, Feuillée published his observations in a dense, massive two-volume work filled with maps, tables of planetary and star sightings, and drawings of plants. Feuillée himself had died in 1732, so regrettably the members of the Geodesic Mission could not get any more firsthand details.[34]

Unlike Feuillée, the other Frenchman to have systematically explored South America was still very much alive, and had a much different perspective on Peru than the aged and abstemious Feuillée. Amedée-François Frézier had been a twenty-nine-year-old military engineer when he was selected to canvass the South Pacific from 1712 to 1714 on the heels of Feuillée's voyage. Frézier's mission was to improve French trade by gathering military and commercial intelligence—charting the coastlines, ocean environment, and above all fortifications, ports, and docks of the cities of Peru and Chile. In addition to his careful surveys and maps, he took copious notes on the plants, animals, and people and their customs. Frézier fully understood that (then as now) sex sells, and in a best-selling book published two years after his return, he detailed the sexual mores of Peruvian women, contrasting their daytime appearance of chastity and virtue with their supposed nighttime excursions to visit their lovers. "In matters of love they yield to no nation," he wrote. "They freely sacrifice most of what they have to that passion." Frézier claimed that in Peruvian culture, premarital sex and adultery were the norm, not the exception—a claim that may have colored the views of at least one of the members of the Geodesic Mission, with tragic consequences.[35]

Frézier returned to France in 1714 with more than just descriptions of how to overcome the defenses of Peruvian cities and their women. In a rare historical coincidence, Frézier most famously brought back specimens of the Chilean variety of the strawberry, which in French bore his own family name, *fraise*. Those South American fruits, much larger and more succulent than the puny European ones, became the primary source of the ones we know today. Frézier took several of the plants back to his home in Brittany, where they grew successfully in the maritime climate, and gave others to Antoine de Jussieu to cultivate in the Royal Garden of Paris, from whence they soon became fixtures in the gardens of Europe. Given their connections to the Royal Garden, it is likely that Jussieu, Godin, and La Condamine (and perhaps Bouguer as well) would have spoken with or written to Frézier to learn all they could about Peru once the mission had been approved.[36]

The drawback to Feuillée's and Frézier's intelligence was that they had only gone as far as the coastal city of Lima, whereas the scientists in the Geodesic Mission would be working near Quito, eight hundred miles to the north and far inland. Therefore, the two travelers' observations would only serve as starting points for understanding the country. Nevertheless, Frézier had become acquainted with Pedro de Peralta y Barnuevo, a noteworthy Peruvian author, mathematician, and astronomer at the University of San Marcos in Lima, who subsequently sent copies of his astronomical observations to the French Academy. Peralta would become a close confidant and correspondent of the members of the Geodesic Mission once they arrived in Peru.

In spring 1735 the final arrangements for the Geodesic Mission were well under way. Both the French and the Spanish contingents were preparing for a departure in mid- to late May, with a rendezvous in Cartagena de Indias in late summer. Unlike the two Spanish team members, who would sail directly to the Spanish colony in the fleet carrying the new viceroy of Peru, the French scientists would make several stops on their maritime voyage.

The agreement with Spain required the Geodesic Mission to debark in the French Caribbean colonies before progressing to Spanish territories. Maurepas gave orders for the forty-four-gun warship *Portefaix* to carry the scientists from the naval port of Rochefort, on France's southwestern Atlantic coast, to Martinique in the eastern Caribbean and then to Saint Domingue (modern-day Haiti). At Saint Domingue, they were supposed to find a Spanish vessel to take them to Cartagena de Indias, since foreign ships were prohibited in the Spanish colonies. But the normally careful Maurepas committed his first major blunder by ordering the governor-general of Saint Domingue, Pierre Marquis de Fayet, to persuade the Spanish authorities to allow a French ship to take them instead.

Maurepas gave Fayet the cover story that it would be quicker for the scientists to get to Cartagena de Indias on a French vessel instead of waiting for a Spanish one—but this ruse concealed the minister's predatory intentions. The real reason for sending the expedition on a French ship, as he told Fayet, was "to try to engage in some commerce with the Spanish, or lay the foundations for it, to which end it is necessary that a capable man [*homme de tête*] be placed in charge of its conduct." Maurepas knew the risks if Spain discovered this illicit trade, but such was the lure of its wealth that he was willing to renounce his explicit promises to Patiño, jeopardizing the Geodesic Mission as well as France's fledgling relationship with Spain for the chance to open the door, even if just a crack, to its colonies.[37]

As Maurepas dreamed of illicit trade with Peru, the preparations for the expedition were unfolding in Rochefort, as planned. The warship *Portefaix* arrived in Rochefort on March 23, 1735, to await the expedition's arrival. *Portefaix* was a type of ship called a *flûte*, similar to a frigate but designed to carry cargo and troops; it had been in service eighteen years, making it old for a wooden warship, and this would turn out to be its last voyage. The commander of *Portefaix*, Lieutenant Guillaume de Meschin, was just six years older than his ship. He carefully supervised the loading of cargo—primarily cannon and grain—destined for the colonies of Martinique and Saint Domingue and the fortress of Louisbourg.[38]

The intendant of Rochefort, François de Beauharnais, was busy making preparations for the arrival of the scientists, their aides, and four servants. Beauharnais would provide guns and swords for each man, as well as ammunition, powder, tents, blankets, surgical instruments, and utensils for their cook, a man named Saint-Laurent who "had spent five or six years in Peru and could pass as a Spaniard," according to Godin. Beauharnais also made arrangements for the expedition members' lodging at one of the best *auberges* in Rochefort, the *Trois Marchands*, with an impressive wine cellar and hearty fare that was popular with the navy crowd.[39]

The members of the Geodesic Mission arrived at Rochefort in dribs and drabs from mid-April to early May 1735. As they came from all over France, it would have been impossible to time their arrivals any more closely. Bouguer and La Condamine left Paris in late March, taking three weeks to traverse a route that the spring rains had made into a muddy soup. They arrived in Rochefort in mid-April, with Jussieu and Seniergues arriving a week later. Verguin's trip from Toulon probably took over a month. Godin (presumably with Godin des Odonais and Couplet) did not leave Paris until April 14, arriving by express carriage on May 7. Maurepas, eager to get the expedition moving, wrote the Spanish court in early April telling them that the French half of the mission was already in port and ready to depart—not quite the truth, but necessary to get the Spanish half of the mission to sea.[40]

It was perhaps Maurepas' impatience, coupled with his lack of experience in such scientific adventures, that caused him to make his second, more catastrophic blunder—this time regarding the mission's leadership. Maurepas had named Godin the sole leader of the Geodesic Mission, granting him absolute power over the other members of the expedition. To some extent, the minister's hands were tied in Godin's selection; he had proposed the original expedition, and of the three scientists he was the most senior member of the Academy of Sciences. Godin, however, was neither the oldest nor the most experienced member of the group, nor had he ever led a team of men in any activity. This alone was not cause for alarm; after all, he could have fallen back on the knowledge of Bouguer, who had been a professor since his teenage years, when he was accustomed to directing men twice his age. Godin could also have leaned on the military experience of La Condamine to ensure that the expedition members followed orders. As it turned out, once the mission began, Godin would arrogantly assume that he could take any action he deemed necessary without consultation and give any order without being questioned. His imperious style of leadership would have disastrous consequences for the entire enterprise.

The fact that Maurepas allowed Godin unchecked power might have been excusable, if only for the reason that nothing like the Geodesic Mission had ever been attempted before. The Geodesic Mission would be the largest group of scientists and observers ever embarked for the sole purpose of carrying out scientific measurements, as well as the world's first international scientific expedition, involving both diplomatic agreements between two nations and participation by members of both countries.

No previous scientific expedition of any country had involved international collaboration—let alone much collaboration of any sort. All prior expeditions under the banner of the French Academy of Sciences had been carried out by individual scientists, often on their own initiative. The astronomical voyages of Nicolas Couplet, Jean Richer, Louis Feuillée, and Amedée Frézier (as well as the Caribbean botanical expeditions of Charles Plumier in the 1690s) were effectively one-man shows that required a minimum of support and logistical planning and whose observations were in many cases carried out on French colonial soil.[41]

Strong leadership was a critical part of any unprecedented, complex expedition such as the one that the Geodesic Mission would soon attempt. The success of the mission would depend not on individual brilliance but on the close-knit cooperation of an entire team, all of whom were needed to carry out the extensive long-distance survey. It would require both personal leadership and diplomatic astuteness—neither of which was particularly encouraged in the culture of the French Academy. Maurepas, accustomed to having his orders carried out by military men, probably expected that his scientists—Godin included—would act in the best interest of both science and their nation, and put aside any personal aspirations for the good of all.

If Maurepas had confidence in Godin's leadership abilities, it was despite two warnings he received just before the mission began. The first was an incident that suggested Godin was guilty of mishandling finances. As expedition leader, Godin was responsible for both drawing and disbursing the expedition's funds from the treasury; he had already been given letters of credit to cover their expenses when they arrived in Spanish America, and all costs on French soil should have been taken care of by the French authorities. Yet Godin inexplicably demanded an additional 2,300 livres (roughly $20,000 today) from Beauharnais, the intendant of Rochefort, just before sailing; when Beauharnais complained to Maurepas, the minister simply brushed him aside, stating that he should give Godin any sum he asked.

Several weeks before the expedition was under way, Maurepas had received a more blatant warning from Bouguer about Godin's behavior. This time, the issue was Godin's treatment of his fellow scientists—a grievance that many of the team members would come to share. Although Bouguer's specific complaints are not known, Maurepas' impatient response suggests what they were:

> Versailles, April 30, 1735
>
> The worries you seem to have, Sir, about the manner in which Godin treats you are without foundation, other than it appears to me that he has profound feelings of esteem and friendship for you. I have recommended to him to get along [*vivre en bonne intelligence*] with you and your companions, and I am certain you will do all you must to maintain the union which is absolutely vital to complete your mission. It was unavoidable that Godin was specifically charged with the conduct of the mission, being the most senior; but regarding the endeavor, he could not and I believe he will never intend to refuse to communicate with you his observations, as it is vital that your [observations] become part of his during the course of the work. Although His Majesty intends that you economize [your costs], M. Godin nevertheless has orders to underwrite all necessary expenses, and you should have no worries on that subject. . . .
>
> [signed] Maurepas[42]

Bouguer was already calling into doubt Godin's intentions to share information with his fellow scientists and questioning his ability to handle the expedition's money. The other team members would soon begin to chafe under Godin's inept leadership as well.

These fractures in the French team were apparent when the scientists' cargo arrived at Rochefort by boat from La Rochelle on May 10. In addition to various scientific instruments, the men had packed clothes, books, sabers, munitions, tents, cooking utensils, and other supplies needed to sustain them in Peru. Lieutenant Meschin unhappily counted over sixty crates at dockside, all of which had somehow to be squeezed into *Portefaix*. "I was obliged to offload 140 barrels of grain for the colonies in order to load [their effects]," he petulantly noted in his log.

On May 12, two days after the scientists' cargo had arrived in Rochefort, the French team set out on *Portefaix.* The wind rose in the predawn hours, and at 10 AM *Portefaix* cleared Rochefort, warping out of the Charente River into the bay of La Rochelle. Shortly after noon the wind died, adding to Meschin's irritation, so the ship remained moored off Ile d'Aix for several days. At 10 AM on May 16, the wind freshened again, and under Bouguer's professorial guidance, *Portefaix* sailed out of the bay of La Rochelle and into the Atlantic, anonymous among the dozens of ships making their way to various ports around the globe. By 6 PM on May 16, when the *flûte* was twenty-four miles to sea, the crew members sighted a whale—the first good omen they'd had in weeks.[43]

As the French members of the expedition put to sea, their Spanish counterparts were still preparing for their voyage in Cadiz. The imminent departure of the new viceroy of Peru made the warships on which the Spaniards would be traveling, *Nuevo Conquistador* and *Incendio*, the center of much attention and to-do in the port city; Villagarcía was accompanied by his son Mauro de Mendoza and forty members of his court, all of whom had to be accommodated in appropriate style. The newly minted lieutenants Jorge Juan and Ulloa, meanwhile, brought clothes, instruments, other goods, and servants for their voyage. Unlike Ulloa, who came from a well-to-do family, Jorge Juan was of modest means and took yet another loan from his uncles to pay for these accoutrements, leaving behind a total debt of almost 1,600 pesos (about $72,000), which he fully expected to repay when he returned.

Early on May 26, 1735, just as *Portefaix* was passing the islands of Madeira southwest of Portugal, Jorge Juan and Ulloa boarded the two warships now anchored on a sandy bottom in the bay of Cadiz. Jorge Juan boarded *Nuevo Conquistador*, on which Villagarcía and his entourage were traveling; Ulloa was assigned to the smaller *Incendio*. The departure of the viceroy and his court was attended with the requisite pomp and circumstance, as Ulloa noted in his log: "At 9AM Villagarcía, the Viceroy of Peru, boarded *Conquistador*, at which he was saluted with 21 cannons. At 9:30 the commandant of the ship was greeted with a salvo of 11 cannons. . . . At 11 AM the wind brought the frigate to a heading of south-southwest." The two ships spent that day and the next maneuvering in the bay of Cadiz, until on May 28 a northwest wind came up, allowing them to point their bows southwest toward America.[44]

In the spring of 1735, three warships were sailing westward on the Atlantic, bearing the members of the Geodesic Mission to their rendezvous in Cartagena de Indias. Godin, Morainville, and Verguin had undoubtedly comforted their wives and children by telling them they would only be gone three or at most four years; the other members must have told each other the same. The dangers, the men believed, were few. Peru must be a civilized place; after all, the Spanish had been there for exactly two hundred years. The scientists knew their work would be difficult, but they did not foresee any major impediments to carrying out their survey and returning home with the single number they sought: the length of a degree of latitude at the equator. The fact that the expedition members were well armed and had set sail on warships was thought of as prudence; in reality, it was a harbinger of things to come.

III

Finding Quito

Mariners have heard the call of the sea for millennia, knowing that each one answers it in a different way. Jorge Juan y Santacilia and Antonio de Ulloa had answered the call early in their youth, even before they were naval cadets. The sea had become their home, their school, their church, their workplace. On many voyages, under the gaze of experienced sailing masters, they had learned its rhythms, its subtle moods and sudden tempers. In spite of their youth they were now full lieutenants, a fairly high rank on Spanish warships, and were in their turn guiding less-experienced sailors in the ways of the sea.

The official duties of both Jorge Juan and Ulloa kept them busy every hour on *Nuevo Conquistador* and *Incendio*, scrutinizing the sails and helm, continuously ordering adjustments to maintain course and speed. Ulloa had recently made the passage to Cartagena de Indias and knew these waters well; both he and Jorge Juan were already skilled navigators, daily taking sightings of the sun and the stars to determine latitude, using a combination of astronomy and dead reckoning to estimate their longitude, and noting them daily in the ships' logbook. Ironically, it was their training for a life at sea that had led to their selection for a mission in the Andes, hundreds of miles from the ocean.

Unlike Jorge Juan and Ulloa, the scientists aboard *Portefaix* were merely passengers with no official duties, and the sea affected them in unexpected ways. La Condamine and Verguin were both seasoned mariners accustomed to life aboard a warship; indeed, Verguin had surveyed the Caribbean fifteen years earlier and knew the route well. The two men busied themselves with the regular shipboard routine, each one keeping a personal journal of the ships' position and winds. For the other expedition members, who had never been on a long sea voyage, the combination of crowding, constant motion, and sheer boredom quickly took its toll. *Portefaix* was not a large ship—at roughly a hundred twenty feet long by thirty feet wide, it was smaller than a town church—but it nevertheless carried five officers, a hundred thirty crew, and over a hundred passengers on top of the heavy load of cargo, making the ship extremely cramped. Passengers slept in shifts belowdecks, in hard little berths (only crewmen slept in hammocks) that were tucked into every nook and cranny.

Portefaix's first day or so at sea was pleasant enough, with land still in sight and just a gentle roll to lull the passengers. But as the ship made its way into the open ocean, the incessant pitch and heave would often drive them belowdecks, where they were engulfed in the fetid atmosphere and constant moaning of ship's timbers and sick passengers. Seasickness affected Bouguer most of all; despite being one of the most renowned professors of hydrography in the country, he had never made a transatlantic voyage. Still, he put on a brave face and did his utmost to stay busy during the month-long journey. He, Godin, and La Condamine spent much of their time with an assortment of nautical instruments that the Academy of Sciences had given them to test. Hadley's octant—which Godin had bought in London—received particular attention from the three. Sighting the sun through finely ground mirrors using carefully calibrated brass fittings, it gave considerably more accurate and repeatable results for calculating latitude than did the traditional backstaff, which was little more than a wooden cross that marked the sun's shadow on a series of whittled marks. Bouguer, ever the professor, also spent time training a vice admiral aboard the ship for the officer's pilot certificate.[1]

Jussieu, the meek and melancholy doctor, found the sea air to be a powerful new medicine that invigorated him and lifted his mood considerably. While ashore in Rochefort he had been in a morose state, awaiting letters from his older brothers, to whom he was deeply attached. Now, out of sight of land, the family's youngest son experienced something like a sense of liberation as he began to make his own way in the world. He had his close friend Seniergues with him and was not even troubled when Seniergues' dog bit him on the hand, causing it to redden and inflame. Jussieu applied a compress soaked in a mixture of water, alcohol, and herbal medicines, noting that "in less than three days, thanks to God, my hand was better and I was doing very well."[2]

The long voyage gave the expedition members time to take the measure of each other. The minor members banded together against the condescending treatment by Godin, who according to Seniergues, "wished to erect himself as the Grand Master."[3] Bouguer and La Condamine, already wary of Godin, also banded together in an unlikely friendship. They had barely known each other at the Academy of Sciences and were polar opposites in almost every way. Bouguer was a serious, almost somber man, passionate about science and mathematics but with a limited worldview, having spent much of his life teaching hydrography in a small coastal town. La Condamine, who had traveled around the ancient world as well as the intellectual circles of Paris, saw math and science as part of a much broader canvas, not taking them at face value but rather looking always at the human side of the equation.

While neither Bouguer nor La Condamine mentioned it in their journals, it is likely that they would have compared notes about the recent scandal involving their playwright friends Paul Desforges-Maillard and Voltaire, which was just being resolved when they set sail from Rochefort. Bouguer's childhood friend Desforges-Maillard, trying to break into the literary scene after a career in law, had been unable to get his poems recognized or even published, so he began submitting them to newspapers under the pseudonym Mademoiselle Malcrais de La Vigne. In a short while "she" became known as "The Breton Muse" and plaudits came from across the literary hemisphere, especially from La Condamine's good friend Voltaire. When the Muse's identity was revealed, it caused great embarrassment among those who had previously rebuffed Desforges-Maillard. Voltaire, amused by the clever deception, took it all in stride and even graciously offered to help Desforges-Maillard in his dealings with the tax authorities.[4] If the two playwrights could see past their differences, then perhaps their friends Bouguer and La Condamine could do so as well.

The crossing of the Atlantic was uneventful. The warship was becalmed just a few days, and those aboard saw only a handful of other ships on their journey. On June 11, *Portefaix* crossed the Tropic of Cancer, where the scientists were subjected to the time-honored ritual of the equator-crossing baptism, in which experienced crew members initiate newcomers into the close-knit world of seafarers; for even though the scientists had not yet reached the equator, they "were destined to do so."[5] On June 20 the crew sighted numerous frigate birds, a sure harbinger of landfall. At 4:30 AM on June 22, as the island of Martinique hove into view across the predawn mist, it must have seemed to the relieved passengers like Columbus's first moonlit view of the New World.

The warship sailed around the island to the capital Fort Royal, where it would moor for two weeks while Lieutenant Meschin offloaded cargo and took on supplies. While *Portefaix* was being readied for the next leg of the voyage, the expedition members were lodged in various locations around the island according to their rank. Bouguer stayed at the grand home of the intendant of Martinique; Jussieu, on the other hand, opted to stay with the Frères de la Charité, a church-run hospital that ministered to the poor, where he turned his medical talents to good use.[6]

The scientists had no official duties during their stay on the island, and they used the opportunity to adjust to the climate of the New World and take in the sights. Martinique was one of France's oldest Caribbean colonies, founded exactly a century earlier to grow the coffee and sugar that supplied France's burgeoning café society. Bouguer and La Condamine inspected the island's sugar-making industry and coffee plantations, the latter of which owed their existence to several coffee plants sent to Martinique by Antoine de Jussieu, Joseph's brother, in 1722.

Joseph de Jussieu, meanwhile, was enthralled by the bounteous variety of tropical fruits and plants in their native setting, which he had previously seen only within the sterile confines of the Royal Garden. He was pleased to find himself unaffected by the oppressive heat, which made the others flag and weaken: "The high temperature of this land, though quite intense, is not at all intolerable and I believe I shall become quite at ease in the tropical climate," he wrote his brother Antoine. Wearying of sightseeing, the scientists began creating their own make-work projects, climbing and measuring the already-measured Mount Pelée, recalculating the well-established latitude and longitude of Fort Royal, and sending back samples of plants that the Royal Garden already possessed.[7]

By the end of June, Meschin was ready to make sail from Martinique for Saint-Domingue, where the expedition would have to find another ship to take them to the coast of South America. The scientists' departure from Martinique would not go as smoothly as their arrival, however. Godin, in a manner that was to become irritatingly common, requested an additional 600 livres from the intendant to pay for "expenses." Then, on the eve of their departure, one of *Portefaix*'s passengers, a Swiss Guard, came down with yellow fever, and La Condamine showed the symptoms the next day. The ship would not wait for him to recuperate fully, so he was "bled, purged, cured and placed on board within twenty-four hours."

Portefaix departed Fort Royal the evening of July 4, wearing north-west for the short hop to Saint Domingue, on the island of Hispaniola. A few days later Meschin noted, "I found myself on the coast of Saint Domingue which was covered in fog," concealing the harbors and navigation hazards. The ship spent three days coasting before finally making landfall on the 11th at Fort Saint Louis. This was not the scientists' final destination on the island; that was Petit Goâve, at the time the colonial capital of Saint Domingue, on the other side of the southern peninsula of Hispaniola.

As *Portefaix* waited out the rare summer fog at Fort Saint Louis, the expedition members began what would prove to become a dangerous habit: splitting up, even in the face of peril, simply to satisfy individual whims. *Portefaix* left its temporary harbor on the 21st without Godin and La Condamine, who had decided to trek overland across the peninsula to make some astronomical observations.[8] It was a minor excursion over well-traveled territory, but it was proof of Godin's haphazard leadership; already deep in a strange new land, he had sacrificed the cohesion of his team to pursue a superfluous goal.

Godin and La Condamine reunited with the rest of the French expedition members in Petit Goâve on July 29, to wait for a ship to take them to their rendezvous in Cartagena de Indias. The governor-general of Saint-Domingue, the Marquis de Fayet, had been expecting them, having received orders from Maurepas to assist them in every way possible. Fayet gave orders to offload the scientists' effects, find the men appropriate housing, and provision them with servants, slaves, and additional equipment for the rest of their journey.

As Fayet was preparing the expedition for their arrival on the South American continent, he also searched for a ship to take them there. He wrote to the captain-general of Santo Domingo (the Spanish half of the island of Hispaniola), asking for a ship to transport the scientists to Cartagena. The captain-general replied that he did not have a ship available but would allow a French ship to take them. This fit nicely with Maurepas' instructions to secretly trade with the Spanish using a French ship, but Fayet soon discovered he could not find a vessel large enough for the scientists and their equipment. "In spite of all the measures I have taken for their voyage," he complained to Maurepas, "nothing is more difficult than finding a ship big enough to carry all their baggage."[9] *Portefaix* could not be pressed back into service, as it was scheduled to depart on August 11 for Louisbourg, and the other ships in port were simply too small. The scientists would have to wait.

During the scientists' enforced sojourn in Saint Domingue, they set about selecting servants and slaves to accompany them to the continent. Bouguer, accustomed to being waited on, as were all gentlemen of that era, had lost his servant to disease, so to replace him he hired a clever man named Grangier, whom he would teach surveying and who would later return to the colony of Saint Domingue as a royal surveyor. The scientists also bought three slaves to carry their equipment and perform other tasks. In that era, buying human beings was little different from buying cattle. In Saint Domingue, the process was especially easy, as the colony had the largest slave population in the Caribbean, well outnumbering the whites on the island; Petit Goâve alone had six hundred whites and two thousand slaves.[10] Even after a century of plantation farming, France still had to import huge numbers of slaves every year from West Africa, since they died on those plantations—through sickness, overwork, and beatings—far faster than they could give birth.

None of the scientists left any record of their thoughts toward the indescribable cruelty meted out by plantation owners for even the slightest infraction. They had almost certainly never seen this level of brutality; indeed, while living in France they probably had never seen slavery firsthand. In theory, any slave coming from the colonies to mainland France was immediately set free—"there are no slaves in France" went the dictum. Nevertheless, some slaves brought to the major slaving ports of Nantes and Bordeaux were not emancipated but rather put to work behind closed doors in the role of domestic servants. For the most part, the French government preferred to keep blacks within the colonies and away from the eyes of the French people, effectively disguising the real source of their country's wealth and power.[11]

Not all Africans on Saint Domingue were slaves. One in particular, a free woman named Bastienne Josèphe, owned a brothel where Louis Godin soon found himself a privileged guest. He favored one prostitute named Guzan, whose tastes ran to the extravagant. It was hardly surprising that Godin took the very first opportunity to set on adultery; one out of seven nonslave women in the colony was a prostitute, and as a contemporary observer put it, across the island "one is surrounded by a most dangerous seduction, that which cannot turn away desire."[12]

Of itself Godin's adultery was no real cause for alarm, but he soon began spending the expedition's money on his philandering. He contributed to the cost of Guzan's upkeep, even ordering his draftsman, Morainville, to paint portraits of Guzan and Bastienne Josèphe. The last straw came when Godin used the expedition's rapidly diminishing funds to buy a diamond for his paramour worth 1,000 ecus (3,000 livres or about $27,000 today). This shocked even the mild-mannered Jussieu, who until that point had been somewhat well disposed toward Godin. In a letter to his brother Antoine, Jussieu wrote, "Mr. Godin our chief . . . has for some time set astronomy aside to take care of more pressing affairs. He's completely occupied with love; I hope his wife doesn't learn of the infidelity of her Adonis, she will perhaps have a mind to take revenge. . . . A lot of money is being spent to satisfy the appetite of the young lady and her knight refuses her nothing. . . . We all see this and suffer, and as for me I shrug my shoulders and think of botany."[13] Godin later denied buying the diamond, claiming that the 1,000 ecus was simply the amount of money in his hands as he was paying off the expedition's lodging. It was a transparent lie, but the other members of the expedition had already made up their minds about him.

Godin's behavior had become increasingly outlandish since the expedition reached the New World, polarizing the expedition even more than it had on the Atlantic passage. While still in Martinique, Seniergues had written a letter back to France stating that he, Couplet, Morainville, and Hugo were already on the point of quitting the expedition as a result of foul treatment from Godin. The mission's leader also denied the minor members the funds they needed for basic expenses, forcing them to turn to their trades to eke out a living instead of attending to their mission. Verguin, an experienced military officer, was the only junior member of the expedition who stood up to Godin and did not jump when he wagged his finger. La Condamine and Bouguer, ostensibly Godin's peers, saw what was going on but did nothing to rectify the situation.[14]

Rather than taking steps to ameliorate the ill will spreading among his team, Godin turned the accusations back onto his subordinates. He defended himself in letters back to France, claiming that he had good reasons for taking the actions he did but without ever specifying what those reasons were. He also petulantly complained that the suspicions and gossip were insulting to him.

While he was alienating himself from the other members of the expedition, Godin was also making a poor impression on the colonial rulers. One aristocrat noted that the scientists "are taking a lot of money with them to South America, [and] spent a fair bit in Cap François before leaving. . . . If they do not leave off with their haughty airs they run the risk of finding themselves in some unpleasant situations."[15]

As their three-month stay in Saint Domingue wound down, Fayet himself complained to Maurepas. He told the minister that the colony had spent almost 30,000 livres on the expedition, including 17,000 paid directly to Godin, and if one counted food, housing, and so forth, Fayet claimed, the government of Saint Domingue had been set back 150,000 livres, well over a million dollars today. Upon receiving these letters and other news about the goings-on of the expedition, Maurepas realized that it was going to cost a fortune to measure the Earth. This was due in large part, he must have realized, to his decision to allow Godin such unfettered powers, and the concomitant failure to take the simple precaution of having one person hold the money as treasurer, with a separate person accountable for payments. But it was far too late to do anything about this; Maurepas would have to trust to providence that the expedition would not turn into a fiasco.[16]

The members of the Geodesic Mission whiled away their time as best they could as they waited for their ship to be readied. Godin, La Condamine, and Bouguer sent separate letters to France on their experiments with Graham's pendulum, reports that were printed in the 1735 memoirs of the Academy of Sciences. Jussieu spent his time botanizing and collecting seeds to be sent back to his brothers. His mood, which had markedly improved during his ocean voyage and stay in Martinique, had descended into depression; Godin wrote that Jussieu had a "somewhat black humor," and Seniergues, increasingly solicitous of his friend, worried that he had "entered into melancholy."[17] Today Jussieu might have been diagnosed with bipolar disorder, a serious mental illness marked by alternating periods of elation and well-being punctuated by bouts of depression and anxiety. Mood swings can be triggered by changes in environment, stress, or unknown causes; in this instance, Jussieu's depression may have been brought on by the increasing friction with Godin that was affecting the rest of the expedition as well. For Jussieu and the others, the initially pleasant sojourn on Saint Domingue was turning into a seemingly endless ordeal.

Relief for the strained expedition members came on September 30, when the sixteen-gun brigantine *Vautour* arrived from Rochefort under the command of Lieutenant Louis du Trousset, Comte d'Héricourt. It was quickly contracted for the scientists' voyage. *Vautour* was half the size of *Portefaix*, barely large enough to carry the scientists and their equipment, along with a dozen other passengers and a garrison of Swiss soldiers, but, given the circumstances, it would have to do. Héricourt, on confidential instructions from Fayet, spent the month of October loading *Vautour* not just with the scientists' baggage, but also with textiles and other contraband merchandise, secreted in the lower holds, to sell in Spanish America.

On October 31, the expedition left Petit Goâve on what would turn out to be a tedious two-week voyage to Cartagena de Indias and the rendezvous with the Spanish lieutenants Jorge Juan and Ulloa, who had now been waiting since July for their arrival. La Condamine was spared from monotony when he recognized the ship's second officer, Jean-Baptiste Sinetti, as an amateur poet of some repute, as well as a mutual friend of Voltaire. La Condamine later wrote Voltaire that he and Sinetti "spoke incessantly of you during the entire trip. It was our only pastime, apart from reciting passages from [your plays], to relieve our boredom."[18]

In contrast to the amateurish exploits of the French scientists, the voyage of the two Spanish officers had been professional, efficient, and uneventful. *Nuevo Conquistador* and *Incendio* had sailed in close company across the Atlantic, averaging five knots or around a hundred twenty nautical miles per day. They were becalmed several days; Ulloa noted on June 7 that they had traveled only sixteen miles in twenty-four hours, the pace of a leisurely stroll. On July 2 the blue of the open ocean gave way to the emerald green of coastal waters, and on July 7 the two ships passed through the Boca Chica entrance to Cartagena Bay, the roadstead of the great city of Cartagena de Indias.[19]

Nuevo Conquistador and *Incendio* remained at anchor for two days while the local government prepared for the official arrival of the new viceroy of Peru. Finally, on July 9, all was ready ashore. First, at 6 AM, the commander of the city fired a nine-cannon salute that was returned by a seven-cannon salute from *Nuevo Conquistador*. The ships slowly moored to the quay and began disembarking officers, crew, and baggage. At 4 PM the Marques de Villagarcía majestically set foot on the ground of his new kingdom, to shouts of "*Viva el Rey*" and cannonades from the city and the two ships.[20]

While Villagarcía and his entourage were being shown to their lodgings, Jorge Juan and Ulloa found the governor of Cartagena de Indias and asked if "he had any notice of the Parisian scientists." With no word of them yet, the two officers proffered their services to map the city and establish its longitude, as required by their instructions from Patiño. After being lodged in a splendid seafront house, they set to work surveying the city during the day, while making astronomical observations at night.[21]

Ulloa acted as Jorge Juan's guide to Cartagena, assisted by memories of his 1730 visit to the magnificent walled city. It was now over two hundred years old, fortified and rebuilt several times at enormous expense. In the 1580s, King Felipe II was said to have complained that he should be able to see the city ramparts all the way from Spain, since they had cost so much. In its heyday Cartagena had been at the heart of the gold and silver trade from Peru to Spain, but since 1700, the arrival of treasure galleons had slowed to a trickle, as the mines were becoming depleted. Although it remained the Caribbean hub of the Viceroyalty of Peru, Cartagena's reduced flow of official colonial trade (that is, ships carrying treasure as well as supplies) meant that much of the commerce was contraband, the smugglers almost always in collusion with the local administrators.[22] Still, the city's importance both strategically and commercially made it a likely target for invasion. The last major attack had been in 1697 by French privateers who had launched their assault from Saint Domingue, and foreign warships were still eyed warily when they entered the harbor.

Cartagena's checkered past was well known to Villagarcía, who fully understood that his foremost preoccupations as the new viceroy would be clamping down on smuggling and defending his kingdom against invaders. His layover in Cartagena was necessarily brief but crucial. The city was, in effect, the capital of the northern region of Peru, comprising modern-day Colombia and Venezuela, and was still only loosely controlled from Lima (in fact, for a brief period from 1717 to 1723 the Spanish crown had split off the northern region of Peru to form the Viceroyalty of New Granada). Given the distance and dangers of the voyage, Villagarcía was not certain to come this way again, so this visit gave him the opportunity to take stock of the situation in the city, at the same time firmly establishing his leadership in the minds of the local government officials.

After two weeks Villagarcía and his fleet were ready to depart on the next phase of his voyage to Lima. On July 25, he boarded *Nuevo Conquistador* to make the trip to Portobelo for the crossing of the Panama isthmus and then to travel down the coast of his new kingdom, stopping at various ports and trekking overland until he arrived in his capital city, Lima. The trip would take him half a year. As *Nuevo Conquistador* prepared to depart, Jorge Juan and Ulloa came aboard one last time to pay the viceroy their respects. By all appearances, they had formed a close bond with him during the long voyage and the brief stay in Cartagena, and they had succeeded in winning his support for their mission. They did not know if they would see him again—their plans did not include a trip to Lima—so they said their final good-byes.[23]

In September, Jorge Juan and Ulloa finished their survey of Cartagena and presented their maps to the city's engineers. Two months had passed without a sign of the French scientists, so the Spaniards now had no official duties, apart from writing occasional letters to Patiño to keep him informed of their wait.[24]

It was at this time that Jorge Juan and Ulloa began keeping personal journals of their voyage through the New World. Their original instructions from Patiño asked only for "a logbook of their navigation . . . plans of cities and ports, their latitude and longitude . . . and observations of various plants or herbs." But from early on the officers decided to also record "descriptions of the cities and customs of their inhabitants; . . . descriptions of civil government and politics, especially concerning the Indians or indigenous people . . . and in general, what is done with the goods from Europe, distinguishing between legal and illicit commerce."[25]

Jorge Juan and Ulloa's notes were to be far more than the usual travelogues, marking the two young lieutenants as exceptional observers, with a perceptiveness far beyond their age. This did not mean, however, that they were unbiased. They had been molded as naval officers in the Academy of Naval Guards, one of the premier educational institutions of the Enlightenment, and they had been trained to believe that scientific, rational thought was the means to solve problems in the social, as well as mathematical, realm. They had been raised in an era, moreover, in which political reform was in the air, especially as concerned the colonies, where the Indian populations were subjugated by both the church and commercial landowners. Antonio's father, Bernard de Ulloa, was one of the most vocal advocates of reforming the system, and this had naturally colored his son's views.[26] As they took in the sights of Cartagena, then, the two officers were on the lookout for corruption, dishonesty, and oppression of indigenous peoples.

The Spanish officers' enforced leisure in Cartagena gave them the opportunity to closely observe the inhabitants of Peru, and the two men were struck by what they found. Although ruled by the Spanish crown, the population of Peru had undergone two hundred years of cultural evolution that had made them as different from Spaniards as modern-day Americans are from their former British rulers. The first thing Ulloa and Jorge Juan noticed was the complex social hierarchy that had evolved among the different races of people, layered onto the usual differentiations of nobility, region, and profession.

The distinction between white, black, and Indian was actually quite finely grained in colonial Peru. Whites were at the top of the order, and at the pinnacle were the "Spanish Spaniards" or *chapetónes*, who arrived from the home country, usually to fill some administrative post or "acquire a moderate fortune and return into Spain," as Ulloa noted. They were also the least numerous.[27]

On the next rung down were the *criollos*, native-born Spaniards, the landed gentry who "owned property and estates, and were highly respected since their ancestors came to these parts with noble titles."[28] They held themselves to be superior to the *chapetónes*, though at first encounter, the only thing that distinguished them was their accents. Often highly educated and fiercely proud, their relationship with the Spanish immigrants was often marked by a combination of admiration, jealousy, and disdain. Even though *criollos* were endowed with solid classical formation as well as local knowledge, the Spanish crown almost without exception chose *chapetónes* to occupy the high political posts.

Blacks were on the tier below whites in the social hierarchy of Peru. Most (though not all) were slaves, and their presence would have been familiar to Jorge Juan and Ulloa. Unlike in France, slavery was openly practiced in Spain, especially in Ulloa's home region of Andalusia, which, by some accounts, had as many blacks as whites. Although intermarriage between blacks and whites was unheard of in Spain, it was occasionally practiced in Spanish America, where slaves did not necessarily work the fields but rather formed part of the whites' households. Children of such unions were called *mulattos*, with specific designations for the amount of "African blood" assumed to be in their veins.

Indians were at the very bottom of the Peruvian social hierarchy. Even though they were not technically slaves, in fact they were press-ganged by whites to work in mines, mills, and plantations, often under unspeakable conditions and subject to horrific physical abuse. The *cholos* or *mestizos*, mixed white and Indian, were somewhat higher on the social ladder than pure-blooded Indians, but still ostracized by whites. At the time the Geodesic Mission was in Peru, they made up a third of the total population and were already the majority in many urban areas.[29]

The social stratification between the races in colonial Peru was almost incomprehensible to an outsider, as was the exceptional cruelty to the Indians on the lower rungs. To the early Spanish conquistadores in search of fortune, Indians were either helpers or impediments to their quest and no different from domestic animals; to starve or beat an Indian into obedience was the same as breaking a horse or bringing a dog to heel.

In the 1500s the Dominican friar Bartolomé de las Casas, working around the Caribbean, Mexico, and modern-day Central America, had been the outspoken critic of the treatment of the Indians, leading a royal commission that attested to the destruction of the Indian nations and the enslavement of their peoples. In 1552 he published a famous and widely translated book, *A Short Account of the Destruction of the Indies*, which made public the horrors—most real, some imagined—that accompanied the rule of the conquistadores in the region from Mexico to Chile. Both his private testimonies and public chronicle (illustrated with graphic depictions of Indians being beaten, hunted and slaughtered) were instrumental in ending the slavery of the Indians, although this was eventually replaced by a system that was just as brutal.[30] Jorge Juan and Ulloa no doubt saw themselves as the natural heirs to Casas in their passion for reformation, which would drive them to write their most scathing criticisms of Spanish American culture.

Jorge Juan and Ulloa's social observations were interrupted the night of November 15, 1735, when the French members of the expedition finally arrived aboard *Vautour*, which anchored off Boca Chica. The following morning the officers went out to the ship to greet their new teammates. As they came aboard, they found a completely different group from what they had expected. The enumeration Jorge Juan and Ulloa had been given by the Spanish government was woefully out of date, listing only five members: Godin, La Condamine, Grandjean de Fouchy, La Grive, and Pimodan. Only two of the originals were now aboard, and the full number of people was double the original. Very well—let the authorities sort it out. This was a moment for science and not bureaucracy. With the team finally assembled, the men made their acquaintances.

First impressions of the Spaniards were mixed. Jorge Juan and Ulloa spoke passable French, as was expected of all educated men in Europe, and Jussieu found the two men "friendly gentlemen, very engaging personalities, quite sociable, well-born and very knowledgeable of mathematics." La Condamine, a former soldier against the Spanish, was more condescending: "We are arriving in Peru with men who carry their love of Physics as trinkets." For their part, Jorge Juan and Ulloa did not record their impressions, simply announcing the Frenchmen's arrival in a letter to Patiño.[31]

Now that the entire company was assembled, the men discussed the next phase of their voyage. It was not immediately apparent that they would take the sea route to Quito, as originally planned; there were many arguments in favor of the overland route down the Andes. Eventually, the group decided to stick with the original plan and proceeded to collect the additional funds needed for the journey. Godin sought out the Cartagena correspondent of the French banking firm Casaubon, Béhic et Compagnie, with whom Maurepas had arranged a 4,000 peso line of credit. The team was down to 9,000 livres after Godin's profligate spending in

Saint Domingue, so these new letters of credit would give the expedition an equivalent of $260,000 to see them through the foreseeable future. For their part, Jorge Juan and Ulloa received their back pay from the Spanish treasury.

The group reassembled aboard *Vautour* on November 24, eight days after the scientists had arrived, and departed for Panama, arriving in Portobelo on the 29th. From there, they would portage across the isthmus before making their way south along the west coast of the continent. Six months had passed since the men had departed from Europe, and they were not even in the Pacific.[32]

Having arrived in Panama, the expedition had to wait yet again as they prepared for the next leg of their voyage. Only forty-five miles separate the Atlantic and the Pacific at the narrowest portion of the Isthmus of Panama, but the steep mountains and incredibly dense jungle made the passage between the oceans a grueling two-week journey by riverboat and mule. After debarking the baggage in Portobelo, the group was once more delayed for several weeks until boats could be found and loaded. Meanwhile the Panama authorities, obeying to the letter the original instructions of the Spanish court, measured each crate to the exact *vara* and *pulgada* (yard and inch) and then completely unpacked and inventoried the entire lot. They checked for contraband cargo and noted the presence of six slaves and seven servants—five for the French scientists, two for the Spanish officers.[33]

As the expedition waited for their passage across the isthmus, they found little to please them in the wretched port town. Portobelo was justly famous for its spectacular trade fairs, held every few years when the treasure fleets arrived to transship gold and silver to Spain. For two months the city came alive with Spanish, British, and Dutch merchants, selling the latest in European clothes, textiles, foodstuffs, and other goods to eager buyers who would resell them at a handsome profit to the fashion-starved colonists in Peru. The rest of the time, however, Portobelo was just a squalid market town at the edge of the jungle, where snakes, pigs, and the occasional panther roamed the streets. Jussieu complained of the interminable wait, stating that although the city had a good harbor, it was "the most unseemly and unhealthy place in the universe."

Morainville became ill, the food was horrible, and the group was spending the equivalent of $1,000 per day on subsistence.[34]

A month after their arrival, the group finally found two large, flat-bottomed boats and began the journey up the Chagres River, partly tracing the route of the present-day Panama Canal. At the town of Las Cruces, they switched to the well-worn trail normally used to transport Peru's silver and gold from the Pacific to the waiting treasure fleet in the Caribbean. Arriving on December 29 in Panama, the colony's capital city on the Pacific coast, the travelers discovered they would have to wait again, this time for a ship to take them to Peru. Meanwhile, the Frenchmen began learning Spanish. Jussieu in particular was a quick study, soon able to "speak well enough to be understood."[35]

As the expedition waited for passage down the western coast of South America, a blunder by Maurepas—this time out of greed rather than haste—almost scuttled the Geodesic Mission for the second time. Unbeknownst to the French scientists, *Vautour* was still in Portobelo, where Héricourt sent a landing party ashore to sell his contraband cargo of expensive textiles. The men were quickly caught and imprisoned, leading Héricourt to capture several Spanish sailors in reprisal. After a hastily arranged prisoner exchange, Héricourt left Portobelo and sailed back to Cartagena de Indias, where he began selling his merchandise to the local Indians. The authorities there were not fooled when he claimed that the various crates and boxes were simply additional baggage for the scientists; they sent word down to Quito to intercept the expedition en route, so as to make certain the team members' baggage did not contain smuggled goods.

The scientists knew how critical it was to be on good terms with the colonial government, but now, even before they had arrived in the city, the officials in Quito suspected the Frenchmen of smuggling. The expedition suddenly had a huge black mark against it, without the members even suspecting what had happened. Héricourt was blissfully unaware of the damage he had done. He returned to Saint Domingue, where word of his success was sent by the commissioner of the colony to Maurepas: "If the British or Dutch had found themselves in a similar position, they would have seized the opportunity for such easy access to such a rich land. . . . The brave Indians who have been subjugated by the Spanish would support such an advantageous trade with France."[36]

The scientists waited in Panama for two months before a merchant ship was found to take them to Guayaquil, the only coastal gateway to their inland destination of Quito. On February 22, 1736, they set sail in *San Cristobal*, under the command of Juan-Manuel Morel. By now all the expedition members were seasoned *matelots*, the French term for sailor (originally meaning "bedmate," since one berth was shared between two men who stood alternate watches on deck while the other slept). They saw that the ship's master, though experienced, could barely take star sightings and was careless in the handling of the ship. Fearing for their safety, they took matters into their own hands. "We kept port and starboard watches," remarked Ulloa, "and our servants manned the helm when the helmsman fell asleep at the tiller."[37]

The unease aboard *San Cristobal* was exacerbated by the tension between Godin and the other team members, which had only gotten worse since Saint Domingue. There were interminable arguments over how they would carry out their mission. In particular, La Condamine and Bouguer insisted, over Godin's objections, on putting ashore at the port of Manta, on the northern Pacific coast of Peru close to the equator, to make astronomical observations. Godin, on the other hand, wanted to continue straight to Guayaquil, the main port on the northern coast of Peru and the entrance to the only passable route to Quito. In a letter to Jussieu's brothers Antoine and Bernard, Seniergues recounted La Condamine and Bouguer's plan to stop at Manta, and he went on to reveal just how disastrous the state of relations had become: "Master La Condamine has already said in front of everybody that if no one else wants to stop there he'll stop on his own and if so Master Bouguer will surely stay with him. Master Godin and the others have not spoken for some time. They fight like cats and dogs and hide their observations from one another. It is not possible that they can finish this journey together." Before their mission had even begun in earnest, the expedition already seemed about to come apart at the seams.[38]

The bickering continued as *San Cristobal* sailed south along the coast toward Manta. The ship crossed the equator on the evening of

March 7–8, and on the afternoon of March 10 it dropped anchor in the port of Manta. At this point, the expedition completely broke in two, a rupture from which it never fully recovered.

Upon reaching Manta, La Condamine and Bouguer insisted on remaining in the region for an extended period, whereas Godin—who had reluctantly agreed to a layover—was determined to immediately sail farther south to Guayaquil, where they would take the only open route inland to Quito. Bouguer and La Condamine believed the stay was necessary to carry out their agreed-on orders from the Spanish government to "determine the precise position of the coast of Peru," a task they hoped to complete as close to the equator as possible.[39] They were also convinced that the whole geodesic survey—laying out a long baseline and triangulating for a distance of over two hundred miles—would be easier on the coastline than in the mountains. Godin, for his part, did not want to deviate from the itinerary that all the members, as well as the governments of France and Spain, had originally agreed to.

The split between Godin on one side and Bouguer and La Condamine on the other had been long in coming. While still in France, it had been apparent that the expedition could not function under Godin's incompetent leadership and ham-handed treatment of the other members. They had almost revolted while on Martinique, and Godin's escapades on Saint Domingue, in particular his lavish treatment of Guzan at the expedition's expense, lost him all trace of credibility. Now the issue of coastal observations was simply the last wedge in the fissure.

The other members of the expedition found themselves straddling the fault line between Godin and his two Academy colleagues. Now that they were in Peru, the minor members had little choice but to follow Godin (his name and not theirs was on the passports, and he still held the expedition's purse), but Bouguer and La Condamine felt they could separate from the group and strike off on their own. Jorge Juan and Ulloa, undoubtedly shocked at the amateur behavior of such distinguished scientists, attempted to patch things up between the three men. Although the Spaniards' intervention failed, their impartiality nevertheless won them the confidence of all the French members of the expedition.

The three French scientists were now so livid that they could only communicate by letter. After only two days reconnoitering the area around Manta, Bouguer wrote a memoir to Godin outlining the advantages of performing their geodesic survey on the coast instead of down the Andes. The land was flat and easily traversed, he argued, and using boats instead of mules "affords us various means of simplifying our work, and more specifically, our transportation becomes much easier." It would appear that Bouguer's appraisal was too optimistic and more of a rebellion against Godin than an objective assessment of the team's options. Jorge Juan and Ulloa, who were comparatively unbiased, looked at the same territory and concluded that "any geometrical operations would be impracticable there, the country being everywhere so extremely mountainous and almost covered with massive trees, that it made the enterprise impossible."[40]

It did not really matter whether the idea was good or not; Bouguer and La Condamine's insubordination spoke for itself. Jorge Juan and Ulloa delivered Bouguer's communiqué to Godin, still aboard ship. Godin was furious and immediately laid out his plans in a two-page missive, which he then handed back to the Spanish officers for delivery to the men ashore. "Seeing as Messieurs Bouguer and La Condamine wish to remain in the Port of Manta without my consent," he imperiously decreed, "and having refused to obey orders . . . I find myself obliged to go as soon as possible to Guayaquil." Godin gave the Spanish officers the option to stay behind, but of course they were duty-bound to follow the expedition's leader. The next day, March 13, Juan-Manuel Morel raised anchor, and *San Cristobal* sailed south. Bouguer and La Condamine, the officers noted in a letter back to the Spanish authorities, "embarked [ashore] with some of their instruments," along with a servant and two slaves.[41] The two scientists were now effectively cut off from the expedition and would have to make their own way to Quito.

Storming out of Manta without two of his team's members was the worst possible decision Godin could make, and yet it was also a continuation of the practice he had begun in Saint Domingue, when he had separated from the group to travel overland. Bouguer and La Condamine had made an equally bad decision to stay behind, for without guides, they had no concept of what plants and animals were dangerous or poisonous, or which were edible. The two groups now made their separate ways down the western coast of Peru, in a completely foreign and therefore potentially hostile territory, where they did not know the geography, the people, or the many languages. By allowing their team to split up in the face of these great unknowns, the three scientists had almost fatally weakened their ability to survive the journey, much less carry out their mission. Finding Quito had now become far more treacherous than they ever imagined.

Lighter by five men, *San Cristobal* worked its way south along the coast toward Guayaquil. Godin was in a rush to reach the port town in time for the lunar eclipse of March 26. In that era before the perfection of the marine chronometer, which would keep accurate time over long ocean voyages, astronomers established the longitude of a distant place by precisely recording the local time of an astronomical event—a lunar or solar eclipse, or the passage of the moons of Jupiter behind the planet—using a land-based clock calibrated to that location, as Richer did in Cayenne. They would compare it with the exact time recorded at a home observatory in, say, Paris or Cadiz. Every hour of difference between the local time and the observatory time meant a difference of 15° longitude. The expedition hoped to observe the precise hour, minute, and second of the beginning and end of the eclipse, so that when the scientists returned to Europe, they could compare their measurements with similar observations in Paris and Cadiz, thus establishing exactly how far west lay the coast of Peru. Unfortunately, as *San Cristobal* hurried into the bay of Guayaquil in time for the eclipse, Godin found it covered in clouds and mist, thwarting any possible observations.[42]

Although cloudy skies had ruined the hoped-for observations, the weather had caused the expedition more immediate problems. The accompanying rains had turned the city into a marsh and swollen the Guayas River to the point that it was impassable by the large canoes needed to transport the scientists' baggage on the first leg of their journey to Quito. After a winding eighty-mile paddle upstream through the jungle to the town of Caracol, the team would have to traverse forty miles of mule trail over the western cordillera of the Andes to the town of Guaranda, in the foothills on the other side of the range. From there the land flattened out to a wide valley that separated the western and eastern cordilleras, and through this valley ran a road leading north for another hundred and twenty miles, around the volcano of Chimborazo and on to Quito. The route between Guayaquil and Quito was heavily trafficked with travelers, imports, and exports, and an unencumbered, veteran traveler could make the journey in just over a week. For these inexperienced, heavily laden scientists it could easily take a month.[43] Even worse, they would have to wait for weeks until the river became passable, before they could set out for Quito.

The expedition's first order of business in Guayaquil was to get funds from the local treasury using the letters of credit Godin had received in Cartagena de Indias. The authorities were careful to record the payments to Godin separately from those to Jorge Juan and Ulloa; the former payments were considered loans that the colonial government would remit to France for reimbursement, while the latter payments would come from the Spanish treasury. Godin received 2,128 pesos on top of the 1,500 pesos he had received in Panama, an equivalent total of $160,000 from which he paid $68,000 to Morel for the charter of *San Cristobal*. Jorge Juan and Ulloa received their back pay plus expenses. Seniergues also filled his coffers by successfully performing cataract surgery on a wealthy resident of the port town.[44]

Throughout the month of April the Europeans remained in the waterlogged city, burning through their available funds while waiting for the flood to abate so that they could depart. The governor of Guayaquil, in the meantime, sent word to his counterpart in Guaranda, requesting a train of eighty mules be brought down to Caracol, and also sent a missive to the president of Quito informing him of the imminent arrival of the expedition. Finally, the skies cleared, and the river lowered, and the expedition was surely heartened to catch sight of the brilliant snow-covered peaks of the Andes, visible from the port town, which they hoped they would soon be surveying.

On May 3, with La Condamine and Bouguer still absent, the expedition began the last leg of its journey to Quito. Whatever eagerness they may have felt would have quickly disappeared in the clouds of mosquitoes that harried them day and night as they rode riverboats inland to Caracol. As Ulloa noted plaintively in his journal, "No expedient was of any use against their numbers; the smoke of the trees we burnt all night to disperse them, seemed to augment rather than diminish their numbers. At daybreak we could not look at each other without concern; our faces were swollen and our bodies ablaze and covered in welts." The men lived on bananas and game, carefully avoiding "the huge numbers of Caymans [a type of crocodile] up to 18 and 20 feet long and proportionally wide" that sunned themselves on the banks.[45]

After eight days of battling the currents of the Guayas River, the expedition arrived in Caracol. There the men painstakingly loaded the mules with their crates, boxes, and barrels, but they were forced to leave a dozen boxes behind until more mules could be found. The carefully crated twelve-foot zenith sector, being especially fragile, would have to be hand-carried the whole distance by six Indians.

The road from Caracol to Guaranda would turn out to be the most treacherous part of the journey from Guayaquil to Quito. The route over the western cordillera was extremely narrow (sometimes as little as seven inches wide), steep, and muddy; one false step by the mules could pitch a rider or a precious box of astronomical instruments into a deep ravine. But the mules were surprisingly sure-footed and had developed their own way of tobogganing down sheer descents, with the rider simply hanging on to the saddle. Nevertheless, the travelers still had to do their part; the path was frequently blocked by fallen trees that could not be simply cut and pushed aside, and at each obstacle the baggage had to be unslung from the mules so that they could clamber over the trunks. At night the team slept in the open, even though the clouds of mosquitoes had not diminished.[46]

After five days of navigating the steep mountain trails, the team arrived in Guaranda, the provincial capital of Chimbo. A small town of around 1,000 inhabitants located in the heights of the western cordillera of the Andes, it was nevertheless an important way station on the route between Guayaquil and Quito. The well-publicized approach of the French scientists gave the townspeople ample opportunity to meet them while still on the road. "We continued our march towards the village with a dozen horsemen leading the procession," gushed Verguin, haggard after so many months of hardship. "Four young Indians joined us, dressed in blue with white sashes, shouting with joy, while bells and trumpets sounded. We were thus led to the house of the Governor, where we drank iced drinks, and while we supped, listened to a symphony composed of harps and violins." Ulloa, bewildered, asked the governor why they had been singled out for such an honor. He replied, "This was not particularly unusual, as it is how we greet people of any rank, as is the custom of all villages in these parts."[47]

On May 21, the team began its final push north toward Quito. On the entire journey they had seen nothing but coastal flats, steep mountain trails, and dense forests, but as they descended the western cordillera, an achingly familiar vista lay before them that evoked memories of the pastoral European landscapes they had left behind. "We thought we were in another world," recalled Verguin in a letter, "by the agreeable aspect of the countryside, varied from one side to the other by the verdure of the newly planted grain, to the other side with thickets and fields ready for sowing, legumes in full flower . . . herds of sheep, cows and other beasts on the plains, with habitations scattered from one horizon to the other."[48]

The expedition followed a long road that left the fertile plains and rose into the cold, dry *páramos* (high mountain prairies) around the great volcano of Chimborazo. The Europeans kept careful note of their passage, as this would be the territory they would soon be surveying in their geodesic work. Away in the distance was the city of Riobamba, which would mark the halfway point of the survey, between Quito and the southernmost extremity of Cuenca. They stopped at farms and villages to rest, including the Hacienda San Augustin de Callo near Cotopaxi, built atop the ruins of an Inca palace that they examined with some haste before moving on.

On the afternoon of May 29, having reached the outskirts of Quito, the travelers were met by two official horse-mounted guards, who accompanied them into the city and to the presidential palace. Once again they were fêted with a grand dinner attended by the local officials, although to their disappointment none of the "very young and very pleasant" women in the palace were allowed in. The local population peered through the windows at the foreigners as they ate. Meanwhile, the authorities took careful account of their baggage, comparing it with the inventory they had carried all the way from Portobelo and noting the presence of six servants and four slaves. The visitors were then shown to their rooms in the palace, where they would be lodged until suitable accommodations could be found in the city.[49]

Far to the west of Quito, Bouguer and La Condamine were finally making their way to rejoin the expedition. The past two months had been gratifying but exhausting for the two scientists, and it was clear that they would need to rejoin Godin's troupe if they were to continue with the most important objective of their mission: the precise measure of the length of a degree of latitude.

After thankfully seeing off *San Cristobal* from Manta two months earlier, Bouguer and La Condamine had busied themselves with astronomical observations along the Peruvian coast. They had taken measurements of the solar equinox on March 21 and the lunar eclipse on March 26, which Godin had singularly failed to observe. This latter observation the two men had taken near Cape San Lorenzo, in order to fix the longitude of what they believed to be the westernmost point of South America.[50] This would allow the scientists to accurately determine the position of the coast of Peru, a critical goal of the mission and one certain to ingratiate them with the Peruvian authorities.

By canoe and on horseback, Bouguer and La Condamine had continued their excursions up and down the coast for another month, working side by side to determine whether the region would in fact be suitable for their geodesic survey. One particular object of study was the atmospheric refraction near the equator, where the bending of light altered the apparent position of distant objects and stars. The scientists would need to correct for this refraction when making both their survey and their astronomical observations while in the Andes.[51] On April 13, Bouguer witnessed an unusual occurrence of such refraction: a rare "inferior mirage" of two suns setting, one atop the other. The two men continued to observe the sun and stars with their quadrant until they determined the point at which the equator crossed the coastline, at a promontory called Palmar.

La Condamine had marked his and Bouguer's discovery at Palmar by painstakingly carving a message in Latin onto a large boulder. In his diary, he recorded the inscription:

OBSERVATIONIBUS ASTRONOMICIS
REGIE PARIS SCIENTIAR ACADEMICA
HOCCE PROMONTORIUM PALMAR
AD AEQUATEORI SUBJACERE
COMPERTUM EST.
ANN. CHR.1736

"Confirmed by astronomical observations of the Royal Academy of Sciences Paris, in the Year of Our Lord 1736," La Condamine wrote, "the equator is located at this promontory Palmar." The inscription was far more than a simple astronomical marker; it was an overt political statement. La Condamine, a former army man, understood the military and political concept of a nation's "presence" in a foreign land. He knew the expedition would only be in Spanish territory a short time, but that this inscription, pointedly identifying the French Academy of Sciences as the source of geographical knowledge, would stand as a permanent reminder to the population that the French were once in Peru and might be there again.[52]

Relieved at being out from under Godin's reign, the two scientists had enjoyed a relaxed time in the area around Manta. They took their time to explore the area, visiting local villages and learning some of the history of the Inca and the Spanish colonists. Though Bouguer and La Condamine had been thrown together principally in joint opposition to Godin, they worked inseparably during their sojourn on the coast.

By late April, it had become obvious that the two French scientists needed to make their way to Quito. Bouguer's health was flagging, and

Figure 3.1 La Condamine's equatorial inscription at Palmar. From Charles-Marie de La Condamine, *Journal du voyage fait à l'Équateur* (1751). Courtesy of Robert Whitaker.

they were running out of funds. They had received no promise from Godin that he would wait for them in Guayaquil before setting out to Quito, so they understood they would need to find their own way to the capital. The safest course of action would have been to travel together, overland to Guayaquil, and then proceed to Quito. Yet at this stage, La Condamine once again showed the combination of curiosity, bravery, and sheer idiocy he had demonstrated at the siege of Rosas. Rather than accompany Bouguer, he decided to strike out alone across the uncharted jungle, without the certainty of a guide or any knowledge of the region, leaving his ailing friend and colleague to make his own way into the unknown.

On April 23, Bouguer and a single slave set off on horseback toward Guayaquil, taking coastal paths and riding along the beach. They arrived on May 19 and, learning of Godin's departure several weeks earlier, set off the same day up the Guayas River. Delayed at Caracol by exhaustion and a lack of mules—Godin had taken almost every last animal in the province, and they could only scrounge up enough to carry a few more crates—they took the path through Guaranda and around the slopes of Chimborazo to Riobamba, turning north and entering Quito on June 10, the last of the expedition to arrive.[53] Bouguer was undoubtedly relieved to see that La Condamine had already arrived safely and would have listened intently to his remarkable story.

La Condamine had tried and failed to find a guide to take him on the direct route across the jungle to Quito, so he, his servant, and his slave had canoed 150 miles north to the mouth of the Esmeraldas River, continuing some ways upstream before debarking and finding other Indian guides and porters to take him upriver. It was a hard slog through the dense rainforest, made worse by La Condamine's cumbersome baggage, including, the scientist noted, "several instruments and a large quadrant, which two Indians had a hard time carrying." By various paths they worked their way southeast toward Quito.[54]

As he had worked his way through the jungle, La Condamine observed something curious: Indians tapping native trees for a white, resinous sap that they would collect in banana leaves. The sap was worked into long-burning torches that rain could not extinguish. The substance would turn brown and elastic when hard, and the Indians molded it into unbreakable bottles to carry water and other liquids. At first, La Condamine simply used the sap to light his way along the trail and noted that a ball of it would bounce elastically when dropped. Soon after arriving in Quito he would send a description and some samples of the hardened material to his friend Charles du Fay at the Royal Garden, marking the first European discovery of rubber. During their stay in Peru, the scientists would discover many more uses for rubber, working the liquid resin into the cloth of their capes and tarpaulins, thus making waterproof coverings for themselves and their instruments.[55]

La Condamine's fascination with the discovery of the strange new sap had quickly turned to fear for his very survival. He and his companions were abandoned by their guides along the trail, and for eight days they wandered alone as best they could, guided only by the Frenchman's compass. They soon ran out of gunpowder, so they subsisted on bananas and the few fruits they could identify. La Condamine "suffered a fever which I treated by diet," although his claim should be treated with some skepticism, since it is doubtful he knew much about the medicinal plants growing in the region.[56] With the Indian porters gone, La Condamine's servant and slave no doubt took their turns wrestling with the cumbersome instruments and baggage that the scientist had opted to bring with him on the journey.

La Condamine, his slave, and his servant had finally emerged from the jungle along the crest of a mountain, and four days later happened on the Indian village of Niguas, where the locals pointed them along the route to Quito. At another village the travelers rented mules and an Indian guide, leaving the quadrant and baggage as a guarantee of payment. At the village of Nono, just fifteen miles northwest of Quito, a Franciscan monk gave La Condamine clothes and other goods on credit.

To reach Quito, La Condamine and his companions had to climb the volcano Rucu Pichincha, on whose eastern slopes the city lay. La Condamine wrote down his impressions as he made that ascent, in a memorable mix of scientific narrative and poetry that would have made his friend Voltaire proud:

> The higher I climbed the more the jungle cleared; soon I saw only sand, and higher up naked and calcified boulders which bordered the northern edge of the volcano of Pichincha. Having reached the highest point of the edge I was seized by a sense of wonder mixed with admiration, at the appearance of a large valley five to six leagues wide, interspersed with streams which joined together: I saw, as far as my eye could see, cultivated lands, divided between plains and prairies, green spaces, villages, and towns surrounded by hedges and gardens: the city of Quito, far off, was at the end of this beautiful view. I felt as if I had been transported to the most beautiful of provinces in France, and as I descended I felt the imperceptible change in climate going from extreme cold to the temperature of the most beautiful days in May. Each instant I added to my surprise; I saw, for the first time, flowers, buds and fruit in the middle of the countryside on all of the trees. I saw people planting, laboring and harvesting all on the same day and in the same place. I have let myself become carried away by the memory of the first impression I had then, quite forgetting that this [journal] is only a place for recounting our academic work.[57]

Descending the eastern slope of Pichincha, La Condamine and his companions had finally reached Quito on June 4 after six weeks of difficult—and occasionally directionless—travel. There they found Godin and the others who had accompanied him over the Andes. Six days later, Bouguer strode into Quito, bringing the team back to its full strength.

After more than a year of unexpectedly arduous travel, of near mutiny and rebellion, the expedition had united once again. If they took any pleasure in their reunion, it would be short-lived. The team was about to receive a rude awakening. They had come to Peru with shockingly few resources of their own, relying instead on a colonial power for their subsistence. Despite its apparent wealth, Peru would soon show that it was unable—or unwilling—to provide for even the most basic needs of the Geodesic Mission.

IV

Degree of Difficulty

Dionisio de Alsedo y Herrera was not at all impressed with the ragtag group of Frenchmen before him, especially given their reputation as learned scientists. As president of the *audiencia* of Quito, one of the most important provinces within the Viceroyalty of Peru, Alsedo was a formidable and efficient administrator. This was due in no small part to his deep mistrust of foreigners, which he had acquired during his time as a British prisoner of war at the age of eighteen. After his release, Alsedo had risen rapidly through government ranks in Spain and its colonies, making a name for himself by clamping down hard on smuggling and foreign contraband. He had been appointed as president in 1728; when the Geodesic Mission arrived in Quito, his term was drawing to a close after eight years.[1]

The *audiencia* of Quito was one of six administrative departments in the Viceroyalty of Peru, an enormous Spanish jurisdiction that extended from Panama to present-day Chile and Argentina, spanning almost the entire continent of South America except eastern Brazil. Each *audiencia*—a roughly equivalent term would be "presidency"—had its own court and president who governed more or less independently from the viceroy, who in turn oversaw taxation, silver exports, and the defense of the realm.

The *audiencia* of Quito, encompassing all of modern-day Ecuador and large portions of Peru and western Brazil, had been established by the Spanish thirty years after they had overthrown the Inca nation and claimed the lands for their own. The Spanish had arrived just as the leader of the Inca nation, the emperor Atahualpa, was emerging victorious from a destructive and costly civil war against his brother. During one battle, Atahualpa sacked and razed the golden city of Tomebamba, the site of modern-day Cuenca and a city that was said to have rivaled Cuzco in splendor, giving rise to one of the many legends of El Dorado. The Spanish, upon hearing these legends, mounted a fierce campaign to conquer the Inca and plunder their gold.[2]

The Spanish conquest of Peru had been rapid and violent. From 1532 to 1534, the conquistadors Francisco Pizarro, Diego de Almagro, and Sebastián de Benalcázar led a two-pronged attack; Pizarro, striking south toward Cuzco, captured and killed Atahualpa fairly early in the campaign, while Almagro and Benalcázar went north to capture the Inca city of Quito, which had been Atahualpa's ancestral home. Although the Spanish handily defeated the Inca forces, the survivors retreated to Quito and burned it to the ground to prevent the conquerors from looting it. Atop the ashes of the Inca stronghold, Benalcázar founded in 1534 the Spanish city of San Francisco de Quito, which became the capital of the *audiencia* when it was formed in 1563.[3]

At first glance it seemed that Quito was hardly an ideal location for a capital city, either Inca or Spanish. It was precariously squeezed onto a narrow and inaccessible bluff at the foot of an active volcano, which rose dramatically above the city. The surrounding hills further confined its growth, and the steep ravines and river gullies called *quebradas* (literally "fractures") that cut through the landscape made it extremely difficult to move freely across the city. But the sun-worshipping Inca had established the site as their northern capital—the name "Quito" referred to the Quitu tribe that originally inhabited the location—in part because of its position at the equator, the place where the sun casts no shadow at the equinoxes, but also because its inaccessibility made it an eminently defensible stronghold without the need for walled fortifications. Its natural defenses, however, could not protect it from being razed by its own defenders before even being attacked.

The climate of Quito was ideal, with mild temperatures that varied little throughout the year. Quito really had only two seasons, wet (November through May) and dry (June through October). The fertile volcanic soil, a result of the continuous activity of Mount Pichincha, ensured that farms almost never lay fallow, inspiring the Spanish to dub the region "*Quito siempre verde*" (evergreen Quito). But they also described it ruefully as *muy accidentado*, very rough, with streets that abruptly rose and fell, and great fissures that split buildings and even whole neighborhoods, all the result of the ongoing series of natural disasters that left their mark on the cityscape. In addition to occasional eruptions and ash flows from Pichincha and other nearby volcanoes, the city suffered frequent earthquakes, which heaved up the roads and destroyed buildings, and torrential rains that triggered great *huaycos* (mudflows), which would sweep down the slopes of the mountain and through the *quebradas*, washing away the bridges in their path. Quito's 30,000 inhabitants—it was quite small for a capital city—simply took all of this in stride as they went about their business, convinced they were living in the Eden of the New World.[4]

Godin and his expedition had come into Quito flanked by ceremonial guards, to what appeared to be a heroes' welcome—but President Alsedo was carefully observing them as they entered his capital, trying to square what he saw with the news he had received. The previous September, in 1735, he had received instructions from the king informing him that the French astronomers would be making observations in his realm and ordering him to "provide every assistance they may ask for in this endeavor." Alsedo was excited by the promise that these men could help map his territory: He understood the strategic and economic importance of accurate cartography, having himself recently commissioned one of the first comprehensive maps of Quito to assist military planning and tax collection. At the same time, he was wary of foreigners and smugglers, so he wrote to the incoming viceroy to "instruct me as to what else Your

Excellency would consider advisable to do." Villagarcía responded in January 1736, soon after his arrival in Lima, telling Alsedo what the president already knew: "Be attentive that the aforementioned Astronomers do not use their royal permissions to another, less desirable purpose."[5] In other words, Alsedo was on his own.

In mid-April of 1736, word had come from Guayaquil that the astronomers had debarked and were requesting mules for their overland trip to Quito. Alsedo, following royal orders to "provide every assistance" to the expedition, quickly fulfilled the request. He was waiting patiently for the group's arrival when, on May 22, he received the disturbing news from Cartagena de Indias. The French warship *Vautour* had returned from Panama, supposedly with additional baggage for the scientists that would be carried directly overland to Quito on the nine-hundred-mile trail along the Andes. In fact, its captain Héricourt "had hidden in the shadows of [the vessel], a great quantity of expensive textiles, for which they were looking for buyers in the port."

For Alsedo, the news that the French scientists may have been part of a smuggling plot presented an intractable problem. He had spent much of his career combating British and Dutch merchants who, while operating legally in the regular Portobelo trade fairs that coincided with the arrival of the Spanish treasure fleets, openly traded in contraband up and down the Peruvian coasts. Now Alsedo had a potential French incursion on his hands, in the very heart of his territory, complicated by the fact that the suspicious party had royal permission to travel anywhere in his jurisdiction. He would have to see for himself what sort of threat these men posed to the fragile order he had imposed on his realm.[6]

Not only had the French scientists come under the scrutiny of the president, but they had also inadvertently chosen to arrive at the worst possible time for Alsedo. Now on his way out of office and with an uncertain political future, he was struggling to hold his power base together, in the face of both economic and political tensions, to bolster his chances of finding a suitable position in the Spanish government. For the past two centuries, the Spanish-born *chapetónes* who came to Peru had held sway against the native-born *criollos*. The Madrid-born Alsedo had, from the beginning of his term as president, allied himself with the *chapetón* political hierarchy, which included most local officials as well as the Jesuits, by far the most powerful order in the country. Alsedo burnished his image as a law-and-order politician by carefully staging several public banishments and executions of criminals. Even with this fearsome (if exaggerated) reputation, Alsedo had been able do little against the flow of contraband goods into the *audiencia*.

Under Alsedo's watch, smuggling had critically eroded the power of his *chapetón* constituency. The economic foundation of the province—and the source of the *chapetónes*' political power—was found in the hundred-odd textile mills that produced the beautiful blue woolen cloth, called *paño azul*, that formed a major portion of Quito's export to the rest of Peru. These textile mills were the primary source of income for their owners—a few powerful Spanish families and the Jesuit order—but the *paño azul* trade had been in decline since the previous century, as cut-price European cloth was smuggled into the viceroyalty in increasing quantities. The problem now was that, with the downturn in the production of *paño azul*, these mills were generating fewer profits, making the Spaniards vulnerable for the first time to a shift in political power.[7]

The first evidence of a power shift had occurred the previous year, in 1735, when the city council came under the control of an opponent of Alsedo. The *criollo* treasurer of Quito, Fernando García Aguado, had placed his family and partisans on the council immediately following his ascension. This challenge to the *chapetón* power base had been short-lived, for early in 1736 a group of Alsedo's partisans, led by his son-in-law (and crown attorney) Juan de Valparda and the head of the local merchant guild, Lorenzo Nates, took back the council. Their grip on power was tenuous, since García Aguado still controlled the treasury.[8] And now, as the Geodesic Mission trickled into Quito, the balance of power was poised to shift once again.

With Alsedo soon to leave office, a new president from the *criollo* faction was set to take power—a development that seemed likely to rob Alsedo of any vestige of power he might otherwise have held onto. José de Araujo y Río, a rich merchant from Lima, had been appointed the next president of the *audiencia* and was bringing in a new native-born faction to rule over Quito, threatening the political structure that Alsedo had so carefully constructed—the *chapetón* power base that was the key to his political future. The appearance of a group of French scientists, possibly carrying smuggled goods, represented a further threat to the Spanish-born elite; any misstep by the Frenchmen could potentially stir up the long-simmering hostility of the local population toward all Europeans (including Spaniards) and would further endanger Alsedo's own standing.

Alsedo's suspicions about the Frenchmen quickly faded once they reached the capital. Now that he was face to face with Godin and his expedition, it was quickly obvious to Alsedo that this dispirited and fractious band posed no threat of smuggling, and their actions to date certainly suggested that they were incapable of any other organized activity. No illicit goods were present in their carefully inspected baggage. Did they know about the contraband trade carried out in Cartagena? No, they were shocked to learn of this; Bouguer would later state that they "could not even think about the illicit introduction of merchandise" while on their expedition.[9]

The Frenchmen certainly had not thought about their finances very carefully, Alsedo soon discovered. The increasingly perplexed president learned that the scientists had spent all but 372 of the 4,000 pesos they had been credited, along with the other funds they had received from the French authorities. In one year, the expedition had burned through a quarter million dollars simply to get to Quito, and now they were nearly broke.

As if the president had not already taken a dim view of the expedition, Godin made matters worse by demanding carte blanche from Alsedo, undiplomatically waving the Spanish king's instructions in his face. "Monsieur Godin showed me another dispatch of His [Spanish] Majesty," Alsedo reported to Villagarcía, "that orders the Viceroy and the Presidents of the *Audiencias* to assist the Astronomers with all the amenities that they ask." Godin asked for horses, houses, and spending money, all to be paid out of the Treasury of Quito on the thin promise of repayment by the French government. Godin's uncouth behavior certainly did not encourage Alsedo to go out of his way to help the Frenchmen. Nevertheless, Alsedo asked for the viceroy's help to pay for all that the scientists requested, since the Quito government had none of the resources needed to provide for their upkeep. The most Alsedo could do was to lodge them in the presidential palace for a few days until they found proper accommodations, pay them their last 372 pesos, and await Villagarcía's reply.[10] Meanwhile, he was left to wonder how these scientists, so highly learned and yet so disorganized, could ever hope to carry out their royal mission to measure the Earth.

As the scientists were shown to their temporary quarters, their hearts must have sunk. The low, elegant façade of the presidential palace, located on the west side of the main square, hid the wretched state of the once-luxurious building. The palace had been damaged by a combination of recent earthquakes and incessant leaks from poor construction, and it had also been subject to the continuous theft of furniture and decorations by corrupt officials such as the *audiencia* treasurer García Aguado. Despite repairs, the building was now almost uninhabitable, and even Alsedo had recently moved out, retaining only his offices there.[11]

The scientists stayed in the ramshackle palace for three days until they could find a house to lodge them all, but after six days even that house proved unsuitable for their astronomical work. They asked for, and were provided with, two houses in the Santa Barbara parish just north of the main plaza, each with a large garden from which the men could conduct their astronomical observations. It was here that Bouguer had joined them upon his arrival. La Condamine, meanwhile, had found alternate arrangements. Having left most of his baggage behind in Nono, he had arrived in Quito carrying little more than the clothes on his back. He immediately sought out the Jesuits, who were not only the most powerful ecclesiastical order in the region, but also considered the most learned. The priests in the seminary next door to the presidential palace graciously accommodated him, providing the weary scientist with clothes as well as lodging.[12]

With the entire expedition now assembled in Quito, it was time to get to work. The two months' separation seemed to have repaired the rift between Godin and the other two scientists; all were more or less on speaking terms again. For the moment, the biggest challenge would be a practical one. Most of the team's baggage was still en route from Caracol,

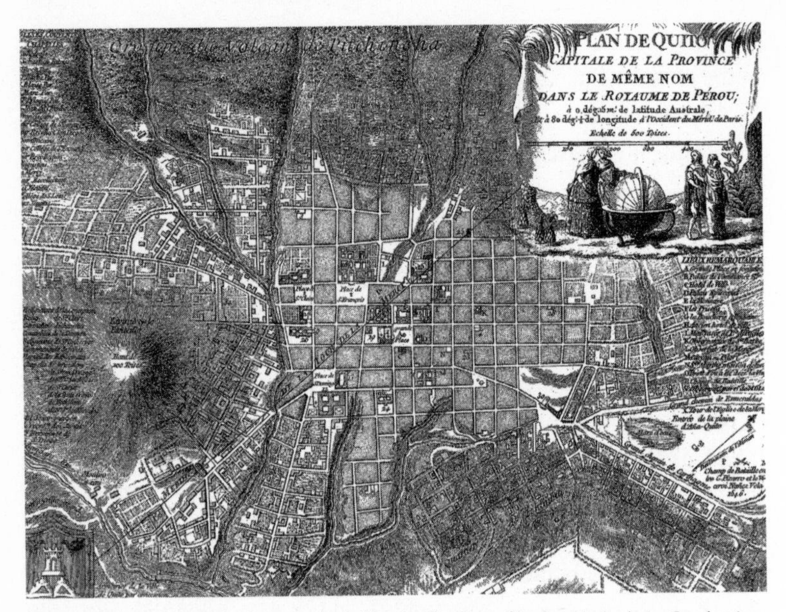

Figure 4.1 Map of Quito, by Jean-Louis de Morainville. From Charles-Marie de La Condamine, *Journal du voyage fait à l'Équateur* (1751). Courtesy of Robert Whitaker.

so they had to make do with the few instruments they had on hand. Hugo set about calibrating them for the surveys that lay ahead.

The expedition's first order of business was to verify the exact position of Quito and draw a map of the city, which would go some way toward pleasing Alsedo. The task was relatively straightforward; Bouguer, Godin, and the Spanish officers established the latitude of the city by fixing the heights of various stars and determined its longitude by observing several eclipses of Jupiter's moons. With that accomplished, and the president—they hoped—placated, the team could turn to the work that they had traveled so far to perform.

In order to lay out the baseline for the triangulation and begin the geodesic survey, the team searched for a large, flat plain precisely at the equator. The engineer Verguin, with young Couplet in tow, spent two weeks reconnoitering the terrain: "On June 11 Couplet and Verguin departed for the plains of Malchingui [eighteen miles from Quito] and Cayambe [thirty miles] that were located right on the equator, to determine whether they were suitable for a baseline of up to 4,000 *toises* [five miles]," reported Jorge Juan and Ulloa. "On the 22nd they found that neither location was suitable."[13]

While the expedition busied itself in searching for a baseline location, La Condamine was nowhere to be seen. He remained in the Jesuit seminary, later remembering it as the happiest time of his entire stay, transcribing his observations from his trip through the jungle, building a sundial in the courtyard, and entertaining a variety of guests. The gregarious Frenchman quickly befriended Ramón Joaquín Maldonado y Sotomayor, a *criollo* magistrate from one of the richest and most powerful families in the *audiencia*. Ramón Maldonado, upon learning of La Condamine's trip along the Esmeraldas River, excitedly informed him that this was the very route he was proposing for a new northern road between Quito and the Pacific. On June 21 Maldonado took La Condamine to the town of Nono, a few miles from Quito and almost directly on the equator, to view the project firsthand.[14]

During his social whirlwind, La Condamine had tactlessly neglected to pay his respects to Alsedo, even though the presidential palace was adjacent to the seminary. Alsedo was annoyed at this slight, though he accepted La Condamine's excuse that he "did not even have decent clothes" to present himself, since his baggage was still en route. The president was furious, however, when he learned that La Condamine had separated from his colleagues and taken another route through his realm without authorization; even worse, the Frenchman continued to live apart from his countrymen after arriving in Quito. To continue to do so, Alsedo scolded, would be "construed as being disunited in carrying out your commission to measure the exact figure of the Earth."[15] La Condamine, chastened, apologized in writing and moved in with his colleagues, returning himself to Alsedo's good graces.

La Condamine was not the only member of the expedition to put aside the measurement of the Earth as their stay in Quito dragged on. By the end of June, just three weeks after the team had begun to search for a location for the baseline, all geodesic work had stopped; the money to pay for the mules, food, and shelter necessary to carry on fieldwork had simply run out. Alsedo's request for additional funds was en route to Lima, but it would be at least two months before Villagarcía's reply could make it back to Quito.

With a long period of astronomical inactivity awaiting them, the French scientists turned their observational skills toward their home for the next few years. Like all travelers, the expedition members experienced a new country through their stomachs. In the more than two hundred years since Columbus's voyage, a bewildering variety of foods had crossed the oceans between the Old World and the New, greatly enriching the lives and diets of people on both continents. Tomatoes, beans, potatoes, and corn traveled westward from the Americas, while African rice and European grains such as wheat and barley were artfully and deliciously incorporated into the local Peruvian cuisine, the combination of which became known as *comida criolla*. Chicken and livestock had been transplanted from Spain to provide meat for the Spanish population, although the Indians still preferred the *cuy*, a domesticated guinea pig that could grow to the size of a small rabbit. French wines were cultivated in Chile, but La Condamine reported that they were "transported in jars coated with a tar that gives them an unpleasant taste. . . . The wealthiest hardly drink it, and shy away from it." The explorers gushed over the cornucopia of fruits found in the New World: "There is an abundance of peas, apples, peaches, apricots, strawberries. . . . A continuous springtime, joined with an eternal autumn, allows flowers, buds and fruits to grow all at the same time on the trees," Bouguer noted in a letter to his friend Jean-Paul Bignon back in France. The travelers reserved their most effusive praise for the chirimoya, a tantalizing, almost cloying fruit that was found nowhere else in the world. "The chirimoya is universally acknowledged to be the most delicious of any known fruit either of the Indies or Europe," exulted Ulloa. "I cannot compare it to our other fruits, and I am tempted to place it above them," said Bouguer, rather more cautiously.[16]

The people of Peru, like their diet, greatly fascinated the travelers. In both dress and customs, the Peruvians appeared to be almost a century behind the French. Culture shock was palpable for both groups; it was as if a band of twentieth-century Americans had suddenly stepped into Victorian Britain. "The people here dress flamboyantly; not infrequently using gold and silver fabrics and fine cloths of silk and wool," noted Ulloa.[17] The women's dresses were ornate and complicated affairs, the likes of which had not been seen in France since the early 1600s. Their skirts were embroidered with several rows of silk lace, above which they wore a doublet (a snug-fitting, short jacket) of gold cloth or linen, sporting puffy sleeves and even more lace. A mantilla of black lace traditionally covered their head and shoulders. On going into town or to church, women would cover their dresses with a *saya* or stiff overskirt and a *manto* or sleeveless hood of black silk so as to modestly veil their faces. Men's clothes were equally ornate; their long jackets and breeches were made of velvet, often accessorized with numerous buttons and fittings, a plush blue cape, and silk stockings.

The French scientists, in contrast to the locals, were clothed in relatively simple pleated jackets and linen breeches, but this was far from the only difference between French and Spanish colonial fashion. The Frenchmen explained these differences to the Peruvians, so that they would "know a little of our world, our French manners, so different from the Spanish . . . [to know] about the current fashion of the French women and gentlemen," as Bouguer related to Bignon.[18] Velvet had become rather passé in Paris, and the Frenchmen would likely have explained to the attentive locals that French women were now wearing less ornate skirts that were hooped and had long sleeves and pleated cuffs, adopted from Britain. French women's coiffures were also much simpler; hairdos were usually short, fixed close to the scalp, and decorated with ribbons. "Peruvian women," Ulloa observed, "wear their hair in braids at the back of their neck, flowing down to form crossed loops, the ends of which are adorned in diamonds and flowers, and very light and airy to the touch."[19] The young, unmarried men on the expedition were evidently not blinded by their scientific endeavors to the charms of the women of Peru.

To the Frenchmen especially, the Peruvian attitudes toward women would have appeared rooted in the previous century. In Peru, women received little schooling apart from their catechisms; most of their education was devoted to learning manual skills like sewing and other household duties. Apart from beauty and a proper dowry, these were all the skills a woman was thought to require to capture a man. Although

French society in the mid-eighteenth century was hardly a model of equality between the sexes, women's formal education was by now seen as an important part of their upbringing, and well-to-do families went to great lengths to ensure their daughters were knowledgeable in the ways of the world.[20] These highly educated women had become society leaders in Enlightenment Paris; they ran the afternoon salons where public opinion was created, and it was in those salons that the resolution of the Earth's shape had been first mooted.

French progress in the social status of women had translated to a more relaxed attitude toward sex and marriage. Girls were no longer strictly separated from boys in the household or at church, so they could court more or less in public. Premarital and extramarital sex was still frowned upon, but already the tendency in France was to look the other way. In Peru, by contrast, girls were kept strictly separated from any boy who was not part of the family, and the "honor" of young women was closely guarded by brothers, uncles, and fathers. Consequently, an intricate series of rituals had developed to maintain that "honor." An unmarried woman could not be left alone with an unrelated man, nor could she flirt openly; it was absolutely unheard of, moreover, for a single woman to promenade publicly with a man, as she might on the streets of Paris. Frézier, the previous Frenchman to officially visit Peru, had been correct that premarital sex and adultery were quite common in Peruvian society, but these affairs were invariably carried out away from public view. Peruvian women had learned to use their *mantos* to full advantage; if they unveiled one or both eyes a certain way, it was a signal to pass a secret note or to rendezvous later that night. Such rituals would have seemed almost quaint to the French, had the displays not been imbued with the deadly serious code of honor that the men continued to follow.[21]

The mutual fascination between the French and the Peruvians was edged with serious misgivings—not only Alsedo's abiding worries about smuggling, but also the generally held suspicion that smugglers were also Jews. Spanish colonists had long conflated the two, believing that the Portuguese and Dutch merchants who plied the Caribbean and the Pacific were secretly Jewish smugglers who sold contraband goods to the unsuspecting locals. Since the reconquest of Spain from the Moors and simultaneous expulsion of the Jews in 1492, the Spanish crown had relentlessly persecuted suspected Jews as a means to maintain firm economic and religious control over the kingdom. The Inquisition was subsequently brought to the Spanish colonies, where prosecutions were carried out with equal fervor. "Judaizing" (being a practicing Jew) was a deadly serious charge under the Tribunal of the Holy Office of the Inquisition of Lima, which had prosecuted over 1,400 suspected heretics and executed 32 since its inception in 1570.

In 1736, the same year the Geodesic Mission arrived in Peru, a suspected Jew named María Francisca Ana de Castro was burned at the stake in a celebrated auto-da-fé in Lima under the eyes of Viceroy Villagarcía. Anyone who did not regularly carry a rosary was branded a heretic and came under immediate suspicion. The French scientists, not particularly overt in their faith, found themselves under close scrutiny from the Tribunal. Bouguer described the gravity of the threat in the same letter in which he rhapsodized about the local fruit to Bignon: "The Inquisition mistook us for Jews, but they were unable to arrest us as we are the subjects of the French king. We were easily absolved of the accusation, since we told the Inquisitor that if he wished to dine with us, we would only serve pork with all its sauces. To this there was no reply, and they must certainly now view us as Catholics."[22]

By mid-August 1736 Alsedo received his reply from Lima, and it was not a happy one. In a short note, Villagarcía said, "I have directed that all remittances should pass through the *fiscal* [crown attorney]" of the *audiencia* of Quito, and that no payments would be coming from the viceroyal coffers.[23] Once again, Villagarcía was telling Alsedo that he was on his own. But the *fiscal* had no money to offer, and Alsedo was certainly not going to move heaven and earth to find more; the scientists were also on their own.

With no money forthcoming from the colonial government, La Condamine came to the rescue. Having foreseen precisely this sort of difficulty, he revealed to the other members of the expedition that he had obtained personal letters of credit from the Castanier bank, worth about 20,000 pesos at the exchange rate, which should tide the group over until they could get more money from France. On August 17 the three scientists signed a formal "treaty," as they termed it, agreeing that La Condamine would use his own money to fund the expedition, with the expectation that he would later be reimbursed by the French crown. La Condamine proposed to go to Lima to exchange the notes after they had finished surveying the baseline. As the journey would take him many months to complete, the French scientists "asked all the merchants [in Quito] to loan them working capital" as the Spanish officers reported to Patiño, but the most they could obtain was 2,000 pesos, apparently from a French doctor living in Quito.[24] With this loan (worth about $90,000 today) the team could finally begin its survey, but the scientists would have to work fast to establish the baseline for triangulation before the rains came in November.

Back in June, Verguin and Couplet had reported that neither the plains at Malchingui nor those at Cayambe were suitable for the baseline. Their measurements, however, had been done without the benefit of several of their important survey instruments, which had been held up en route from Caracol because of a lack of mules. New mule trains carrying those instruments had finally arrived in the interim. With the dry season quickly drawing to a close, the scientists decided to reexamine the area around Cayambe. On September 10 they departed Quito, with La Condamine (as increasingly was his wont) taking a different path from the others', a more easterly route through Puembo and Yaruquí. On the 12th, the main party arrived at the Hacienda Guachalá, a large farm estate on the edge of the plain, where they resided for several days. They soon became friends with the owner, Antonio de Ormaza, with whom several of the members would develop lifelong relationships.

While La Condamine was still making his own way toward the group, Bouguer and Godin began measuring the plain, and—even before they completed their survey—they had to agree with Verguin's original conclusion. Although it was precisely on the equator, the plain's uneven terrain, with the six-hundred-foot-wide Rio Pisque ravine that ran across it, made it an unsuitable site for their baseline. Once again, the expedition would have to rethink its plans.

The day after the scientists had rejected Cayambe as a baseline site, La Condamine rejoined the group and brought with him some good news.

He told his colleagues of the plateau at Yaruquí, which he had just traversed, and suggested that it might be more suitable for the baseline than Cayambe. Although Yaruquí was several miles south of the equator, it was much closer to Quito than Cayambe and, apparently, much flatter as well. Bouguer, La Condamine, and Jorge Juan left that same day to examine the plateau, leaving Godin, Ulloa, Hugo, Verguin, Godin des Odonais, and Couplet to complete the survey of Cayambe.[25]

While Bouguer, La Condamine, and Jorge Juan were off scouting Yaruquí, tragedy struck the main body of the expedition. On September 17, young Couplet, who had left Quito feeling "slightly indisposed," was now attacked by a "malignant fever" that no one at Cayambe could identify. Reports of Couplet's illness suggest that he was most likely suffering from malaria, whose symptoms (fevers, sweating) flare up, subside, and then worsen every few days, as the plasmodium, a parasite, alternately invades red blood cells and is then released into the bloodstream in progressively larger quantities. Jussieu and Seniergues, likely still back in Quito, were not in Cayambe, where they might have been able to identify and cure the disease.[26]

At the time of the Geodesic Mission, the only cure for malaria was a powder called quinine, which is made from the bark of the cinchona tree, native to Peru. Scientists now understand that the bark causes the plasmodium to poison itself. The curative powers of the bark had long been known to small groups of Peruvian Indians, who introduced quinine to Jesuit missionaries in the 1620s, after which it was regularly exported to European doctors. Jussieu specifically was intent on learning more about the native cinchona tree to improve the drug's effectiveness; La Condamine had brought some quinine with him from Europe and in fact had used it to cure a *criollo* of a recurring fever during his stay on the coast in Portoviejo, expressing surprise that the man had not heard of the remedy, as it came from his own country. In fact, though quinine was known to the Spanish Peruvians, many did not believe it to be useful to cure malaria.[27] Had La Condamine been at Cayambe, he might have provided a cure for Couplet.

On September 19, after being "confined to bed just two days," Jacques Couplet died. He was buried in a church at Cayambe, his tomb covered by a stone inscribed in Latin.[28] That evening, the members of the Geodesic Mission gathered at their two separate locations in Cayambe and Yaruquí to observe a lunar eclipse, in order to establish the longitude of each site. A series of earthquakes ominously interrupted the sightings.

The same eclipse that the scientists on the Geodesic Mission observed was also recorded by Jacques Cassini on his Thury estate outside Paris, where it was then the morning of September 20. It was probably observed by most of the French intellectual elite, likely including Couplet's family, who may have witnessed the darkening of the moon with some foreboding about their young son. They had entrusted him to Louis Godin, who now had the burden of informing the family of his death.[29] He also faced the challenge of carrying on the mission short one man. In the meantime, he had determined Cayambe to be unsuitable for the expedition's purposes, so after interring Couplet and making some final astronomical observations on September 23, Godin and the remaining members departed for Yaruquí.

As Couplet had lain dying, Bouguer, La Condamine, and Jorge Juan had been busy examining Yaruquí. The long, thumb-shaped plateau of grassy terrain was oriented approximately north-south, situated close to the little pueblo of Yaruquí, from which it derived its name. The plateau gave the expedition over seven miles of clear run for a baseline, better than the six miles Jacques Cassini used for his surveys back in France. At the southern end, near the hacienda at Oyambaro, the land climbed to a wide terrace, a perfect location for one endpoint of the baseline, with a clear view down the flat plateau. At the northern end, near the Caraburo hacienda, the land dropped abruptly into the deep Guayllabamba valley, forming a four-hundred-foot cliff at whose edge the team could place the second endpoint. In between the two points were mostly cattle and sheep pastures, dotted with small farmhouses, churches, and the large complex of a wool mill.

The expedition had been searching for a flat expanse on which to measure their baseline, but "flat" was a relative term in the Andean landscape, which had been shaped by volcanoes, floods, and earthquakes. Yaruquí's nominally level surface dropped almost eight hundred feet from the plateau's southern to the northern end, with many terraces and undulations that the scientists would have to account for. The plateau had been carved into ravines—one, near Oyambaro, was nearly fifty feet wide—and lifted into hillocks by *huaycos* that had dumped tons of earth, rock, and mud across the plain. Frequent earthquakes had twisted those features into jagged shapes, made more uneven by farm terraces that had been carved out of the earth since before the time of the first Incas. The expedition would have to clear, excavate, or bridge each of these obstacles in order to measure the baseline with near-perfect accuracy. Unlike the organizational and fiscal calamities that had haunted the team over the past year, *these* were precisely the sorts of difficulties for which the scientists had prepared.

Upon arriving at Yaruquí, Bouguer, La Condamine, and Jorge Juan had first spent several days trekking up and down the plateau to establish the path of the prospective baseline. They buried large circular millstones near Caraburo and Oyambaro to mark the endpoints, above which they erected tall pyramids of wood and thatch that could be easily seen at long distances. They then laid out a series of tall, cross-shaped markers about every mile or so between the endpoints, to align the baseline for straightness.

Just as the three men completed the alignment for the baseline, the rest of the expedition arrived on the scene, in time for the most difficult phase of the operation. In order to create the baseline, an absolutely straight path, seven miles long and just eighteen inches wide, had to be dug into, ripped up from, and scraped out of the landscape. For the scientists, who had been accustomed to a largely sedentary life back in Europe, this would involve eight days of backbreaking labor and struggling for breath in the rarefied air. "We worked at felling trees," Bouguer explained in his letter to Bignon, "breaking through walls and filling in ravines to align [a baseline] of more than two leagues." They employed several Indians to help transport equipment, though Bouguer felt it necessary that someone "keep an eye on them."[30]

For their part, the Indians were also observing the scientists, but to them "all was confusion" regarding the scientists' motives for this arduous work. The long, straight baseline they had scratched out of the ground certainly resembled the sacred linear pathways that Peruvian cultures, since long before the Incas, had been constructing. These pathways,

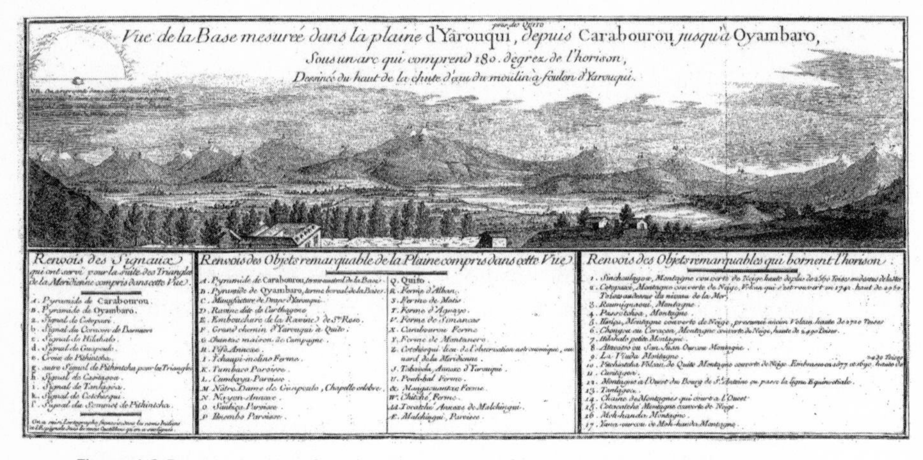

Figure 4.2 Panoramic view of the baseline at Yaruquí, by Jean-Louis de Morainville. From Charles-Marie de La Condamine, *Journal du voyage fait à l'Équateur* (1751). Courtesy of Robert Whitaker.

called *ceques*, connected distant sites that possessed religious or ceremonial importance. Indians would still routinely walk the *ceques*, making offerings at the sacred sites along the way. All along the Andean landscape were hundreds of *ceques* that ran arrow-straight for many dozens of miles, directly up steep hills and down sheer valleys, their builders impervious to changes in terrain. A thousand miles to the south, their ancestors had created the intricate *ceques* that radiated among the giant animal figures now only barely visible on the barren plains of Nazca.[31] This fairly short line that the scientists were laboriously creating on level terrain would have been mere child's play for the Indians who had created the *ceques*.

By the end of September, the baseline had been completed. A narrow, dusty-brown ribbon, it stretched across the plain, invisible in the distance but for the orderly formation of crosses that rose above it. The scientists' next task was to measure this line, which would be the first leg of the first triangle in their survey.

In order to ensure that the endpoints of the baseline would be visible from the triangle's intended vertex, La Condamine and Verguin returned to the summit of Rucu Pichincha, some sixteen miles away. There, at the summit where La Condamine had first viewed Quito, they erected a whitewashed stone signal, the first of many such vertex markers they would erect during their survey. The two men also constructed a shed to protect them during their observations. Meanwhile, back on Yaruquí, the other members prepared the measuring poles and made final adjustments to their instruments.[32]

Godin and Bouguer agreed to conduct the measurement of the baseline with two teams. Godin's team of four men would start at the southern end near Oyambaro, while Bouguer's team of five would begin at the northern end, near Caraburo. The scientists had convinced themselves that such duplication was necessary to achieve the incredible accuracy they sought in this measurement; Cassini, after all, had deduced a prolate Earth based on a perceived difference of just nine hundred feet in the sixty-plus miles of a degree of latitude he had measured in his survey of France. Splitting into two groups would help to ensure the accuracy of the critical baseline measurement, from which the dimensions of the rest of the chain of triangles would be calculated. The separation also solved the problem of the continuing friction between Bouguer and Godin, whose relationship never fully recovered from their quarrel on the shores of the Pacific earlier that year. Returning from the summit of Pichincha, La Condamine and Verguin—neither of whom had fully reconciled with Godin, either—joined with Bouguer, who also pressed his servant Grangier into service as a replacement for Couplet. The two Spanish officers now separated for the first time in the voyage, Jorge Juan joining Godin and Ulloa going with Bouguer.

Bouguer's team began their measurements first. On October 3, they set to work on the northern end of the baseline. Godin and Jorge Juan, assisted by Godin des Odonais and Hugo, began five days later. To ensure accuracy over the uneven ground, the teams divided their measurements into seven sections, each partitioned by two of the cross-shaped alignment markers. In conducting the measurements, both teams used three stiff twenty-foot poles of dry wood, three inches square in cross-section and tipped with copper plates. Two men would carry the rearmost pole forward and carefully lay it in front of the frontmost pole, which was held in place by another man, forming a slowly advancing chain along the brown strip of land. Bouguer's team took the precaution of color-coding each pole so they were always laid in the same order.

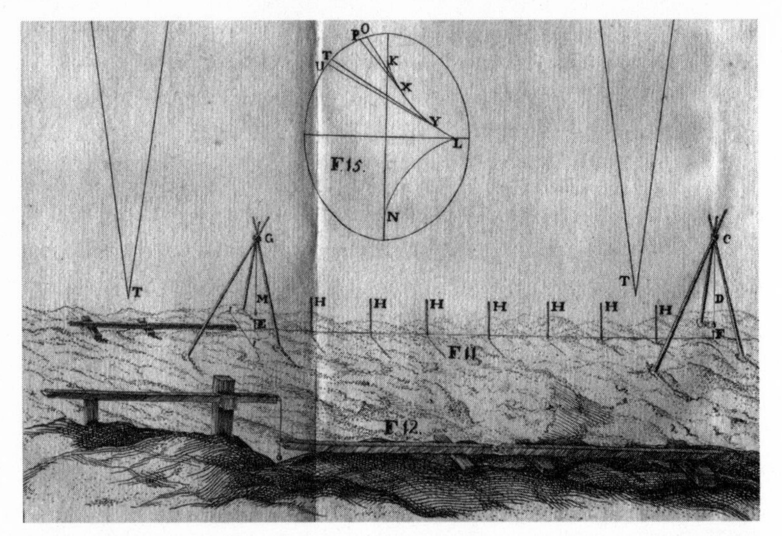

Figure 4.3 Measuring the baseline using wooden poles. From Jorge Juan and Antonio de Ulloa, *Observaciones astronómicas* (1748). Credit: Special Collections & Archives, Nimitz Library, U.S. Naval Academy.

Each team regularly verified the length of the poles against a six-foot iron *toise*; Godin retained the one fabricated in Paris, while Bouguer's team used a duplicate that La Condamine had instructed Hugo to make. Both teams aligned the direction of the poles with a horizontal cord (which they also checked for level with a plumb bob and right angle), but they used slightly different methods to support the poles during the actual measurements. Godin and Jorge Juan followed Cassini's practice of using adjustable wooden easels; Bouguer found them too prone to wind gusts, so his team carefully leveled each pole close to the ground with wooden wedges. Bouguer carried out the procedures for his team during the first week, but as his confidence in Verguin grew, he ceded the operation to him.

The two teams started out quite slowly, measuring less than 250 feet the first day. As the weeks wore on, the men worked faster and faster, covering half a mile a day toward the end of the process. They "were obliged to sleep on the ground" while in the field so as not to waste the cooler daylight hours of early morning and late evening, only retreating to their tents for a short midday break to escape the oppressive heat.[33] The teams passed each other in mid-October, comparing the measurements they had carefully penned in their little five-inch-by-seven-inch notebooks, each page crammed with figures front and back, since paper was often in short supply.

Bouguer's team completed their measurements first, on November 3, with Godin finishing two days later. Retiring to their lodgings at the hacienda in Tababela, near Oyambaro at the southern end of the baseline, each team began adding up their measurements to calculate the overall distance between the endpoints. They were soon astonished at how closely the two teams' measurements matched up; over the seven miles they differed by less than three inches. The scientists split the difference and took the distance as 6,272 *toises*, 4 *pieds*, 3 *puces*, and 7 *lignes*. They next set about the arduous and complex calculations of reducing the baseline to a single horizontal level, accounting for the slope of the terrain by reestimating all the measurements as if they were taken at the lowest point of the baseline, the northern end. Each of the three scientists employed a different method of calculation, but toward the end of November they all agreed on the figure of 6,274 *toises* (40,154 feet) as the corrected horizontal length of the baseline, which they would use for all subsequent trigonometric calculations.[34] It was an exhilarating moment; by establishing and measuring the baseline, the men had accomplished the critical first step in the triangulation, with both extraordinary precision and a minimum of conflict. Yet almost immediately after this achievement, they received news that plunged the future of the expedition into doubt.

On November 21, the men received their first packet from France, after more than a year's absence. Not only was it devoid of additional letters of credit from their patron Maurepas, but it also contained a heart-stopping letter to La Condamine from his friend (and Bouguer's enemy) Maupertuis, head of the Paris Academy Newtonians. Maupertuis informed La Condamine that he was undertaking his own expedition to determine the figure of the Earth. He was headed in the opposite direction: to the Gulf of Bothnia on the Arctic Circle in Lapland, the site of today's Finnish/Swedish border.[35]

Back in France, on May 25, 1735—just ten days after *Portefaix* had departed La Rochelle and entered the Atlantic—Maupertuis had read an essay before the Academy of Sciences with a new mathematical analysis of the shape of the Earth. He quickly followed his first presentation with a second that restated the advantages of conducting geodesic measurements far from Paris, which the Peru mission was created to do. Maupertuis' goal, however, was to build support for a *second* geodesic mission, which would go to the Arctic Circle. It would be much shorter and would return the results far more quickly than the Peru mission, even if it did start a year later. It was clear from Maupertuis' memoirs, as well as from private conversations with his friends, that the esteemed scientist—filled with the same wanderlust as his disciple La Condamine—was angling to lead this expedition to the far north. The story went that he finally persuaded Maurepas by playing guitar and singing to the minister while he was convalescing from an illness.

By the end of summer the approvals had been granted. Maupertuis would be accompanied by the mathematician Alexis-Claude de Clairaut, who, although twenty-two years old and a genius since childhood, was myopic—hardly a desirable trait for an astronomer. Maupertuis and Clairaut holed up in Jacques Cassini's estate at Thury to learn practical field survey and astronomy, a set of skills these normally deskbound mathematicians had not yet developed. On September 8, 1735, while the Peru expedition was still waiting for a ship in Saint Domingue, Maupertuis had penned the lines that La Condamine would read at Oyambaro a year later:

> You will perhaps be surprised when you learn that there will be a trip to the North so as not to miss anything in the determination of the figure of the Earth. France, by its two voyages, will accomplish the greatest feat ever made for the sciences. We (for I am on this voyage) are counting on departing next March for Bothnia where we will die of cold while you die of heat. In spite of all the gold in Peru I'm annoyed that you will not be in our caravan. . . . If you have some ideas on this matter we ask that you send them to us in Paris where they will be forwarded to the Polar Circle.[36]

Maupertuis, now embarked on his geodesic mission to the Arctic Circle, would almost certainly be the first to determine the shape of the Earth.

With this letter, the whole future of the Peru expedition was thrown into doubt. It was as if a great battle had been fought and lost before they even knew they were at war. The Arctic Circle was only 1,200 miles from Paris, just over a month's journey overland and through European waters. The notice of Maupertuis' imminent departure was now over a year old; for all they knew, he'd already carried out the measurements and returned to Paris wreathed in glory. With Maupertuis certain to succeed in his quest, there stood a real chance that the Peru expedition would be seen as redundant and recalled before they had completed their work. They had now depleted their funds, no further money had come from France, and it was a throw of the dice whether La Condamine would succeed in cashing in his letters of credit at Lima. In short, they did not known if they could continue with the mission or even have the money to return home.

The Spanish officers Jorge Juan and Ulloa were in just as morose a state as the French voyagers, but for entirely different reasons. Since their arrival in Cartagena de Indias over a year earlier, the officers had supplemented their astronomical work with their self-appointed task of finding and describing the oppression of the Indians, which was little changed since the time of the conquistadores two centuries before. So far in the expedition, the two men had had no difficulty filling their notebooks with examples of oppression and brutality. On their arrival at the hacienda Guachalá near Cayambe, they had witnessed the owner, Antonio de Ormaza, dupe a neighboring Indian into selling him his plot of land for a pittance. "The owner told us that he had tried to get the Indian to sell the land for a long time," the officers noted, but had been unsuccessful thus far. When the French scientists arrived at his hacienda, Ormaza told the man that they had come "to seize his land and return it to its rightful owners . . . but out of charity the owner agreed to give the Indian something for the plot. The Indian believed the outrageous lie and sold his land [to Ormaza]."[37] Such duplicity, Jorge Juan and Ulloa noted, was a depressingly common feature of the relationship between the Spanish landowners and the Indians.

Jorge Juan and Ulloa had witnessed more than mere duplicity since leaving Quito. They reserved their harshest critique for the sinister practices of the *mita*, which had replaced outright slavery with a system of forced labor of the Indians. Based on an Incan system of communal labor for public projects (like repairing bridges), it had been perverted beyond recognition under Spanish colonial law, which required Indian villages to provide essentially free, dawn-to-dusk labor to any Spaniard who requested it.

While measuring the baseline at Yaruquí, they had witnessed firsthand the harsh working conditions of the *mita*. Near the midpoint of the baseline was a large wool mill, which the two officers thoroughly explored. Such self-contained mill complexes (called *obrajes*) were the economic backbone of the *audiencia,* producing *paño azul* for export, and the Yaruquí mill was one of the largest, containing 10,000 head of sheep and employing over two hundred Indians at a time. At these facilities, raw wool was sheared, washed, carded and spun, soaked in hot copper kettles filled with indigo dye, and then spread in the sun to dry before being woven into long bolts of blue cloth.[38] Each operation required a large crew of Indians, who were rounded up before dawn each day and locked in workrooms with no ventilation and only little light. Those working in the dyeing vats had the hardest time of it, with the heat and smell adding to the already stultifying conditions.

The Spanish officers railed against the brutal treatment of the Indians at Yaruquí: "In the evening when it is too dark to work any longer, the overseer collects the piecework he distributed in the morning. Those who have been unable to finish are punished so brutally that it is incredible. . . . They whip the poor Indians by the hundreds of lashes." They went on to note that the owners fed the workers leftover grain and meat that had already spoiled, sickening people who were already weakened by the terrible conditions in the mill. "The majority die in the *obrajes* with their piecework still in their hands," the officers wrote.[39] These atrocities were the human price paid for the beautiful blue cloth that the men and women of Quito wore daily. Although the officers were able to bear witness to the sickening reality of the industry, they could do little to ameliorate the suffering of the Indians they encountered.

It was in a melancholy humor that the scientists and officers finished their observations and recalibrated their instruments, sighting on the baseline endpoints and various landmarks to adjust the degree markings on their quadrants. On December 5, they began the two-day journey back to Quito. That night, between midnight and 1 o'clock, a strong earthquake shook the ground for almost a minute before subsiding, leveling homes and churches up to fifty miles distant from the capital.[40] The sinister omens, it must have appeared to the expedition, were accumulating rapidly.

Now back in Quito, as La Condamine was preparing for his trip to Lima in an attempt to exchange his letters of credit, the scientists had one last astronomical observation to perform. As the first European astronomers to operate at the equator with modern instruments, they were ideally situated to accurately measure the vertical projection of the equator into the sky, known as the ecliptic, which describes the Earth's angle of tilt relative to its orbit around the sun. It is this tilt that gives the Earth its seasons. Modern techniques have revealed that the ecliptic is tilted by 23° 26', but in 1736 this angle had not been precisely measured, since, as Bouguer put it, "the obliquity of the ecliptic can hardly be observed anywhere to a sufficient degree of accuracy but near the equator."[41]

The precise measure of the ecliptic not only was of immediate importance to the French scientists for their own survey but would be useful to astronomers worldwide as a means of calibrating stellar observations. On the December 21 solstice (the date when the Earth's northern hemisphere is tilted farthest away from the sun), they used the zenith sector to measure several stars that gave them the precise angle of the ecliptic. They repeated these observations over several days and later at the June 1737 solstice. Their observations were remarkably accurate, less than 2 minutes (1/30 of a degree) different from today's accepted figure. Bouguer and La Condamine, having forwarded the results to Paris, ensured that this newfound knowledge would be available to other astronomers as well. They sent a copy of all their ecliptic observations to Edmond Halley, which Halley subsequently had translated and published in English.[42]

The covering letter to Halley that accompanied Bouguer's and La Condamine's measurements revealed their frustrations with the unanticipated degree of difficulty in carrying out their work. The two scientists explained that they "had not seen Godin's work," since he had not given it to them for their correspondence—despite the fact that Godin knew Halley personally. Just as Bouguer had originally feared before the mission was even begun, Godin was beginning to act independently of his colleagues, leaving them to carry out their work on their own.

They also speculated resentfully that Maupertuis had likely already returned from what they saw as a much easier expedition to the Arctic Circle, where he had "not found there the obstacles of ignorance, prejudice, barbarity and the impossibility of following the king's orders which we endure at every step."[43] It is telling that they had chosen to reveal their irritation to a British astronomer, given the rising tension between the two nations. They were cut off from any assistance from France and were apparently concerned that their strenuous efforts would be seen as irrelevant in light of Maupertuis' certain success. Their situation appeared hopeless, and it was about to get far, far worse.

V

City of Kings

The advent of the December solstice in 1737 coincided with another stellar event: the arrival of the new president of the *audiencia* of Quito, José de Araujo y Río. Araujo was a lawyer from a wealthy family of merchants in Lima and a member of a close-knit group of *criollo* nobles who had attended the best schools in the capital and used their social connections to further their political power in the court of Spain. In 1732, Araujo had bought the position of president of Quito for 26,000 pesos, going all the way to Madrid to ensure his petition was granted. Although most applicants paid a "benefit" to the court to become a president, Araujo had paid a very high price (over a million dollars today) for this particular *audiencia*, as it gave him control over the highly lucrative trade between Quito and other parts of the colonial Spanish Empire. His extended family and childhood friends controlled other ports of entry into Spanish America, from northern Peru all the way to Mexico, forming a network that could effectively control most of the contraband trade (and therefore the enormous profits) coming into the Pacific as a result of the Portobelo trade fairs, the next of which was scheduled for 1739. Buying the presidency was an excellent investment.[1]

Araujo and several of his colleagues traveled from Madrid to Mexico in early 1736 and then took the warship *San Fermín* from Acapulco to the northern port of Paita in Peru. From there Araujo slowly trekked northward to Quito, with a long layover in Riobamba. When he finally entered the capital on December 26, the reason for his delay was readily apparent. Trailing behind the incoming president was a mule train with over a hundred thirty crates containing wines, silver wares, Chinese silks, and porcelain. These were not Alsedo's personal effects; while in Acapulco, he had bought a large shipment of Asian goods from the annual Manila galleon in order to sell them in Quito. He had no problem illegally stowing them aboard a royal warship, since it was under the authority of Araujo's friends in Lima. Nor did he face any customs duties, since the port of Paita was controlled by his brother-in-law. Araujo had stopped in Riobamba to drop off another sixty-six crates of goods to be sold by his confederates there.[2]

On December 28, Araujo assumed the title of captain-general of Quito, the military commander of the militia. The following day he assumed the presidency from Alsedo, who, along with his *chapetón* partisans, had been dreading the entrance of Araujo and the shift in power to the native-born *criollo* faction. They now had reason to be furious; Araujo's huge cargo of smuggled goods was sure to drive down prices and ruin the businesses of local merchants, many of whom belonged to the guild led by Alsedo's friend Lorenzo Nates. On December 29, the day Araujo assumed office, Alsedo wrote scathing letters to the king and the Council of the Indies, demanding the removal of Araujo from office for having illicitly bought the presidency and for illegally smuggling goods into the capital.[3]

Within days the conflict between the factions deepened even further. The selection for the city council was at hand, and Araujo collaborated with the *criollo* treasurer Fernando García Aguado to place their candidates on the council. The vote taken on January 1, 1737, however, favored the *chapetón* faction led by Juan de Valparda. The following week,

Araujo disqualified the results "in order to remove dangerous illusions and establish peace and goodwill," thereby allowing him to place his *criollo* candidates on the council. This led to howls of protest by the *chapetón* faction. There was now a general stalemate as both sides frantically wrote to the viceroy for support.[4] The French scientists and Spanish officers had been detachedly observing these events even as they completed their observations of the December solstice. They did not imagine that they, too, would soon be explosively thrust into the middle of this struggle for political power.

The expedition had first met with Araujo around December 30, just before the contentious council elections, and had been no more pleased with the new president than were the *chapetónes*. The encounter had been a complete disaster, resulting from serious missteps on both sides. Right from the beginning, Araujo had treated the scientists and officers "with particular disdain," as a witness later reported; this was one of his first opportunities to demonstrate to the foreign-born *chapetón* faction that a new order was in place and that these Europeans were to be regarded as interlopers. For their part, the scientists and officers had addressed Araujo as *vuesa merced* (your grace), the term used between equals, instead of *vuesa señoría* (your lordship), the term used for addressing those of high rank. They may have been spoiling for a fight with Araujo, as they had begun the meeting by addressing him as *vuesa señoría* only to revert to *vuesa merced* when Araujo did not address them equally. They had not done this out of ignorance; rather, the men fully understood the Spanish hierarchy of salutations and were making it clear that they regarded themselves as envoys of King Felipe, to be treated as equals. Araujo bore this violation of protocol silently throughout the meeting, and in turn violated protocol himself when he did not return the team's visit soon afterward, as he had done with other senior officials. Slighted, the expedition members agreed not to call on Araujo again.[5]

There was now no question of Araujo providing additional funds for the expedition; that door had been slammed firmly shut. La Condamine made the final preparations for his journey to Lima to exchange his letters of credit. During his stay in Quito, he, like his colleagues, had resorted to selling personal items to stay solvent. He had the additional burden of raising enough money to pay for mules, food, and lodging during his trip. La Condamine was by far the wealthiest member of the expedition, however, and he had inexplicably—but fortuitously—brought with him many expensive and superfluous items such as buttons, silks, and jewelry. Through an intermediary he had set up a small shop in the Jesuit seminary where he had first lodged and sold off these items to the fashion-starved members of Quito's elite. Alsedo bought fine linens; Ramón Maldonado bought diamonds and emeralds for his wife, as well as a set of the *Memoirs* of the Academy of Sciences for himself; his brother Pedro Vicente Maldonado bought a set of silver and gold spoons.[6] By January 18, La Condamine had managed to raise enough funds for his trip, so he departed for the two-month overland journey to Lima, presumably with his servant and slave in tow.

To reach Lima, La Condamine chose to travel by land rather than by the faster route of a ship from Guayaquil. He was determined to scout the southern end of the intended chain of triangles, which the expedition planned would end at Cuenca, as well as to investigate some Inca ruins and—most importantly—pass through the region of Loja, where the cinchona tree was found. Joseph de Jussieu specifically intended to examine this plant, its bark being the source of quinine, and he had provided La Condamine with a list of the botanical information required to make a proper assessment. La Condamine spent three days in early February slogging through the mountainous countryside of Loja and gathering information from local collectors on the so-called fever tree before continuing south.[7]

On February 28 La Condamine arrived in Lima under the thick coastal mist, called *garúa* by the locals, which blankets the region for months at a time. The dry and dusty city—it hardly ever rained—had been established in 1535 by Francisco Pizarro, set on a featureless plain on the banks of the Rimac River five miles inland and walled for protection against attack. Its lack of natural scenery was balanced by the ornate architecture of its buildings and the endless display of riches its wealthy population paraded through its narrow streets. Even the lowliest craftsmen in Lima wore clothes embroidered with silver and gold. Instead of walking, most people rode about in mule-drawn coaches, the simplest of which were heavily gilded and covered with rich ornaments. At one point the merchants of the city literally paved several streets with gold to demonstrate their wealth and power. As the seat of the viceroyalty, Lima was the main beneficiary of the riches that came from the mines in Potosí and other locales across Peru, all of which flowed to "the City of Kings" to be minted into coin before being shipped to Spain and its other colonies.[8]

La Condamine had high hopes of quickly trading his letters of credit for hard currency, which appeared to be plentiful in Lima. But when he approached the Spanish businessmen he had been referred to by the Castanier bank, they informed him that all the money from that year's mining operation was now being loaded onto a frigate at the nearby port of Callao, to be sent to Panama. There was not a peso to be had in the whole city. Fortunately, La Condamine chanced upon Thomas Blechynden, the principal agent in Panama of the South Sea Company, which held the British monopoly on trade with Spanish America. Blechynden had traveled to Lima to collect on several debts, but though he was now flush with cash he could not safely send the coin to Europe, for fear of piracy and the ever-present danger of a sudden outbreak of war. La Condamine and Blechynden quickly reached an agreement, and the Frenchman signed over 60,000 livres of credit (out of the 100,000 livres credit he brought) to the British merchant, who could then cash them in later. La Condamine now had 12,000 pesos, the equivalent of half a million dollars, to tide the expedition over until more funds could be sent from France.[9]

The scientist's next port of call was the palace of the viceroy himself. La Condamine had brought letters of introduction from Marguerite-Thérèse Colbert de Croissy, duchess of Saint-Pierre, whose brother, France's foreign minister, knew Villagarcía personally. (Marguerite-Thérèse was also the woman who had introduced Voltaire to Emilie du Châtelet, the friend and lover he shared with Maupertuis.) Villagarcía happily lodged La Condamine in his palace but grew restive when the scientist asked him to approve an extension of their credit from the Spanish government; though he'd already said no to Alsedo, the viceroy reluctantly agreed to put the matter before the Council of Finance, which decided such matters.

In pleading his petition before the Council of Finance, La Condamine had to act as his own solicitor—and in a foreign language, no less. The council heard his case and then deliberated for several weeks, in the course of which La Condamine occupied himself with astronomical and pendulum experiments. He was likely joined by Pedro de Peralta y Barnuevo, the famous astronomer and polymath at the University of San Marcos who had known the expedition's colleague Amedée Frézier when he came to Peru twenty years earlier. La Condamine finally received word that the council had denied his request for unlimited funds but did allow him another 4,000 pesos in credit.[10]

La Condamine did not have much time to bemoan the council's decision, for a new problem had surfaced from an unexpected quarter. While he had been awaiting the council's decision, La Condamine had been astonished to learn that Jorge Juan y Santacilia had just arrived at the viceroy's palace. The expedition's problems with Araujo had considerably worsened soon after La Condamine's departure, and the young Spanish officer had fled to Lima to plead their case to Villagarcía himself.[11]

While La Condamine had been away, four large crates had arrived in Quito containing astronomical instruments, including quadrants that would be critical to the expedition's triangulation work. They had been sent from Paris to Cadiz in 1735, too late for the voyage of Jorge Juan and Ulloa, but had been carried on the next voyage and shipped through Panama to Quito. Trouble had arisen when the crates arrived in Quito. An official had demanded 20 pesos in back pay for the mule transport from Guayaquil, but the treasurer García Aguado, on orders from Araujo, refused to pay. Early on the morning of January 30, Ulloa had sent his slave to Araujo with a letter demanding that the president order his treasurer to pay the fee. Araujo promptly returned the letter, insisting that the lieutenant use the proper form "*vuesa señoría*" in all his correspondence.

Upon receiving Araujo's rebuff, Ulloa impetuously decided to confront the president personally. By all indications, the crown attorney Juan de Valparda was the instigator of this ill-considered action. Valparda had been the host and confidant of the Spanish officers since their arrival, and as leader of the *chapetón* faction he was now engaged in a head-to-head battle with Aguado and Araujo for control of the city council.

Since the expedition was under the protection of the king, Valparda likely reasoned, its members could be safely deployed against the *criollo* faction with little fear of reprisal. The hotheaded Ulloa would have been Valparda's obvious choice to confront the president, as his more reflective friend Jorge Juan—who might have restrained him—was out of the city.

Ulloa barged into Araujo's home around 11 AM on January 30, the same day he had received the president's letter, and demanded an audience. Araujo's secretary explained that the president was convalescing in bed from a recent illness; Ulloa tried to push past the man, but Araujo's mixed-race servant blocked his way. Horrified that "a mulatto would dare to put his hands on my chest," Ulloa shoved him aside, shouting, "How dare you try to stop me from entering!" As he entered Araujo's bedroom, the president's wife tried to calm Ulloa down, further enraging him. "Madam," he shot back at her, "don't meddle in something you don't understand; if you were a man like me you would know what I'm talking about."

Turning to Araujo, Ulloa unleashed a diatribe heard by everyone in the president's household and even on the plaza outside. He did not use the term "*señoría*," Ulloa spat out, because he did not recognize Araujo as his superior. As an envoy of the king, only the viceroy was above him. As a naval officer, moreover, Ulloa was not a member of the militia, so even though Araujo was captain-general of Quito, he did not outrank Ulloa militarily. The lieutenant then pointed to Araujo's baton of office and venomously told him, "you *bought* your military title for 26,000 pesos, and it will expire in eight years," whereas Ulloa had earned his commission on merit. With that final jab, the officer turned on his heel and stormed out into the main plaza, cursing like the sailor he was.

Araujo may have suspected from the "26,000 peso" remark that Juan de Valparda had been behind Ulloa's attack, but the president had a bigger problem on his hands than the exposure of his own corruption. Not only had these foreigners aligned themselves with the *chapetónes* against his *criollo* faction, but now they also seemed to be placing themselves above any authority except the viceroy. For Araujo, this was too much to bear. He immediately called a council of justice to place Ulloa under arrest, a motion that Valparda, as crown attorney, resisted. Ulloa sent word of his predicament to Jorge Juan, who quickly returned to the city and was called to Araujo's home the following day, January 31. After a short meeting, during which Jorge Juan fully supported his friend's position, he found Ulloa, and the two returned to their lodgings at Valparda's house, next door to Araujo's home, to decide their next course of action.

Jorge Juan and Ulloa left the house around 2:30 PM that day and were walking across the main plaza when they were stopped by a motley band of militiamen armed with swords and pistols. The group had been sent to arrest Ulloa. Among the militiamen were Araujo's secretary and his servant, who quickly grabbed Ulloa from behind and threw him against a wall, the officer later recounted. "Then," Ulloa said, "the president's secretary closed on me, and pulling a pistol from his pocket, shoved its barrel against my chest, saying that at the least movement he would fire, as those were his orders." Jorge Juan, a skilled combatant and veteran of many battles against hordes of corsairs, drew his sword and pistol. The militia, ill-trained and with no real fighting experience, stood no chance against him. Jorge Juan wounded two men, including Araujo's secretary (who later died from his wounds); the two officers then ran across the plaza to seek refuge in the Jesuit seminary, where La Condamine had stayed just months earlier.

Both sides were now in a bind. Araujo's militia could not invade the sanctuary of the Jesuit seminary, while Jorge Juan and Ulloa could not even get a letter to the viceroy for fear of interception. The stalemate lasted for a week, until, at 2 AM the morning of February 7, Jorge Juan slipped unnoticed from a small window in the seminary and secretly fled the city. He headed straight to Lima, where he petitioned the viceroy for assistance. They had been shipmates aboard *Nuevo Conquistador*, after all, and that bond would have transcended all rank or privilege.

While Jorge Juan was off in Lima and Ulloa was still locked in the Jesuit seminary, the French members of the expedition who had remained in Quito pleaded with the authorities on behalf of the Spanish officers. The group's testimonials to the viceroy were written in French, for the men had not yet mastered Spanish. The first letter, written by Godin on February 11 and signed by Seniergues, Jussieu, Verguin, Morainville, Hugo, and Godin des Odonais, declared that the officers "had shared our tasks, measures and calculations . . . and faced the same fatigue and dangers" to carry out the Geodesic Mission. A further letter, written by Bouguer on the 17th, added that the Spaniards "brought their enlightenment and all possible vivacity" to the mission. The officers' growing band of supporters among Quito's elite also came to the rescue; José de Zenitagoya, a local judge, wrote to the Spanish court stating that Jorge Juan and Ulloa assisted the French scientists in such a "dedicated, ambitious . . . and Christian way that only a pen tainted in hate and malice could undermine the good name of these loyal gentlemen."[12]

To counter these testimonials, Araujo attempted to place the entire expedition in disrepute by accusing them of selling contraband goods. This charge was easily refuted by those present in Quito, though La Condamine was not in the city to defend himself. When the news of the whole affair reached Lima in late March, so did Araujo's orders to have the crown attorney of Lima search La Condamine's possessions for smuggled goods. They found nothing illegal; La Condamine was more bemused than annoyed, for by now both he and Jorge Juan had discussed their difficulties concerning Araujo with Villagarcía, who issued specific orders to the president to make their problems go away and allow the expedition to continue.[13]

In early May 1737, their work in Lima complete, Jorge Juan and La Condamine boarded a Panama-bound frigate at Callao. They made a leisurely voyage north along the coast, stopping in Paita before disembarking in Guayaquil on May 29 and making their way overland to Quito. The two men arrived in the capital of the *audiencia* on the morning of June 20, just in time to find Bouguer and Godin observing the solstice to confirm the ecliptic they had first measured back in December. That was the extent of the astronomical work accomplished during La Condamine's six-month absence. The expedition had now been away more than two years and had not yet completed a single degree of triangulation.

Although the team had made some powerful enemies in Quito and its scientific progress had been slow, La Condamine's return brought some good news for his colleagues. His voyage to Lima had been a success, netting the expedition 12,000 pesos in hard currency, plus another

4,000 pesos credit and the explicit goodwill of the viceroy; they now had both the funds and the political backing to finally begin their geodesic work in earnest. Ulloa's predicament appeared to have resolved itself; he had departed the Jesuit seminary back in March and now roamed the city without fear of imprisonment. Even the team's instruments had been finally released (though the insistent official never got his payment for transporting them). Araujo, under Villagarcía's orders, dropped the charges of contraband against La Condamine and indeed seemed to hope that everyone would forget that the whole incident had ever happened.

Besides making a tenuous peace with Araujo, the French scientists and Spanish officers also found they had gained new respect in the eyes of the residents of Quito, in part for standing up to the self-important president and his collaborators. The team was now firmly ensconced within the social, political, and intellectual elite of the *audiencia*. Jorge Juan and Ulloa continued to live in the home of crown attorney Valparda, while Bouguer lived in the home of the prominent lawyer and inquisitor of Quito, José Dionisio Sánchez de Orellana, the uncle of the future president of the *audiencia* of Quito and the same official who had once questioned the scientists' Catholic faith before being invited to a pork dinner. La Condamine chose to live independently, purchasing a house on Calle Imbabura near the Church of La Merced, whose tower the team would use for their long-range survey operations. Other members of the team found lodgings in the prosperous Santa Bárbara district immediately to the north. Quito's leading merchant, Lorenzo Nates, acted as the expedition's banker, and the team members counted the Jesuits as close confidants and were welcomed by them wherever they traveled.[14]

The respect that the expedition members were enjoying in Quito was due in part to their singular dedication to their mission, which still remained something of a mystery to most of the inhabitants of the city. Indeed, it was from this lack of comprehension that the visitors earned their unique sobriquet, *los caballeros del punto fijo*, "the gentlemen of the fixed point." José de Zenitagoya, the local judge, had noted in his testimonial to the Spanish court that he understood that the scientists were sent from Paris "for their mathematical enquiries as to the fixed point of the equator." The inhabitants of Peru believed—quite mistakenly—that by determining this fixed point, the scientists would somehow establish the exact size and figure of the Earth. Folklore had it that, for the right price, the expedition could even move the equator all the way to Lima. But if the phrase *caballeros del punto fijo* were said in a certain way, one could hear echoes of Cervantes in its meaning: "knights of the exact." Standing on the plateau of Quito and looking down the avenue of volcanoes that awaited their arrival, they could easily have viewed the peaks as the hulking giants that Don Quixote imagined when he saw the great windmills. These giants, however, could not be conquered with sword and lance but rather with pens, quadrants, and science.[15]

VI

The Triangles of Peru

Bouguer and Godin had not been idle while La Condamine and Jorge Juan were away in Lima. The expedition's instructions from the Academy of Sciences did not specify which geodesic method the scientists should use to determine the figure of the Earth, so Bouguer and Godin had begun the year 1737 squabbling over their options: Bouguer wanted to measure several degrees of latitude by going along the north-south meridian of Quito, while Godin proposed measuring several degrees of longitude east to west along the equator, from Quito to the coast.

The original plans for the expedition had called for comparing the length of a degree of latitude at the equator and one in France, but geodesy was still a relatively new science, and there was little consensus as to the optimum method. The size and shape of the Earth could be also determined by comparing degrees of longitude taken at different points on the globe. Godin argued that by determining the length of a degree of longitude at the equator, they would have a standard reference that any other scientist, at any location on the globe, could use to measure the Earth. Bouguer opposed Godin's plan on the grounds that measuring longitude was not as accurate as measuring latitude; moreover, the terrain from Quito to the coast was dense, flat jungle with few clear lines of sight for a survey.

The argument between Bouguer and Godin went back and forth for months. Bouguer, reaching the limit of his tolerance, threatened to quit if the dispute went on—and then Godin suddenly reversed course and agreed to the latitude survey. Unbeknownst to Bouguer, Godin had received a letter on March 9 from Maurepas, directing him to use the latitude method, as Maupertuis and Clairaut had determined that measuring longitude would be inaccurate. Maurepas' letter had actually been written over a year earlier, but it arrived at a timely moment, fortuitously cutting off a potentially ruinous debate. Still, Godin kept its existence a secret from Bouguer. Once again, the leader was operating independently from the rest of the expedition.[1]

Having finally settled on the latitude method, it was now necessary for the scientists to map out the specific points of triangulation to finalize their route. In March, while La Condamine and Jorge Juan had been petitioning the viceroy in Lima, Bouguer struck out along the road north from Quito, mapping the dense forested terrain for about seventy miles before returning in May. Verguin then went south along the cordillera, returning in June with a map of possible survey points as far south as Riobamba. La Condamine, on his return to Quito in late June, showed the group his map of the area between Riobamba and Cuenca as well as a survey of the region around Cuenca, which could serve as a second baseline for verification. The initial survey maps included the estimated locations of the triangles, as the approximate length of a degree of latitude at the equator was already known to within a mile or so. With this survey, they would know its length to the nearest yard.[2]

As the team gathered all the information together, a detailed plan of attack emerged. The double column of volcanoes to the south of Quito offered the most promising vantage points for triangulation, extending about 3 degrees of latitude (just over two hundred miles) to the city of Cuenca. The men would begin their geodesic work around Quito, using

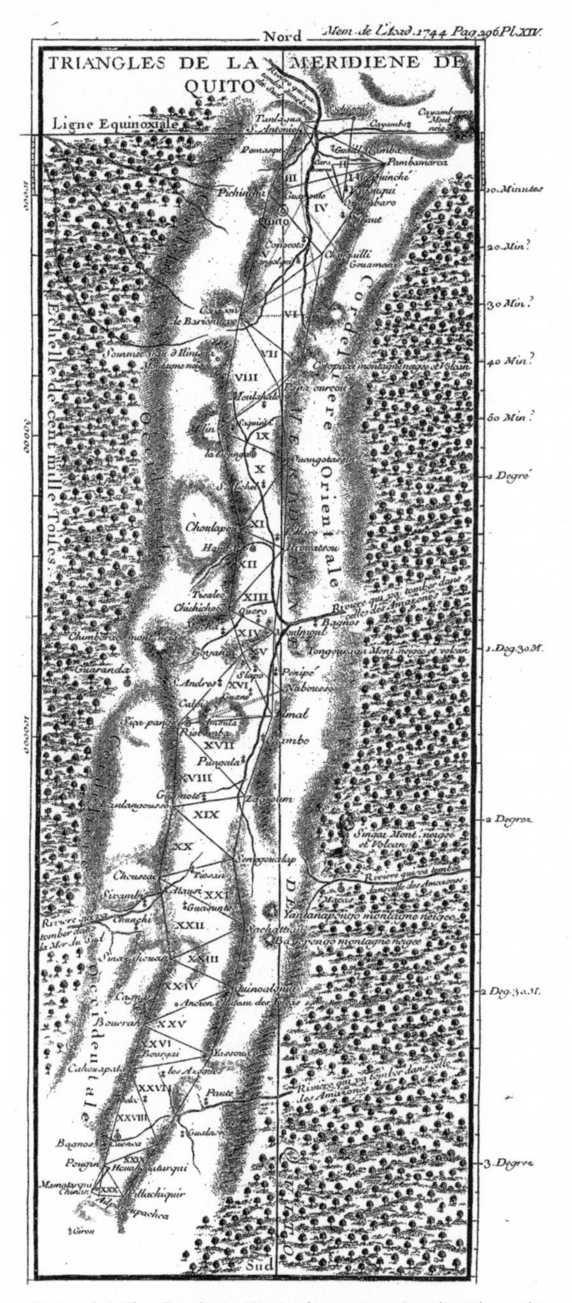

Figure 6.1 The Geodesic Mission's survey triangles along the Andes. From Pierre Bouguer, "Relation abrégée du voyage fair au Pérou" (1744). Credit: Académie des sciences-Institut de France, Paris.

the Pichincha and Pambamarca volcanoes as the vertices of their first triangles, since each had a clear view to the baseline at Yaruquí. They would then move southward among alternating pairs of volcanoes on either side of the great valley and establish stations near the summit of each. To improve accuracy, the survey triangles were laid out to be approximately equilateral, so that each surveyed angle should be around 60 degrees, and the stations were chosen to be roughly the same altitude to minimize corrections for the vertical distance between the observed points.

As at the baseline, the survey would be divided into two parties for both redundancy and speed. Bouguer, La Condamine, and Ulloa would comprise the first party, while Godin and Jorge Juan would be in the second. Advance teams of the junior members (Verguin and Grangier for the first party, Godin des Odonais and Hugo for the second) would go ahead of the expedition leaders to establish the stations and transport the tents, material, and equipment up the mountains; various Indians as well as personal slaves and servants would help carry and maintain these critical items. The two survey parties would follow routes that would keep them out of each others' way, with one party on one range of mountains while the other party was on the opposite range.[3]

From the observation outposts, the two groups would measure the angles that would allow them to create, slowly but surely, an enormous chain of triangles south along the Andes. The baseline the men had measured at Yaruquí would form the leg of the first triangle, which would have its vertex at Pichincha to the west, with a second triangle to the east ending at Pambamarca. Using the length of the baseline and the angles formed between the baseline and the mountain stations, the scientists would calculate the dimensions of these first triangles, which would provide them with one side for each of the next triangles in the chain. The parties would then ascend Pichincha and Pambamarca and measure the angle between two widely spaced stations on the next set of mountains, using Euclidian geometry to fill in the missing sides and angles for each triangle, thereby "closing" it. As each set of triangles was closed, the men would put away the equipment and move on to the next mountain, slowly advancing the chain of triangles southward. With careful coordination, the two parties could enclose the planned thirty triangles with a minimum number of observations, eventually arriving at Cuenca, where they would create and measure a second baseline to double-check their accuracy.

After the expedition confirmed the length of the chain of triangles, routine astronomical observations and mathematical calculations were all it should take to determine the length of a degree of latitude. Once they had the overall length of the chain, the scientists would take simple star sightings to establish the latitude at each end. Dividing the length of the chain by the difference in latitudes would produce a single number, the length of a single degree of latitude at the equator. When this was compared to the length of a degree back in France, they would know for the first time the true figure of the Earth. The process seemed laborious but uncomplicated, and they expected to be finished and on their way home by May of the following year—1738.[4]

During July and early August, the scientists adjusted their quadrants, which would be the workhorses for their survey. Now that the problems with Araujo had been resolved and the shipment from Spain had been released, each scientist had his own quadrant. Roughly two to three feet in radius and constructed of heavy iron to ensure rigidity, the quadrant was the shape of a quarter-circle (i.e., measuring through 90 degrees) with both a stationary arm and a movable arm, which was pinned at the center and swung along the fixed limb (the curved portion, marked with portions of degrees, called "graduations"). To measure the angle between two survey stations, the fixed arm would be aimed at one station, the movable arm swung to align with the second, and the angle between them read from the limb. Both arms were fitted with telescopes to give the greatest possible accuracy; in good conditions they had an angular precision of about 20 seconds of arc—about six inches seen at a mile's distance.[5]

The quadrants suffered damage and distortion during their 7,000-mile trek and had to be recalibrated before they could be used. A typical method for recalibrating the angular measure was to use an extremely long cord to lay out a two-mile-diameter semicircle with the quadrant at the middle, then marking off equal divisions along the circumference of the semicircle. By carefully sighting on the known angles between the markings of the great semicircle, the proper graduations (from a

Figure 6.2 Quadrant of two-foot radius, fabricated by Claude Langlois (1730). Credit: Bibliothèque de l'Observatoire de Paris.

half degree to two degrees apart) could be accurately etched into the quadrant's limb.[6]

While the men were meticulously preparing their instruments, La Condamine also took steps to ensure that the financial preparations for the survey would be handled with the same attention to detail, in contrast to Godin's earlier, profligate spending. La Condamine was, for the foreseeable future, funding the entire expedition from his own personal fortune and was not at all certain of being repaid. By late July, as he carefully noted, he had advanced a sum of 7,988 pesos to expedition members (about half the total available) to carry out the next phase of the work.[7]

On August 14, the two parties departed Quito for the first measurements of angles for the triangulation. The initial angles would be taken at the ends of the baseline at Yaruquí; the first vertex would be the peak of Rucu Pichincha, several miles west of Quito, where Bouguer's party would camp, and the second vertex would be at Mount Pambamarca, twenty-five miles east of Quito, to be occupied by Godin's group.

Climbing over a mile through snow and rock to reach the top of Rucu Pichincha, Bouguer's party moved into the shed erected the previous year by La Condamine and Verguin. It consisted of a few wooden posts covered by reeds and skins, and it offered hardly any protection against the harsh, alpine elements. Despite the spartan conditions, the men set to work, erecting large pyramids of timber, straw, and fabric, whitewashed with lime and lye, to make signals that could be clearly seen through a telescope, even at thirty miles' distance.

The men in both groups managed to make some scientific observations while atop the volcanoes—though neither party was able to obtain the most sought-after measurements. The men managed to measure the altitude and temperature of their stations using the latest instruments: precision barometers to record heights and new thermometers, developed by their Academy of Sciences friend René Antoine de Réaumur, to record temperatures. Nighttime temperatures dropped to −5 degrees Réaumur (about 20 degrees Fahrenheit), and the thin air quickly dissipated any heat. As Bouguer observed, "even with a pan alight with coals in our midst, and candles all around, the water froze in our very drinking glasses." Fog and mist frequently covered the mountain, and worse still, when the weather did clear for a few minutes, Bouguer's party could not sight simultaneously on the Yaruquí baseline's endpoints and on Pambamarca, as one or the other was then obscured by clouds. Godin and Jorge Juan experienced similar hardships on their mountain. For twenty-three days they endured the bitter cold and biting winds without measuring a single angle, so in early September both parties returned to Quito.[8]

Their time on Pichincha and Pambamarca, the men later realized, had baptized them into their next two years of existence. The climb up the mountains had been treacherous, made worse by the wind, cold, and sudden fogs that obscured the narrow, rocky paths; both Jorge Juan and Godin had suffered dangerous falls. But the real danger was far more insidious, for they had also suffered from a condition that few Europeans had ever experienced: altitude sickness, colloquially known as *soroche*, a term derived from the word for a silver ore that locals believed was the source of toxic vapors that caused the condition. Altitude sickness is today understood to be the result of the lower oxygen level, which, at the heights at which the expedition was operating, was about half that of sea level. Lowered oxygen levels can cause swelling in the brain but affect people in unpredictable ways; the young and healthy can be laid low while an older person may have no symptoms. Ulloa, who was just twenty-one, collapsed unconscious during one part of his climb; La Condamine, age thirty-six, had bleeding gums; others vomited. Meanwhile Bouguer, who was approaching forty, frequently ill, and the worst sufferer of seasickness, was unaffected by the heights.[9]

The Indians who accompanied the scientists to the tops of Pichincha, Pambamarca, and subsequent peaks didn't have nearly as difficult a time as the European scientists. For one thing, their ancestors had lived in the Andes for many thousands of years, and they had evolved massive lungs and an incredibly efficient circulatory system that enabled them to endure the altitude and cold with little apparent effort. Yet even they did not stay overnight on the mountain summits, wisely descending instead to a lower altitude and returning in the morning to clear the snow off the scientists' hut and tents. The Indians had also learned to chew coca leaves and drink a tea (called *mate*) brewed from the plant, both of which alleviated the symptoms of altitude sickness; even today, visitors to the region follow these customs. Though they do not state as much in their journals, it is likely that the expedition members would have picked up these practices as they became more experienced mountaineers. They believed, as did all Europeans before the Himalayas were surveyed, that the Andes they were now traversing were the tallest mountains in the world. Indeed, the expedition members had already climbed higher than any European explorer; at over 15,000 feet, Pichincha is as tall as Mont Blanc, the tallest mountain in Western Europe, which would not be ascended for another fifty years. Yet the men would have to climb higher still, and endure even greater hardships, before they could complete the triangles of Peru.

After recuperating in Quito for a few days, the parties reconsidered their plans. Their first action was to erect new signals around Quito, based on the clearest viewing angles from Pichincha and Pambamarca; one signal would be on the plain of Schangailli (nowadays Sangolqui) to the southeast, another at the hacienda of Cochesqui to the northeast, and a third on the treacherous peak of Tanlagua, twenty miles north. Next the men repositioned the station on Pichincha; it was now obvious that the summit was too fogbound, so in mid-September Bouguer's party established a new post some 1,300 feet lower. They remained there for another two months but were again unsuccessful in trying to take sightings on Pambamarca, the baseline, or the newly erected signals.

It was in the new station on Pichincha that Bouguer learned why Godin had suddenly agreed to take latitudinal measurements instead of longitudinal ones. On September 27, Bouguer received his own copy of the letter from Maurepas, delayed in transit for unknown reasons, which had instructed Godin to abandon the longitude measurement. Bouguer and La Condamine, now realizing that Godin had hidden this instruction since March, must have fumed at Godin's subterfuge, but impotently—for the object of their scorn was back in Quito readjusting his quadrant, visible through their telescopes far below.[10]

Both parties returned to Quito on November 7, having failed yet again to take the necessary measurements. The teams spent the rest of November and most of December futilely attempting further observations on Tanlagua and at the Yaruquí baseline. Then, finally, in late December the weather cleared enough for Bouguer's team to measure their first angles from Pichincha. It had taken over four months to get these preliminary measurements.

As 1737 drew to a close, the expedition members ruefully looked back on an entire year spent with almost no survey results to show. Yet there had been some progress: La Condamine's voyage to Lima had shored up their political and financial support, Jorge Juan and Ulloa were somehow managing to keep the warring French scientists together, and the entire expedition was working toward a common goal. They could now look forward to the new year with some hope.

The year 1738 began with a spark of inspiration from the normally recalcitrant Godin. One of the expedition's thorniest problems had been the fact that the wood-and-fabric signals used to guide their measurements were frequently either blown down by the wind or carried off by local Indians, for whom timber was a precious commodity on the treeless highlands. The signal at Pambamarca had been repaired seven times during their stay, and each time it took a week or more for the weather to clear so that the scientists could note the problem, then trek to the site, re-erect the signal, and return to make observations. Now Godin came up with the simple idea of using their tents as the signals, as they were of very sturdy construction and well recognized by the locals as the property of the scientists. In late January, Bouguer's party climbed Pambamarca, this time successfully taking sightings on the new signals set up around Quito, and finally closed the first of their series of triangles.[11]

At the same time that they completed the first set of triangles, the members of the expedition were treated to a sight rare at the time: an atmospheric phenomenon called a glory, fogbow, or cloudbow. Seen quite commonly today by airline passengers and immortalized as "Ulloa's ring" or "Bouguer's halo," the phenomenon occurs when the sun is at one's back, casting a shadow on a cloud or fogbank. Both Ulloa and Bouguer noted its otherworldly beauty: "We saw as if in a mirror an image of each one of us, and centered upon our heads were three concentric rainbows. . . . As the subject moved, the circles moved with him," remarked Ulloa. Bouguer wrote, "We could even distinguish the arms, legs, head and the entire body of each individual."[12]

The glories were a pleasant interlude in what was becoming an excruciatingly long slog. The scientists had planned to create a chain of thirty triangles, but by the end of January 1737 they had managed to close exactly two. During the next three months the parties managed to close another four triangles, successively ascending the volcanoes of Guamaní, Corazón, and Cotopaxi. Though the men were becoming more skilled at surveying, the elements still conspired against them. The various accounts by the members of the expedition tell the same story. There on the equator it was not the heat but the cold that affected them most. The nights were bitter and the parties were often assailed by hailstorms and blizzards whose fury had no equal back in Europe. Ulloa memorably recounted their misery:

> We generally kept within our hut. Indeed, we were obliged to do this, both on account of the intenseness of the cold, and the violence of the wind. We heard the horrid noises of the tempests, which then discharged themselves on Quito and the neighboring country. We saw the lightning issue from the clouds, and heard the thunder roll far below us. . . . We

> suffered from the asperities of such a climate. Our feet were swelled, and so tender we could not even bear the heat, and walking was attended with severe pain. Our hands were covered with chilblains, our lips swelled and chapped, so that any speaking or other movement of the mouth led to immediate bleeding. Our common food in this inhospitable region was a little rice boiled with some meat or fowl; instead of fluid water, our pot was filled with ice; and while we were eating, every one was obliged to keep his plate over a chafing dish of coals, to prevent his provisions from freezing.[13]

It is doubtful that the scientists, back when they were planning the expedition within the comfortable surroundings of the Louvre palace, could have envisioned the piercing cold and harsh winds they now endured. The weeks spent at the Yaruquí baseline laboring dawn to dusk under the equatorial sun must have seemed an almost pleasant memory.

The extreme weather conditions also hampered their observations. The equatorial skies, the men now realized, were almost constantly covered by dense clouds, a phenomenon (unexplained until the nineteenth century) caused by the convergence of trade winds from the Northern and Southern hemispheres, coupled with the intense convection of moist tropical air. The expedition had not counted on how persistent the cloud cover was, having guessed a year before that they would be finished by May 1738. That date was fast approaching, and they had not even completed a quarter of their survey.

Time weighed heavily atop the cold mountain stations, where the teams had to wait anywhere from a week to a month at each station before the skies would clear for the few minutes needed to take a single sighting. Most of the men probably worked at their journals and diaries to pass the time, but Bouguer, still officially a royal professor of hydrography, brought with him the manuscript he had already started at Maurepas' behest, before the expedition was even planned. He was developing a groundbreaking concept in naval architecture: that ships could be designed using mathematical rules based on physics, instead of timeworn rules of thumb.

When Bouguer had been called to join the Geodesic Mission, he brought his manuscript with him to complete during his extended absence.

During the long intervals on the mountains when observations were impossible, he explained to his friend Réamur, he spent his time "meditating on the construction of vessels and writing a Treatise [of naval architecture]. I am trying to finish it here, in order to then weigh up the rules at leisure, for as far as it will be possible, on the voyage that we will make to return to France."[14] High atop the Andes and miles from the ocean, a new science of ship design was being born, one of the many unexpected outcomes of the Geodesic Mission to the Equator.

In April, after successfully closing six triangles, the entire expedition was obliged to return to Quito for *semana santa* (Holy Week) and Easter, or risk the suspicions of Bouguer's host, the head of the Inquisition. The city was now much calmer than when Ulloa had his run-in with Araujo the year before, and the political turmoil of 1736 was all but forgotten. The president had reached a truce with the *chapetón* faction, and Alsedo had departed Quito for Cartagena de Indias the previous October, thus removing one of the principal instigators of the conflict.

For all of its appearance of calm, Quito was still a dangerous place. Duels were a routine occurrence, and most commoners carried daggers, while upper-class men wore swords and pistols. The small standing army that Araujo had organized soon after Jorge Juan's daring escape was now disbanded. It had really been Araujo's personal bodyguard rather than a proper troop of soldiers, but without it the city was once again left with no police force other than a handful of volunteer night watchmen.[15]

Given the lack of a police force in Quito, it would have come as no surprise that in early May, when Bouguer's slave was fatally knifed by a local *mestizo*, the culprit was not prosecuted or even pursued. Killings were often a weekly occurrence in the city. Indeed, the murder of the lower classes by whites and *mestizos* had attained a mythic status through the legend of the *pishtaco*, which had sprung up soon after the Spanish had colonized the country. The terrifying *pishtaco* was invariably thought to be a white or *mestizo* male, often handsome and well dressed, who hunted Indians and blacks at night and killed them for their body fat, which he then reportedly boiled down and sold as grease to keep wool and sugar mills running, or to metal foundries for use in casting church bells. Parents would commonly put mischievous children in their place by warning that the *pishtaco* would come to get them, a practice that continues even to this day.[16] The Frenchmen, however, were men of science; they would have no more believed stories of *pishtacos* than they would have the tale of *Cendrillon* (Cinderella) conceived by their countryman Charles Perrault.

Legends and rumors were rife in Quito—and not just about *pishtacos*. Many of the Peruvian locals refused to believe the tale that these *caballeros del punto fijo* had really traveled all that way just to measure the Earth. One day, Ulloa, dressed in the commoner's garb the team members wore for surveying, was walking to an appointment when a gentleman who knew him as part of the Geodesic Mission stopped him and, taking him for a servant, explained that he was not convinced by their explanation that they were in Peru for scientific reasons. "Neither he nor anyone else would believe," reported Ulloa, "that the ascertaining of the figure and magnitude of the Earth, as we pretended, could ever induce us to lead such a miserable life; that, however we may deny it, we had doubtless discovered many rich minerals on those *páramos* [high prairies]."[17] The comment might have struck Ulloa as particularly ironic, given the continued financial problems the expedition was facing.

The team was planning the next phase of the survey, a yearlong trek to carry the triangulation all the way to the endpoint at Cuenca, but they would have to work fast to make every peso count. As Bouguer noted, "our expenses have doubled when they should have halved; the loan Mr. La Condamine made is already spent; the 4,000 pesos promised by the viceroy is only arriving in dribs and drabs." Back in April, the expedition had received 4,000 pesos from France with a promise of 8,000 in the future. But this aid came with a stern warning from Maurepas: "You will maintain your accounts in order to explain to the king your expenses. . . . We are convinced that this sum will suffice for all further expenses and needs." The royal checkbook had now been firmly closed. To hedge against further debt, Godin took out a loan of 3,400 pesos from Pedro Vicente Maldonado, the younger brother of Ramón Maldonado, the wealthy local magistrate whom La Condamine had quickly befriended soon after his arrival in Quito in 1736. Pedro Vicente's loan was just one of many acts of kindness and support from the Maldonado family that helped keep the mission going during its most grueling days.[18]

Pedro Vicente was born in 1704, one of nine children of the wealthy and well-connected Maldonado family of Riobamba. Trained as a teacher, he had received the finest Jesuit and university education before embarking in 1734 on a project to create a new road from Quito to the Pacific coast along the Esmeraldas River, the same project that his older brother Ramón had described to La Condamine upon the Frenchman's arrival in Quito. In 1737 Pedro Vicente had been named governor of the Esmeraldas province. In April 1738 he had providentially been in Quito to present his plans for the Esmeraldas road to President Araujo when he learned of the expedition's troubles and promptly offered the loan to Godin.[19] The entire expedition would come to see Ramon, Pedro Vicente, and the extended Maldonado clan as hosts and surrogate families during their stay in the *audiencia* of Quito.

With the money from France and the loan from Maldonado, the scientists could now equip themselves to carry out their scientific labors. During their three-month stay in Quito, they purchased mules, saddles, and other supplies in preparation for the long triangulation. They also conducted various experiments, including one on the speed of sound at high altitude, which they tested by dragging a pair of small cannon to hills at opposite ends of the city and timing the interval between the flash and the roar of the discharge.[20]

On July 9, 1738, the entire expedition trooped south out of Quito along the same road they had used to enter the city two years before. Three days later they arrived at the Corazón volcano, where they made several joint observations before dividing into their respective parties and leapfrogging past each other on the way south.[21] The two parties crisscrossed the cordillera in opposite directions, Bouguer's party always on one side of the valley while Godin's was on the other—an arrangement that doubtless pleased both men. Bouguer and his colleagues spent a month measuring angles on Corazón before crossing the wide valley to Papa-Urco, a much easier survey station, where they completed their sightings in just three days.

On August 16 Bouguer and La Condamine set out from a nearby hacienda toward the foothills of Cotopaxi, where they were to make their next sightings. Ulloa had already gone with the advance team to set up the instruments and signals. The two scientists soon found themselves in open countryside with nightfall approaching and could find no shelter against the cold. As La Condamine later explained, "Our saddles served as headrests and M. Bouguer's coat served as mattress and blanket; a varnished taffeta cape that I had luckily brought with me, secured to the ground with our hunting knives, became our tent, which sheltered us from the frost which fell all night." When day broke, the men found that a thick fog had rolled in, causing them to become disoriented. Their mules disappeared, and only La Condamine relocated his. As the fog cleared, he spotted the pair's Indian porter carrying their tent poles and a bit of bread. "I sent him back along the trail to share his provisions with M. Bouguer and to help him find his mount," noted La Condamine. "I shortly thereafter found the rest of our party who had set up camp. . . . My first task was to send a mule and some food to M. Bouguer, and then to use my quadrant to take advantage of the fine weather."[22] This almost tender account of their survival demonstrates how much La Condamine and Bouguer had come to depend on each other.

La Condamine and Bouguer's shared hardships and experiences had deepened the bond between these two dissimilar men, and that bond was becoming the nuclear force at the core of the expedition. Godin's influence in the overall group had waned to a pale insignificance—he now only spoke to the two Spanish officers—and his previous indifference was deteriorating to outright hostility. "Things are getting worse and worse with Mr. Godin," complained Bouguer. "I have not seen him for over four months, and we only write to each other when it is necessary. . . . I do not understand his anger or his fury."[23]

With Godin growing more withdrawn, Bouguer now assumed the leadership of the expedition and issued orders for carrying out the remainder of the survey. Even their banker, Lorenzo Nates, "will no longer authorize payment to Godin without my intervention," explained Bouguer in a letter home.[24] The outcome of the entire Geodesic Mission, it seemed, now rested on the shoulders of the man who had been the least willing to go.

Despite Godin's continued intransigence, the two parties worked quickly during the months of September and October, closing another seven triangles. Fine weather, combined with Bouguer's orders to have each party measure only two angles of a given triangle instead of all three (reducing redundancy but saving time), enabled this acceleration. In one instance, on Mount Nabuso, Bouguer's party was able to set up their station, measure all the angles, and descend in the space of twenty-four hours. On November 8, 1738, the two parties arrived at Riobamba, halfway to their final destination of Cuenca, having now measured a degree and a half of latitude.

Once the expedition had reached the halfway point of their surveying, Godin returned to Quito with Jorge Juan to resolve some unnamed financial matters. While away, Godin fell ill with a form of malarial fever and was delayed in his return. Ulloa also fell gravely ill with an unspecified ailment, remaining in Riobamba. The rest of the expedition took advantage of an invitation by José Dávalos (brother-in-law of Pedro Vicente Maldonado) to stay nearby at his spacious hacienda of Elén, where they would recover from their maladies and recuperate at the nearby thermal springs for the next three months.[25] (Although the hacienda no longer exists, tourists still flock to Los Elenes, as it is now known, to take the curative waters.)

While in Quito, Godin was undoubtedly treated for malaria by the expedition's doctor, Joseph de Jussieu. Yet during this time, Jussieu's name is curiously absent from the published accounts, as are the names of his botanizing assistants, the surgeon Jean Seniergues and the draftsman Jean-Louis de Morainville. Indeed, from the expedition's arrival in Quito in mid-1736 until early 1739, little information exists on their whereabouts. The available evidence points to the likelihood that the expedition's crippling lack of funds and the intense scrutiny of the Spanish authorities had confined them to Quito for many months—or, in the case of Jussieu, for several years.[26]

An official report by Alsedo, who had departed Quito for Cartagena de Indias in October 1737 and encountered Seniergues when the surgeon had arrived there in 1738, gives some indication of the botanists' activities while in Quito.[27] According to Alsedo's testimony, they had occupied themselves "not only with the care of their companions, but with [the treatment] of their ailing neighbors of the city," and in fact Seniergues had come to Cartagena specifically to replenish the group's medical supplies, which had been expended in their humanitarian efforts.

As there were few European-trained doctors or surgeons in Quito, the Frenchmen's services would have been in great demand. Accompanied by their friend, the émigré French doctor Raimundo Dablanc (who, back in 1736, had loaned money to the expedition to carry out its initial geodesic surveys), they would likely have made the rounds dressed in the peaked hats, red coats, silver-buckled shoes, and blue *paño azul* capes that marked the local doctors as distinctively as their gold-handled walking sticks.[28] The men certainly did not lack for work, as the church-run hospitals were frequently overcrowded, understaffed, and in disrepair. Sickness and death were constant neighbors in the eighteenth century, wherever one found oneself; child mortality rates in "backwards" Peru were actually little different from those in "advanced" Europe. On both continents, parents knew that one in four newborns would die in the first year, and half their children would not live to adulthood.[29] The anguish of loss was not diminished by this knowledge, and neighborhoods in both Europe and the Americas regularly resounded with the keening of bereaved mothers and fathers.

Infectious diseases were the primary cause of this suffering. In the eighteenth century, they raced around the world in epidemic cycles, borne between Europe and remote regions such as Peru by caravans and ships carrying goods throughout the already globalized economy. None of the dozen or so great killers—including measles, yellow fever, smallpox, malaria, and diphtheria—originated in South America. Many of these diseases incubate in domesticated herd animals (pigs especially) that live in close proximity to humans, whereas for much of their history the Indians had only guinea pigs and llamas, which made for bad hosts. The diseases came with the first conquistadores and quickly spread far beyond their original contact with the indigenous peoples of the Americas, devastating their populations long before most of the victims ever saw a European. Smallpox, for example, likely killed off much of the Inca population in the late 1520s, including its great ruler Huayna Capac, leaving his kingdom in disarray and making it easy pickings for Pizarro and his conquistadores when they arrived just a few years later. By the beginning of the 1600s, the Indian population around Quito had been cut down by almost 90 percent and was rebounding only very slowly. Every decade or so another epidemic laid the population low; the last great epidemic of 1718–1723, most likely influenza, destroyed a quarter of the Indian population in Peru before extinguishing itself.[30]

The two medical men who accompanied the Geodesic Mission were likely overwhelmed with tending to the city's sick and dying, while their countrymen were off measuring the Earth. Nevertheless, they had different attitudes toward remuneration for their services. Jussieu, frequently depressed at his inability to botanize the medicinal plants he'd come so far to collect, profited little in monetary terms from his medical expertise. From his letters and the descriptions written by his family and colleagues, he appears to have been a humble man, quite uninterested in fortune, and frequently dependent on the charity of his growing circles of friends in Peru to lodge and care for him.

Seniergues, by contrast with Jussieu, benefited handsomely from his surgical skills. He had jump-started his fortune back when the group had first arrived in Guayaquil by successfully treating the cataracts of a wealthy resident of the port city. His practice afforded him enough money to support Jussieu as well to finance his own overland journey to Cartagena de Indias, where, in addition to retrieving medical supplies and some scientific instruments for the expedition, the surgeon bought Spanish garments and other merchandise for resale and acquired two personal slaves. While in Cartagena, Seniergues also befriended the Spanish admiral Blas de Lezo, who had arrived in March 1737 to take over the defenses of the city, now under increasing threat from the renascent British navy. Lezo, who had commanded Jorge Juan as a cadet and likely had a hand in his selection to the Geodesic Mission, was incensed to hear the surgeon's tale of the expedition's financial woes and promptly loaned the group 4,000 pesos.[31]

Seniergues' return to Quito at the beginning of 1739 apparently roused Jussieu from his melancholy. The doctor of medicine now decided it was time to see with his own eyes the famous cinchona tree, which until then he knew only through La Condamine's enthusiastic but unbotanical descriptions. Together with Morainville, Jussieu and Seniergues planned their first botanizing expedition to the mountainous Loja region. Seniergues, now flush with money from his profitable trip to Cartagena, would fund their botanical expedition as well as provide a substantial loan to Godin to carry on the triangulation.[32]

Jussieu, Seniergues, and Morainville departed Quito for an initial rendezvous with the rest of the company in Riobamba in February 1739 and then left their companions on March 22, traveling several weeks before arriving in the forests of Loja. They spent three months botanizing in the region, at last making detailed observations of the only known source of quinine: the cinchona tree. The scientists were not the first Europeans to have explored the tree's healing properties; quinine, in fact, was called "Jesuit's bark," after the priests who first learned of it from the local Indians and who now controlled its export. European importers awaited every shipment of the dried, pulverized bark as eagerly as those of gold and silver. Scientific understanding of the tree's medicinal value, however, lagged far behind its application. Although quinine's use was by now widespread, its effectiveness was very much hit-or-miss: Sometimes the bitter powder cured malaria; sometimes it didn't. There was a pervasive suspicion of fraud (such as mixing cinchona bark with that of other trees), but since no botanist had carefully studied the tree in its natural state, no one knew how to detect a forgery.[33]

La Condamine's earlier observations of the cinchona tree had been a milestone in the study of quinine but were not particularly accurate. When La Condamine had passed through Loja on his way to Lima in 1737, he saw only a few trees in their natural state, getting most of his information secondhand from bark collectors. During his stay in the capital and on his return voyage to Quito, he wrote his observations into a short memoir, which he had sent off to the Paris Academy of Sciences. The memoir, read before the Academy in 1738, provided Europe its first eyewitness report of the miraculous fever tree.[34]

La Condamine's short memoir on the cinchona tree had all but ensured his steady rise as one of France's most celebrated explorers, but its anecdotal stories left much room for improvement. La Condamine said that the plant was known locally as the *quinquina* (bark of barks) and came in three varieties distinguished by the color of the inside of the bark. Red bark, La Condamine had been told, was the most potent, followed by yellow and then white. When dried, the barks look very similar, which partially accounted for the wide range of effectiveness of the medicine. After setting out a history of the cinchona's use, La Condamine explained how Indians collected and dried the bark, and he described the tree's physical appearance with rough sketches of its leaves and fruit. His observations were rife with errors, but they nevertheless gained traction in Europe; the Swedish botanist Carl Linnaeus used La Condamine's faulty information to describe the tree in his famous taxonomy of plants *Genera plantarum*, naming it the cinchona after the legend (later proved to be completely fabricated) of the Countess of Chinchón, the wife of the Peruvian viceroy who in the 1630s was supposedly cured of malaria with its powdered bark.

Finally energized and with access to funds, Joseph de Jussieu now intended to improve on La Condamine's flawed account by intensively studying the trees in their habitat. From April through June 1739 he and his small team journeyed through the Loja countryside. He and Seniergues carefully collected and cataloged the plants, while Morainville sketched and painted beautiful aquarelles of the leaves and flowers, as well as of the birds and wildlife of the region.

The team carefully laid out the fruits of its labors in an eighteen-page memoir handwritten in Latin, *Descripto arboris Kina Kina*. Compared with La Condamine's almost chatty report, this was clearly the work of a well-trained botanist. Jussieu carefully delineated the differences between the various species of cinchona tree (he identified seven) and explained the plant's unusual mechanism of propagation, the fruit swelling and then literally exploding, casting the seeds far into the surrounding jungle. Most importantly, he described how a trained person could distinguish between the different types of barks by taste and smell to ensure that only the most potent variety was used. *Descripto arboris Kina Kina* was a work of seminal importance, one that would have marked Jussieu as an equal to his more famous brothers. As the group finished their observations, he was eager to send plant samples to his brothers in the Royal Garden of Paris and to send his manuscript and Morainville's accompanying drawings to the Academy of Sciences for publication. The

Figure 6.3 Cinchona tree, watercolor by Jean-Louis de Morainville (1739). Credit: Bibliotheque Centrale du Museum National d'histoire Naturelle, Paris.

men could not have known that their work would not be seen publicly not just during their lifetime, but for almost two centuries.[35]

While Jussieu and his team were cataloging their findings about the cinchona tree, the astronomers had been recuperating from their labors and preparing for the final phase of their survey. Bouguer, La Condamine, and Ulloa remained at the Elén hacienda through the beginning of 1739, enjoying the hospitality of the Dávalos family, as they would many times in the future. The warm springs near the river no doubt aided their recovery, but the household itself also provided a much-needed diversion, in the form of the three daughters of the family: María Estefanía, Magdalena (married the year before), and Josefa, who although just ten years old was already fluent in French and could translate the family's French encyclopedia into Spanish as quickly as if she were reading her native language. La Condamine was entranced by the eldest daughter, the unmarried María Estefanía, noting that "she could play any instrument she cast her eyes upon, and painted in miniature and in oils without ever having had a teacher." He wistfully lamented that María, with all her talents, only had desires to become a Carmelite nun.[36]

The scientists' three-month distraction from geodesy had been sorely needed. Back in September 1738, they had received the news from France they'd long been dreading: Maupertuis and his team had indeed finished their survey at the Arctic Circle and had triumphantly returned to Paris with their findings.

The letters from Europe had recounted the story of the second Geodesic Mission in painful detail. Maupertuis and Clairaut had been accompanied on their expedition by two other Academy members and the young Swedish scientist Anders Celsius, later famous for his eponymous temperature scale. In May 1736 they had left France for Stockholm, where they had been fêted by the king of Sweden before voyaging north to the Gulf of Bothnia. In July 1736, they had begun surveying and triangulating north along the Torneå River. Unlike the equator mission, Maupertuis had decided to leave the measurement of the baseline until the end of his survey, when the rivers had frozen and the team could carry out the survey across flat ice. The scientists had decided to survey about fifty-seven miles, just under a degree of latitude, as compared with the two-hundred-odd miles that Godin, Bouguer, and La Condamine were tracing. Under Maupertuis' firm leadership and aided by a platoon of Swedish soldiers, they took just three months to finish their triangulation. By December they had established a nine-mile baseline across the frozen river and made their astronomical sightings at each end of the chain of triangles. They conducted additional observations the following spring.

By August 1737 Maupertuis was back in Paris, having spent little more than a year away but able to claim that the Earth was officially oblate. He had reported to the Academy of Sciences that his team had measured a degree of latitude at the Arctic Circle as 57,437 *toises*, considerably longer than the 57,060 *toises* measured by Picard and Cassini at Paris. For most scientists as well as the public, this was solid proof that the Earth was flattened at the poles and that the scientific acolytes of Descartes had been wrong; Voltaire called his friend Maupertuis the "flattener of the Earth and of Cassini."

Although its members surely felt that Maupertuis and his team had beaten them to the punch, the Geodesic Mission to the Equator still had a role to play in the debate over the figure of the Earth. Jacques Cassini, for one, had not been impressed by Voltaire's barbs; after criticizing the many flaws in the expedition's measurements, he huffed that the matter would not be settled until the equatorial mission returned.[37] For the Cartesians in the French Academy, the expedition now toiling in South America was the last hope to vindicate the theory of an elongated Earth. For Maurepas, the sponsor of both the Arctic Circle and the equator missions, the primary issue of navigation was still unresolved, for both sets of data would be needed to accurately determine the precise dimensions of the planet.

For the scientists laboring in faraway Peru, the situation would have seemed even grimmer. They knew their work was still important, but they would have worried that its impact would be lessened in the eyes of their Academy colleagues by the results from the Arctic Circle. The news of Maupertuis' victory would likely have hit Bouguer hardest of all. Maupertuis was the one figure at the Academy who had consistently attempted to undercut him and derail his rise to the scientific elite. Bouguer had agreed to join the Peru expedition in order to help cement his reputation as a first-rate scientist, and now it would likely have appeared to him that Maupertuis' victory had undermined him yet again.

Soaking away his injuries and disappointments in the thermal springs at the hacienda and facing several months of forced leisure, the restless Bouguer now fixed on an experiment that might, perhaps, thrust him back onto the scientific center stage. He had long imagined finding a direct way to measure, and therefore validate, Newton's gravitational force of attraction. Though the *Principia* was now fifty years old, no one had yet devised a method to prove the theory that gravity attracted bodies as a proportion of their mass and diminished as the square of their distance from each other. Bouguer's great insight, which he had struck upon back in France, was to understand that if a mountain were large enough, its mass should exert enough gravitational pull to move a weight in a measurable way. None of the European mountains were sufficiently massive to facilitate this sort of experiment—but here at the equator, surrounded by what were thought to be the world's largest mountains, Bouguer could hope to find one with enough mass to test his theory.

The obvious candidate for Bouguer's experiment rose into the sky just fifteen miles to the northwest of the Elén hacienda. Chimborazo, then assumed to be the tallest mountain in the world, was the only major volcano in the region that the expedition did not use for its triangulation: a colossal, isolated cone whose base measured twelve miles in diameter and whose peak, at four miles in altitude, dominated the landscape for a hundred miles around. According to Bouguer's calculations, the mountain's mass was negligible compared with the Earth (about seven billion times less), but he could nevertheless bring a highly accurate pendulum bob to within a few thousand yards of its center, where the gravitational pull of the mountain should be enough to measurably deflect the pendulum from true vertical.

Bouguer's gravitational experiment was admirably simple. To establish the true direction of vertical, he would travel far away from the mountain, and use the twelve-foot zenith sector to measure the angle of the plumb bob to fixed stars overhead. Closer to the mountain, he would measure the new angle of those same stars to the plumb bob, now—he predicted—pulled ever so slightly toward the mountain. Subtracting the first angle from the second would give Bouguer the deflection caused by the mountain's gravity, which should enable him to directly calculate Newton's force.

In December 1738, Bouguer, La Condamine, and Ulloa had set up camp on the flanks of Chimborazo to carry out the experiment. There they observed the angle of the plumb bob to ten stars, including Sirius and Aldebaran. They then moved four miles due west, far from the mountain's tug of gravity, and measured the plumb bob against the same ten stars to establish true vertical. Unfortunately, because of the extreme cold, bad weather, and problems with the instruments, the scientists did not achieve the hoped-for precision. Bouguer had initially estimated that the plumb bob would be deflected by 1' 43", but the observed deflection was only around 7", presumed to be far too small a discrepancy given the gravitational attraction that should have resulted from Chimborazo's enormous size. While the results were not close to Bouguer's prediction, they nevertheless suggested that the mass of the volcano was, indeed, exerting a degree of gravitational influence over the pendulum—the first such example of Newton's theory of attraction.

Bouguer was puzzled by the result of his experiment (going so far as to suggest that the volcano was hollow) but was nevertheless pleased that he was the first to prove Newton right. He quickly sent his memoir to Paris, where it was read before the Academy in October 1739. Bouguer's findings didn't have the effect that he had hoped; his memoir apparently evoked yawns from his fellow scientists at the Academy, who after Maupertuis' return may have momentarily forgotten Bouguer was in Peru at all. Bouguer's findings were not even published in the Academy's annual collection of memoirs, forcing Bouguer to append it to his own account of the expedition several years later.[38]

Neither Bouguer nor any other scientist repeated the pendulum experiment in his lifetime, although it was never wholly forgotten. In 1774 the British astronomer Neville Maskelyne, attempting to determine the mass of the Earth, carried out a similar experiment on the Scottish mountain Schiehallion, giving a nod to Bouguer without mentioning him by name. Other scientists were more vocal in their appreciation of the importance of Bouguer's initial findings at Chimborazo. As later scientists carried out more extensive surveys of the planet, they found that the Earth's crust was not a uniform slab, but rather a highly complex structure riddled with areas of greater and lower densities, like raisins and voids in a loaf of bread. By the early 1800s, those variations in density—which, among other things, geologists now use to find oil and minerals—became known as Bouguer gravitational anomalies, or simply Bouguer anomalies, in honor of the man who first attempted to directly measure gravity.

In February 1739, as Bouguer had been puzzling over the results of his gravitational experiments, Godin and Jorge Juan finally returned to Riobamba. They were accompanied by Jussieu, Seniergues, and Morainville, who would shortly depart on their pathbreaking botanical expedition to Loja. The remainder of the team, meanwhile, would continue the surveying project along the peaks of the Andes. Godin, who had been convalescing in Quito since November, was finally well enough to continue the surveying work, and now that the two survey parties were back to full strength, they could complete the chain of triangles to Cuenca, just over a hundred miles distant.

An observer watching the survey parties trek from one mountain to another would have been hard pressed to distinguish them from the multitudes of *chapetónes*, *criollos*, and Indians taking the same roads. The French scientists and Spanish officers, riding horses at the head of the mule train that carried the instruments and supplies, had long ago discarded their jackets and breeches in favor of regional attire: loose shirts and trousers, covered with dark, all-weather ponchos. Their servants and slaves, similarly dressed, walked or rode mules. The Indian workers, wearing rough cloth and open sandals, invariably walked.

The Indians who worked for the astronomers were often forced to do so under the *mita* system, hired for short periods by local Jesuit priests and hacienda owners on behalf of the astronomers. Slightly smaller, on average, than the expedition members—Indian men were around five foot two, compared with five foot five for European men at the time—they nevertheless possessed incredible strength and stamina, carrying surprisingly heavy loads up the mountains when even mules could not. They hauled equipment and supplies, guided the parties through the countryside, cleared terrain for observation posts, carried messages, and performed the countless minor tasks without which the expedition would have long since failed. Most of the Indians spoke only a dialect of Quechua, the native Inca language, but they were most likely directed by a group leader who spoke Spanish and relayed orders from the astronomers. Far ahead of the survey parties, the wives and children of the Indian men would have set up camp and cooked the evening meal.[39]

For all their dependence on the Indian workers, the European accounts of the expedition say almost nothing about their contributions. In fact, the few words the scientists spared on their invaluable helpers were uniformly disparaging. La Condamine claimed they were "barely distinguishable from beasts" and accused them of habitually stealing supplies and equipment. Bouguer considered the Indians to be devoid of imagination, "only capable of slavish imitation, and incapable of creating anything new." Even the presumptive heirs of Bartolomé de las Casas, Jorge Juan and Ulloa, wrote disparagingly of the same people whose mistreatment they were now carefully chronicling: "The Indians are in general remarkably slow. . . . Neither their own interest, nor their duty to their masters, can prevail upon them to undertake any work. . . . [The man] sits squatting on his hams, this being the usual posture of all the Indians, and looks on his wife while she is working . . . but unless to drink, he never moves from the fireside until obliged to come to the table."[40]

The prejudices of the Academy scientists may in fact have contributed to the peculiar behavior of the Indians around them, who tailored their behavior to the Europeans' biases. Juan de Velasco, who would become one of the great historians of colonial South America, was a twelve-year old living in Riobamba when La Condamine came to the city in late 1738. Many years later, Velasco described what happened when La Condamine visited his family's household:

> Supposing an intelligent person, let us say an academician, examines an Indian: he asks various questions to discover the depths of his intelligence, and to what point he exercises his limited potential. The Indian, realizing that his examiner has a prejudicial ulterior motive, acts maliciously. What does he do? If he's seen to be an idiot, and *four* times more stupid, he'll act stupid and an idiot times *twenty*. And what happens? While the academician forms his opinion and decides he is little less than a beast, and writes it in his *Diary of His Voyage*, the Indian laughs with his companions at duping his examiner. This is a known fact, which I can swear to, for it happened in the same house where several times the said academicians were hosted. I was in that house and though I was but a child I took notice of these events and recorded them.[41]

The supposedly careful European scientists, who took such great pains to take the measure of the lands at the equator, seemed almost determined to take the mismeasure of the people who lived there.

Despite their declared contempt for the natives they encountered, the scientists occasionally found cause to marvel at the Indians' historic achievements. In May 1739, while attempting observations—with little success—at the summit of Bueran, La Condamine suggested to Bouguer that they profit from the grey skies to visit an ancient "fortress" several miles to the east near the town of Cañar. Bouguer and La Condamine did not show even a trace of irony when they marveled at the ruins of this imposing complex—one built by the very culture that Bouguer had dismissed as "incapable of creating anything new."

While archaeology was a passion for La Condamine, it was not for Bouguer, but he nevertheless humored his good friend (perhaps repaying La Condamine for his help with the gravitational experiments). Together they spent a week reconnoitering the now-famous Ingapirca ruins, taking measurements and drawing detailed plans of the site. They assumed that the massive structure—almost twelve acres of stonework—was originally an Inca fortress (the name means "Inca wall"), but modern fieldwork indicates that it actually began as a temple complex and was already a thousand years old when the Inca overran its builders, the Cañari, and greatly enlarged the site. La Condamine, later writing the first detailed archaeological description of an Inca ruin, marveled at the craftsmanship of the structure's builders, who did not even have iron tools. As at other Inca sites, the massive blocks (up to sixteen by thirty feet) at Ingapirca were fitted together without mortar, perfectly smooth on their faces and joined so tightly on every side that the seams were barely visible. "My description of the ruins may give an idea of the material, form and perhaps the solidity of the palaces and temples built by the Inca," La Condamine explained, "but not their extent or magnificence."[42] The scientists, awed by the architectural prowess of the Inca, apparently never made the connection that the Indians they disparaged as "beasts" were in fact the direct heirs of those great engineers.

As the year 1739 wore on, the two survey parties—helped by many Indian workers whose names have been lost to history—advanced quickly southward, closing fifteen triangles between February and July. The teams were plagued with the same problems as they had encountered earlier, including clouds, frost, and harsh winds. One night in March, the surveyors witnessed an eruption of Sangay, one of the most active volcanoes in the cordillera; the bright flow of lava illuminated the entire mountainside. In late April both parties reached the summit of Sinasaguán, the highest station from which they would survey during their entire triangulation.

The first dawn on the towering peaks of Sinasaguán showed great promise for a speedy completion of the survey. As the sun rose, La Condamine could see Cotopaxi, some hundred miles away, and noted that "I had a bird's-eye view of the mountains in between, and the neighboring valleys, as if they were laid out on a map below me." But as the morning wore on, clouds rolled up from the valley, and the winds picked up, eliminating any hope of observations. As the days wore on, the conditions only intensified. Beginning the third night and lasting several days, the worst series of storms the scientists had ever seen came roaring across the peak with hurricane-force winds, hail, snow, and lightning, ripping up their tents, dashing one of their horses to its death, and driving their wiser pack animals to seek shelter in the ravines below the campsite. The parish priest far below in Cañar, aware that the scientists were on the mountain now enveloped in black clouds, led a Mass for their safety.[43]

Finally, on the morning of May 7, the weather cleared, and in the space of a few hours, the scientists surveyed all their angles, finishing their observations before noon. Finally they could depart Sinasaguán and proceed with the rest of their observations. La Condamine kept a peculiar memento of his harrowing time on that mountain: While huddled on its summit he'd received a packet of letters from France, in which his friends expressed worry that he was suffering too much from the equatorial heat.

Although they had completed their observations on the highest mountain in their survey, the scientists soon found that the future had other travails in store. In mid-July 1739, with the two parties in sight of Cuenca and the end of the triangulation process, the deep fracture between Godin and the other expedition members resurfaced. Both parties had originally agreed that a second verifying baseline needed to be measured at the southern end of the chain of triangles. Such a baseline would allow the scientists to check the overall precision of their triangulation, by comparing the calculated length of the verifying baseline (which would form the last side of the final triangle) against its actual measured length. La Condamine had originally proposed that the chain end on the plain at Tarqui, about seven miles south of the city, and Bouguer had agreed to the plan. Upon arriving at the site, however, Godin quickly opted for another location called Baños, closer to Cuenca. Despite their protests, he would not be reconciled with the decision of his colleagues.

It was vital that the two parties share a baseline to ensure the accuracy of their triangulation and inconceivable that, having come so far, they could not agree on this final measurement—yet that is exactly what happened as the parties approached Cuenca. Although Bouguer was now de facto leader, he could do little except write impotent missives to Godin, railing at him for wasting time and endangering the whole mission. Godin, for his part, simply refused to work with his colleagues, saying that he would not do so even if ordered by the Academy of Sciences itself.[44]

Once again, the two parties went their separate ways. In late July Godin and Jorge Juan, aided by Godin des Odonais, measured the seven-mile baseline at Baños, a difficult operation since the plain was crossed by three rivers and dropped sharply from one end to the other. In August, Bouguer and La Condamine, assisted by Ulloa and Verguin (and presumably Bouguer's servant Grangier, whom Bouguer had trained as a surveyor), measured the six-mile-long Tarqui baseline. Part of that baseline was covered by a warm, shallow swamp, so to measure it the men waded out in the knee-deep water and floated the measuring poles on the surface. Both parties then checked the calculated lengths of their verifying baselines against their measured length; the Baños baseline showed a difference of six feet, while the Tarqui baseline was just a single foot greater than its calculated length. Satisfied with the results, the scientists marked the ends of both baselines with large millstones, as they had done at Yaruquí, in the event they would need to take the measurements again.[45]

With the survey observations complete, the scientists now began the tedious but straightforward process of computing the overall length of the chain of triangles with repetitive trigonometric calculations. The groups both returned to Cuenca on August 23, where each holed up to separately make their computations. The astronomers had been updating their calculations in the field as they surveyed southward, so now they could finalize their results and compare the numbers.[46] This was not just a matter of plotting the triangles and computing the sides. For every triangle whose dimensions the scientists calculated, they had to apply one set of corrections to account for differences in the elevation of the survey points, another for closing the triangles, and yet a third for the variations from the north-south meridian.

The teams had already taken measurements that would now allow them to adjust their calculations to account for differences in elevation between the observation sites. The scientists had measured the vertical angles at each survey point, using the plumb bobs on their quadrants to fix the local vertical. Because of the curvature of the earth, the local vertical of each station was minutely different from those of the other stations, which the scientists corrected for by employing spherical geometry (the first time it was ever used in geodesy). The scientists "reduced" (brought to the lowest elevation) each observed angle to its equivalent horizontal measurement, so as to compute the corrected horizontal distance between survey points. They then reduced the entire chain of triangles to a single elevation corresponding to the Caraburo end of the Yaruquí baseline, the lowest point of the survey; and because this elevation was well above the curve of the surface of the Earth, they would later have to reduce the entire chain to sea level in order for their measurements to be applicable to the surface of the Earth itself.

After reducing the chain of triangles to a single horizontal level, the scientists had to correct the angles that they had measured within each triangle in their survey. They had to make minute adjustments to the observed angles so they added up to exactly 180 degrees, the sum of the angles in a triangle. These corrections were necessary to account for known inaccuracies in the teams' instruments, as well as for the effects of refraction—the bending of light that altered the apparent position of distant objects, which Bouguer had tediously examined while on the coast of Peru.

Finally, the scientists had to adjust for the fact that the overall chain of triangles was not precisely north-south, but was angled around 14 degrees west from the meridian they had established through the tower of the Church of La Merced in Quito. They therefore had to carry out several more trigonometric calculations to establish the actual north-south distance between the latitudes of the baselines along the meridian.

The complete set of calculations involved many repetitive steps, each of which might contain arithmetical or trigonometric errors that could go undetected until the final results were calculated. Bouguer and La Condamine carried out their calculations independently as a check on each other. When they compared their initial results in September 1739, the results were astonishingly close: Bouguer's calculation of the north-south distance between the baselines was 162,965 *toises*, while La Condamine's was 162,995 *toises*. Over a distance of 215 miles, they differed by just 30 *toises*, or about 64 yards. The two men likely would have felt some relief that their two years on the mountains had yielded such accurate results. Godin and Jorge Juan would likely have made the same type of independent calculations, but they left no record of their work.[47]

As the scientists compared their miraculously accordant calculations, they received even more good news. Seniergues had already returned from Loja to Cuenca back in July, and Jussieu and Morainville arrived in the city on August 22, armed with a complete set of notes, drawings, and samples of the cinchona tree. For the first time in many months, the entire expedition was together. The group looked forward to a few days of rest in Cuenca before beginning the final phase of astronomical observations to establish the latitude of each end of the triangle—a simple set of procedures that should give them, in just a short time, the true figure of the Earth.

With their goal within reach, the expedition members could look forward to returning to Europe to present their findings. In a letter back to Spain, Jorge Juan and Ulloa optimistically predicted the timeline for their departure: "As far as we can tell, we will finish everything and leave for Cartagena in January or February of next year [1740] as long as no accident impedes us."[48] The mission, it seemed, was almost complete.

Although the expedition had reached a significant milestone, none of its achievements were known back in Paris, where news from Peru lagged from six to eighteen months behind. In addition to the usual vagaries of shipments from the Pacific Ocean, from which packages had to cross Panama to the Caribbean and then traverse the Atlantic (a series of journeys in which cargo regularly fell off mules or was lost to theft or shipwrecks), the on-again, off-again hostilities among Britain, France, and Spain also played havoc with communications between the continents. Several of La Condamine's shipments, containing antique silver idols, vases, stuffed animals, and fossilized bones, were either lost, destroyed in battle, or thrown overboard by suspicious seamen. In one instance, the British seized his memoirs on the ruins of Ingapirca after capturing the ship that carried them, but when the secretary of the Royal Society discovered their contents, he graciously forwarded them on to France.[49]

Because of the long, hazardous distance between the Old World and the New, the popular press in Europe had very little information on what was happening to the Peru expedition. In early 1738 a few packets of letters had arrived in France, allowing the *Mercure de France* and *Gazette d'Amsterdam* to print a handful of articles attesting to the fact that the expedition's observations were advancing slowly, because of an excess of clouds and fog as well as a lack of funds. The editors of the *Gazette* went on to predict, without much basis, that the astronomers would soon be home from their toils in Peru.[50] Indeed, speculation was rife in both academic and political circles about the fate of the expedition, but no one, not even Maurepas, whose ear for news and shrewd sense of character gave him an almost preternatural ability to predict events, had any idea whether the mission would be a success or not.

In December 1738, another packet from Quito had arrived at the Academy of Sciences. Maupertuis, who since his return from the Arctic Circle had been waiting for confirmation of his evidence for a flattened Earth, was dismayed by what he read. Writing to Anders Celsius, now back in Sweden from the Arctic Circle expedition, he remarked: "The strife going on between them prevents me from hoping that they'll ever finish anything. After some terrible scenes they've reached the point of not speaking to each other. . . . Their dissentions mean they won't get anything done. I just hope one of them doesn't get his throat cut."[51]

Maupertuis, normally an astute observer, could not imagine how prophetic his words would be—or how far he was from comprehending the reasons behind the death of one of the expedition members.

VII

Death and the Surgeon

The expedition's surgeon, Jean Seniergues, was behaving strangely. Ever since he had returned to the *audiencia* of Quito from Cartagena de Indias in early 1739, it had been apparent that he was not the man who had left for the Caribbean almost two years earlier. It seemed, quite simply, that he had outgrown his previous self.[1]

While the expedition had been foundering for most of its time in Peru, Seniergues had been prospering. By privately ministering to the wealthy citizens of the towns through which the expedition passed, the surgeon had made money while the others had been losing it; his business dealings in the Caribbean city of Cartagena had netted him an even more considerable fortune. Seniergues had returned from his recent travels with two black slaves originally from the French Caribbean colonies, Joseph Cujidón (now serving as Seniergues' cook) and Agustín Congo. Previously, only the expedition leaders had had slaves, so in a sense Seniergues was tacitly asserting that he was now their equal—one of the senior members of the Geodesic Mission.

Seniergues' recent, uncharacteristic arrogance may have had something to do with his good fortune thus far in the mission. Whereas he had once worked side by side with Joseph de Jussieu to heal the sick and wounded they encountered, Seniergues was now imperious and even belligerent in his dealings with the local population. He also began drawing apart from the rest of the expedition, inexplicably departing the forests of Loja to go to Cuenca, leaving behind his good friend Jussieu, who was completing his studies on the cinchona tree. Seniergues came to Cuenca in July 1739, about the same time the astronomers arrived in the city to begin their measurements of the verification baselines at nearby Tarqui and Baños.

Seniergues came from humble beginnings—surgeons in France were still considered manual laborers and often had second jobs to make ends meet[2]—so his sudden wealth and its attendant status may have given him the fatal hubris so common to parvenus. It is also conceivable that his subsequent behavior—social deviation and delusions of persecution and grandeur—may have been the early symptom of a mental illness such as paranoid schizophrenia, which often first appears in adulthood. Across three centuries, it is impossible to know for certain.

The first record of Seniergues' belligerence comes from a report by the local magistrate (*corregidor*) of Cuenca on July 12, 1739. The surgeon was with Antonio de Ulloa near the verification baseline at Tarqui when Ulloa was inexplicably attacked and wounded by a young *mestizo*. Seniergues and Ulloa lodged a complaint with one of the local magistrates, Mathías Dávila y Orduña, who ordered his deputy marshall to put the man in jail. The marshall tracked the attacker to a house in Tarqui, but before the official could arrest him, Seniergues and Ulloa—who had followed behind—burst into the house and dragged him from his hiding place in the attic. The two men dragged the *mestizo* to Seniergues' house, where the surgeon ordered his slave Cujidón to "flog him two hundred times, and then rub pork fat into his wounds" before turning the man over to the marshal.[3]

Seniergues' brutalization of Ulloa's attacker was only the first in a string of controversies in which the surgeon embroiled himself. While locals in colonial Peru would occasionally take justice into their own hands, they hugely resented the fact that two foreigners had brazenly flouted the law and violently punished a member of the community. Rumors of further attacks began circulating, including a persistent one that Seniergues had beaten a local man with one of the measuring poles used on the baseline in Baños—a doubtful claim, since those carefully calibrated poles were inestimably valuable. But Seniergues' actual behavior was not much better: He had a tumultuous and very public run-in with a local Indian leader and another with a city official, in which the surgeon threatened the man, saying that he would "cut off his ears."[4] The patience of the local authorities was wearing thin, and Godin himself was sharply informed that the expedition had lost the respect of the mayor of Cuenca, Sebastián Serrano de Mora y Morillo de Montalban.

Despite Seniergues' mounting unpopularity with many of Cuenca's citizens and officials, he managed to remain on good terms with several of the city leaders. He frequently dined with Mathías Dávila y Orduña, to whom he sold about $30,000 worth of Spanish garments brought from Cartagena, and he also befriended Nicolás de Neyra y Perez de Villamar, a captain in the militia. Seniergues' services as a surgeon were still in great demand, moreover, and despite his mistreatment of many locals he continued to treat the sick and wounded without regard for how well they could pay.

It was, in fact, Seniergues' humanitarian activities that led to the Geodesic Mission's most violent event. One of Seniergues' less affluent patients, Francisco de Quesada, had been suffering from worsening bouts of malaria, and in mid-August Seniergues was called on to treat him. Rather fortunately, the surgeon had brought a supply of cinchona bark back with him from Loja, and within a few days the medicine took effect. Francisco began to regain his health, to the relief of his wife, Gertrudis, and his only child, Manuela. They were too poor to pay Seniergues, but Francisco, noticing a growing attraction between the affluent surgeon and his young and beautiful daughter, hit upon a scheme both to compensate Seniergues and to settle an old debt.

Seniergues was not the first man in Cuenca to have taken notice of Manuela. The previous year, she had been formally engaged to Diego de León y Román, Cuenca's deputy attorney general and a church warden.

The marriage would have been León's second; his first wife, Isabel, the sister of Seniergues' friend Nicolás de Neyra, had recently passed away, leaving León to care single-handedly for their young daughter. Although León at first found Manuela a pleasant distraction (her nickname, La Cusinga, was from a Quechua word meaning "Happy One"), he jilted her in favor of the younger sister of Cuenca's mayor, Josefa Serrano de Mora, whom he married soon after.[5]

Both Francisco and Manuela were furious with León. It was public knowledge that he had taken Manuela's virginity, casting deep shadows over her marriageability. León initially promised to pay the family off (presumably topping up Manuela's dowry to improve her chances of marriage), but after his own wedding he reneged on the deal. Now, with Seniergues taking interest in Manuela, Francisco saw a way to exact the overdue payment from León, while also paying the surgeon his fee.

Francisco's scheme was to have Seniergues approach León for the money, ostensibly to cover Francisco's medical bill, which the foolhardy surgeon agreed to do. Seniergues sent word to León that he was expecting payment, and some days later, on Monday, August 24, León sent a female mixed-race slave to the Quesada household with his reply. She was accompanied by Nicolás de Molina, León's good friend and Francisco's cousin, to watch over her. Upon entering the Quesada household, the slave at first indicated that she had been instructed to retrieve some clothing that León had left there—but then, the slave approached Manuela and slapped her, "saying that she did so," as one eyewitness recounted, "in order that her Frenchman remove the blow." Being struck by a slave was an unimaginable insult and hammered home León's reply to Seniergues and the Quesadas: He had no interest in paying the fee, but rather wanted Manuela's new paramour to defend her wounded honor. Seniergues, now caught up in the affair, beat the slave with a stick, sending her back to León with his own message that "he demanded satisfaction for this, since he had not come to the house for that woman, but to cure Francisco de Quesada."[6] Seniergues, in other words, had called León out to a duel.

Challenging a town official to a duel was by far the surgeon's most pugnacious—and irrational—decision yet. Dueling had been outlawed and out of favor in France for a century, and the surgeon almost certainly did not understand the rituals involved. He had known the family less than two weeks, and to judge by his later actions, he and Manuela were not deeply in love, as his challenge would seem to imply. He certainly did not need León's money, moreover, and he could have just as easily waived his own fee to the Quesadas. Seniergues was entering into a family dispute that was not his, and to boot he was a foreigner who had not shown the slightest respect for Peruvian society or its customs. Now the hotheaded Frenchman had picked one fight too many.

Seniergues' position in Cuenca had never been more precarious. By the time he delivered his challenge to León, Seniergues was under almost constant scrutiny in the town. He rarely left his home without his two slaves, the three of them well armed. This was in part to avoid imprisonment; the vicar of Cuenca, Juan Bernardino Jiménez Crespo, had asked the local marshals to arrest Seniergues on the grounds that he was illicitly involved with Manuela de Quesada. Crespo, a stern, cheerless priest who some years later would bring criminal charges against any couple who danced together,[7] was not simply railing against extramarital sex with a foreigner. His friendship with the now-united León and Serrano families, both of which were prominent in church and government affairs, strengthened his own political position, and now Seniergues was directly threatening those two families.

In the late afternoon of Wednesday, August 26, two days after Seniergues had challenged León, the surgeon and his two slaves were in the main plaza when they spied Diego de León and his wife. The couple was accompanied by Seniergues' friend Nicolás de Neyra and several others. Hiding behind a pillar, Seniergues waited until the party came close, then stepped out in front of the surprised León, demanding that he draw his sword. León instead reached into his cape and drew his pistol, but the flintlock misfired. Seniergues drew his cutlass and attacked, but he lost his footing and fell into the rain gutter that ran down the center of the plaza. Neyra and his companions grabbed the surgeon and separated the two men. Another militia officer appeared, and together they hustled Seniergues back to his house, the Frenchman still cursing and threatening, "if Diego does not give me satisfaction, I'll kill him or cut off his ears."[8]

Neyra was now in an untenable position. He counted Seniergues as a friend, but as an officer of the militia he could not allow the man's belligerence to continue—especially against León, who had once been Neyra's brother-in-law. Seniergues' behavior was an even bigger problem for the members of the Geodesic Mission; the rising ill will toward the French surgeon threatened to spill over and interfere with their astronomical observations, which were now all that stood between them and the true figure of the Earth.

Seniergues' friend Jussieu had arrived in Cuenca a few days earlier, but it does not appear that he was able to dissuade the surgeon from his course of action. A local Jesuit priest attempted to reconcile León and Seniergues, arranging for the two parties to meet at his lodgings two days later, on the afternoon of August 28. At the appointed hour Jorge Juan brought a reluctant Seniergues before the priest, but Neyra and León were nowhere to be seen. For León, the matter had progressed far beyond words, and it seemed that Neyra was slowly coming around to this view as well.

The escalating drama between Seniergues and León took place against the backdrop of a town-wide celebration in Cuenca. The festival of Our Lady of the Snows (Nuestra Señora de las Nieves) commemorated a vision of the Virgin Mary that had appeared before a group of snowbound travelers, saving them from certain death. The five-day festival had begun on August 25, three days before the failed meeting between Seniergues and León, and was marked by extensive drinking of *aguardiente* (a strong alcoholic drink similar to rum) and parading in costume, culminating each afternoon with a bullfight.

The bullfighting ceremonies that accompanied the festival of Our Lady of the Snows were an integral part of the Spanish culture that dominated Peru. According to legend, Francisco Pizarro had brought bulls to Peru almost as soon as he established the city of Lima in 1535. By 1739, the ceremony had evolved more or less into its present-day form: The bull, first weakened by men on foot and on horseback armed with pikes and lances, met face-to-face with the matador, who, cape in hand, drew the bull ever closer, finally dispatching him with a thrust to the heart from a long sword, the *estoque*. In cases in which the bull did not die immediately, the matador would thrust a smaller, razor-sharp sword called a *verduguillo* into the animal's neck, severing the spine and killing the creature instantly.

Eighteenth-century bullfights were public spectacles that took over the towns in which they were held. A makeshift bullring, constructed around a plaza, formed the center stage; in Cuenca, this was near the Tomebamba River at the small plaza San Sebastián, named for the church on its northern side. Sand was spread over the cobblestones, and scaffolding was erected all around the square, forming a series of loges on two levels where up to 4,000 people could watch the spectacle.

For the fifth and final day of the festival of Our Lady of the Snows on Saturday, August 29, the entire Geodesic Mission was invited to attend the bullfight. The expedition members did not sit together but rather split into their various factions. Godin and Godin de Odonais sat in the second-floor loge of Thomás Melgar, on the southeast corner of the plaza, where they were joined by Jorge Juan and Ulloa, who would have been able to explain the bullfight to them. On the west side, at the invitation of the parish priest Gregorio Vicuña, were Bouguer, La Condamine, Morainville, Hugo, and Verguin; Jussieu and Seniergues were nowhere to be seen. The Frenchmen had undoubtedly never seen a bullfight before, with the possible exception of the Toulonnaise engineer Verguin, who may have witnessed the bloodless bullfights in the south of France. There, the object was to snatch a ribbon from between the bull's horns. The event for which the Frenchmen were now assembled was guaranteed to be far bloodier.

Shortly before the spectacle's scheduled start time of 4 PM, Jean Seniergues staggered into the plaza—tipsy, like most of the attendees, from drinking *aguardiente* all afternoon—and made straight for the ground-floor loge on the east side of the plaza where Francisco de Quesada, playing dress-up in a matador's red cape, sat with his daughter, Manuela, and their cousin Antonia Domínguez. By now the entire city knew about Seniergues and Manuela, and also knew that he had called out Diego de León. But this was not Paris; it was an outrage for Seniergues to be seen in public with his "concubine," an affront made even worse by her father's clear consent to the illicit affair.

Diego de León and Juan Jiménez Crespo, seated together by the church, could only stare at Seniergues' display in impotent fury. But Nicolás de Molina, who had witnessed Seniergues beat his friend's slave a few days earlier, was livid at seeing his cousin Francisco de Quesada fêting the surgeon. Molina caught Quesada as he left his table to enter the plaza and, brandishing his sword, began dragging his cousin toward a magistrate. When Manuela saw Molina accosting her father, she cried out to Seniergues, "They're taking him prisoner!" Seniergues pulled out his cutlass and pistol and, closely followed by his equally well-armed slave Cujidón, jumped from the loge, shouting, "I'll kill the rogue and his whole family!" Turning to Cujidón, he ordered, "Kill them all!" Molina retreated, and Seniergues managed to pull Quesada back to the loge, where a relieved Manuela nervously tried to assuage her paramour by laughing off the whole incident as a joke.[9]

During the confrontation, Nicolás de Neyra, one of the bullfight's organizers, had ridden into the plaza on his richly decorated horse and prepared to start the ceremonies. Seeing what had just happened, he trotted to the balcony where Godin, Ulloa, and Jorge Juan sat, asking them to calm Seniergues down and keep him at his table. "We will take care of this," the Spanish officers replied. Neyra then turned and, halting before Seniergues' loge, told him not to worry and that no one was threatening him. Seniergues, angrier by the minute, spat back that he would kill Neyra. Now exasperated at the threats of his erstwhile friend, but also wary of his sudden bursts of violence, Neyra asked, "Why would you do this?" while wheeling his horse around to exit the plaza. "Wait and see!" cried Seniergues as he leapt up, overturning the table before the startled women in the loge.

Neyra had seen enough. He dismounted at the northern barricade, where the bulls were normally led into the ring, and angrily announced that, because Seniergues had threatened to kill him, the bullfight was cancelled and he was going home. He then stormed out of the plaza to find the mayor, Sebastián Serrano, in order to put the surgeon in jail.[10]

The crowd was furious and began murmuring against the French. Jorge Juan and Godin descended from their loge to speak with Seniergues, but not sensing any immediate menace from him, they started back to-

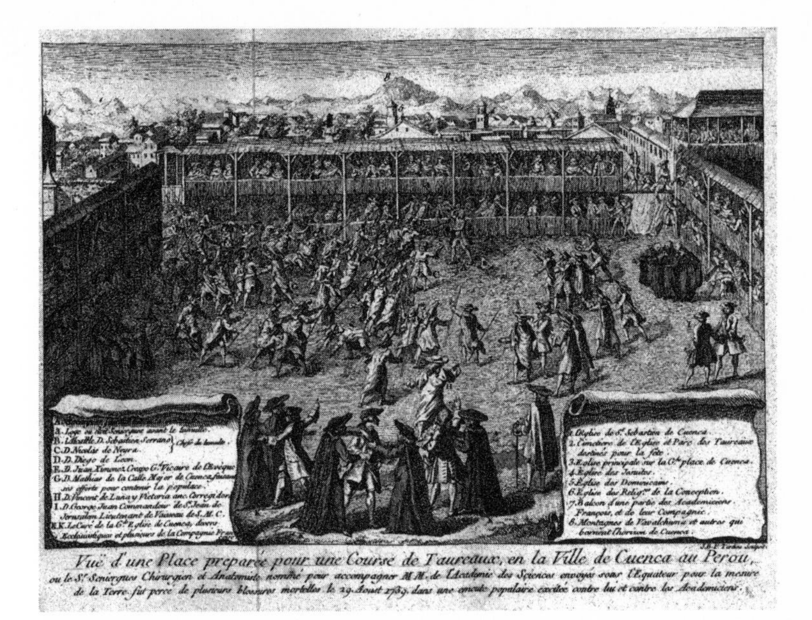

Figure 7.1 The attack on Seniergues at plaza San Sebastián, Cuenca: Saturday August 29, 1739, about 4:30 PM. From Charles-Marie de La Condamine, *Journal du voyage fait à l'Equateur* (1751). Courtesy of Robert Whitaker.

ward their table. Just then, an uproar came from the northern gate, and the crowd, assuming the noise was from the bulls approaching, began to cheer. In fact, the noise marked the arrival of Neyra and Serrano, at the head of over a hundred men armed with the pikes, lances, and swords they had originally intended to use on the bulls.

Marching up to Seniergues' loge, Serrano pointed at the surgeon and said, "Give me your weapons." Instead, Seniergues jumped into the plaza with his cutlass raised over his head and his pistol leveled at Serrano. The Frenchman tried to fire both barrels, but each time the flintlock misfired. His cutlass sliced down at Serrano, but the blow was intercepted by a man named Manuel Armijos. Neyra, standing to Seniergues' left, thrust his *estoque* into the surgeon's left hand, causing him to drop the pistol. Serrano yelled, "Seize him in the name of the king!"[11]

Despite the efforts of a local army officer to contain the tumult, it quickly spilled into the plaza, surrounding Seniergues and pushing him against the loge. Serrano once again called for the crowd to seize the surgeon, but other voices started rising: "Kill the French foreigners! Down with the government! Long live the king!" Men began wrenching up the cobblestones in the plaza and heaving them at Seniergues, who attempted to fend off the onslaught with his wounded arm while brandishing his cutlass with the other. One well-aimed stone smashed into his right arm, knocking him to the ground and sending his sword clattering. Struggling to his feet, he backed toward the southeast barricade, desperately seeking escape from the mob.

The other members of the expedition quickly rallied to Seniergues' defense. La Condamine, mustering his soldier's reflexes, dropped down the ladder to aid his companion. Bouguer, a professor since age sixteen, had never served in an army and almost certainly would have been overwhelmed by the sudden chaos, but he doggedly followed his friend. Jorge Juan, a feared swordsman, was already in the plaza, but as the naval officer attempted to maneuver around the crowd to defend Seniergues, he was grabbed repeatedly to prevent him from also becoming a target. It was in any event too late, for he had lost sight of Seniergues under the press of the mob.

Now at the barricade, the crowd repeatedly jabbed at the surgeon with pikes and lances, wounding him but not fatally, weakening him like a bull before the matador strikes. Seniergues, pleading for his life, had begun forcing his body through the gate, leaving his torso exposed to the mob.

Nicolás de Neyra, fully caught up in the mêlée, was next to the surgeon as he struggled to escape the plaza. Too close to use his full-length *estoque*, the militia captain grabbed a short sword from a young man nearby and with his left hand thrust it into Seniergues' exposed left side. It was a *verduguillo*, the thin blade designed to carve through the massive neck vertebrae of a bull, and it easily sliced between the man's ribs and into his abdomen, piercing his spleen.

At first, Seniergues would not have known what had happened; struck repeatedly by stones and spears, he would almost certainly not have noticed the small tear in his jacket amid all the other wounds he had received. Finally wrenching himself free from the gate, he staggered down the street and around the corner into the house of Thomás Melgar; in the upper story of the house, Ulloa and the Godins were still in the loge, trying to see what was happening. Falling through the door downstairs from them, Seniergues crumpled unconscious onto the central patio.

The other expedition members were still struggling to reach Seniergues. La Condamine and Bouguer met up with the others in the plaza and tried to cross it to rescue their colleague, but they were told that Seniergues had been killed. Still determined to find him but blocked by the crowd at the southeast barricade, they turned back toward the north gate. Exiting the plaza, they ran down a side street toward Melgar's house—only to be confronted by the crowd that had first attacked Seniergues and was now surging forward to finish him off.

The armed mob advanced toward the Frenchmen, crying "Kill the French foreigners!" The scientists backed up, turned a corner, and ran for their lives, pursued by hurled stones and cries of "*Viva el Rey!*" Bouguer was hit by a stone and slowed long enough for a pursuer to thrust his sword at the Frenchman, tearing through his jacket and into the middle of his back. In the nick of time, a priest named Félix Moreno pulled him inside the home of Gregorio Vicuña, who then barred the door. The rest of the company scattered to their homes and waited for the mob to disperse. The Quesada family, meanwhile, hid in the San Sebastián church, fearing for their lives.

As Seniergues lay half-dead on the floor of Thomás Melgar's courtyard, blood seeping from his wounds, the crowd outside began pushing through the entrance of the house and into the courtyard. Sebastián Serrano entered the courtyard and tried to calm the crowd, sending one of the French servants to find a priest. Seniergues, meanwhile, was pulled into another room and laid on a bed, where a local doctor attended him.

At first, even the local clergy seemed unwilling to minister to the stricken Frenchman. When the expedition's servant arrived at a nearby church asking for a confessor, an unrepentant Crespo—who had recently petitioned for Seniergues' arrest because of his dalliance with Manuela de Quesada—snarled, "What good are sacraments to these heretics?"[12] The French servant found another priest and returned with him to Melgar's house, where they found Joseph de Jussieu standing nervously in the doorway, afraid to enter with the mob still milling about the courtyard.

After being assured of his safety, Jussieu was escorted to the room where Seniergues had now regained consciousness and began examining the surgeon's wounds. Most were superficial, but Jussieu soon noted the thin, deep cut on the left side of Seniergues' abdomen, and on further examination found that his friend was hemorrhaging internally. Seniergues immediately understood his fate, crying out, "It's in a very bad place: I am lost!"[13]

Despite Seniergues' fears, his prospects—and those of the rest of the expedition members—initially looked hopeful. As Jussieu was examining Seniergues, the authorities had managed to disperse the mob, allowing the rest of the expedition members to enter Melgar's house. They carried Seniergues to his house, where Jussieu and a local barber (who, like many in his trade at the time, doubled as a surgeon) cared for the Frenchman night and day. Bouguer's wound was serious but not life threatening, requiring just a few day's convalescence, and during that time Seniergues' wound did not appear mortal either. Indeed, Serrano and Crespo even demanded that the surgeon be imprisoned for resisting arrest, which the magistrate Mathías Dávila y Orduña denied.

The officials in Cuenca took immediate action against any further violence after the attack on Seniergues and his comrades. Street patrols were organized to keep another mob from forming, and the authorities in Cuenca also quickly posted an edict prohibiting the gathering of more than two people, on pain of a hundred lashes. The expedition's safety was not the primary motivation for these actions. Rather, the authorities feared that the revolt against the Frenchmen would turn into a general rebellion of the indigenous people against Spanish rule; cries of "Down with the government! Long live the king!" had been heard many times before in Peru. Violent uprisings had occurred almost in every decade since Pizarro's conquest, sometimes against the *mita*, sometimes against taxes, at other times against the general oppressiveness of the Spanish regime. Two years earlier, in 1737, an Indian headman named Ignacio Torote had led a violent rebellion directed against Franciscan missions, which was quickly and brutally put down by the Spaniards.[14]

Seniergues' wound did not at first appear mortal, but as the days progressed, his condition worsened. The spleen itself is not a vital organ, but without a way to prevent infection and control the internal hemorrhaging, Jussieu and his colleagues could have done little except apply leeches to the surgeon's abdomen—now slowly distending as blood seeped in from the damaged tissues—and apply cold cloths to Seniergues' wound as his pain, fever, and vomiting rose. These symptoms were the harbingers of death, and the surgeon would have known that all Jussieu's skills could not save him.

On August 31, Jussieu wrote his brothers to share the news of his predicament. "I've been turned aside by the illness of Seniergues who has been dangerously wounded with a thin cut on the lower left side of his chest," he wrote, which prevented him from telling them of his trip to Loja. "We French were almost overwhelmed by the uprising," he continued, "and barely escaped being exterminated. Seniergues alone has paid for all of us. His state allows me not a moment's rest, and I am at once pharmacist, surgeon and doctor."[15]

Knowing the end was near, Seniergues dictated a court statement and his last will and testament. In the latter, he named Jussieu and La Condamine as the executors of his estate, requested to be interred in the city's Iglesia Matriz cathedral, and listed a series of charities and debts to pay. Seniergues ordered his slaves be left in the care of La Condamine and bequeathed the rest of his considerable fortune to his parents, Guillaume and Françoise Seniergues. His will never once mentioned Manuela de Quesada; if she was ever at the center of Seniergues' thoughts, she was now far from them. In his court statement, Seniergues attempted to cast himself as a victim. He never mentioned Diego de León as the instigator, blaming instead Serrano and Neyra for the attack and asserting that "although I had a cutlass in one hand and a pistol in the other, I harmed no one and only drew them in self defense."[16]

Seniergues hung on for four more days after laying out his will and giving his deposition, but finally the infection and loss of blood proved too much. He died on Wednesday, September 2, 1739, at 10:30 PM, just over a week after the whole affair began.

Even before Seniergues' death, the magistrate Mathías Dávila and the mayor, Sebastián Serrano, had begun criminal proceedings against his attackers. The two officials deposed the first eyewitness the day after the attack, even though it was a Sunday; La Condamine and Bouguer brought a criminal complaint before the judges two days later. In fact, it quickly became apparent that one of the judges was also one of the accused: The Frenchmen's complaint named Serrano, along with Neyra and León, as one of the principal agents of the assault on Seniergues.

The case against Seniergues' attackers took channels markedly different from those of modern court proceedings. The legal process in colonial Peru started with the deposition of witnesses by government scribes, who also took statements from the defendants, who themselves may or may not have been represented by a lawyer. Those testimonies were summarized by the court notaries and sent to the judge, who often was not an independent authority but rather a political figure, such as the city mayor. The judge then reached a verdict and handed down the sentence. There were no courtrooms, juries, or examinations and cross-examinations of witnesses. Decisions were based on an equal mixture of legal precedent, local custom, and biblical law.[17]

In the case of the Seniergues trial, the wheels of colonial justice ground reasonably efficiently despite some initial setbacks. The court records document the political struggles between Serrano, who as mayor initially presided over the case, and Dávila, who wrested it from his control on the grounds that Serrano was one of the accused. Dávila eventually appointed an independent judge.[18] That judge handed down verdicts in August 1740, a year after Seniergues' murder. Diego de León received six years in prison and a fine of 1,000 pesos for having provoked the revenge attack on Seniergues; Sebastián Serrano and Nicolás de Neyra each received eight years and a 2,000 peso fine for having led the attack; two other men received lesser prison sentences for having injured Seniergues during the assault.

The initial verdicts were not the end of the trial. A further complication ensued when control of the *audiencia* of Quito was transferred from Peru to the newly created Viceroyalty of New Granada. Although New Granada was established in August 1739, just before the trial began, its viceroy did not arrive until 1740; so in the end the viceroys of both provinces claimed jurisdiction and had to approve the proceedings.

The case against Seniergues' assailants would ultimately drag on for five years, as the judge's verdict was met with appeal, and control over the proceedings passed from one jurisdiction to another. Of the five defendants, Serrano was the only one to appeal his sentence. His attorney protested the fact that Serrano could not even leave prison to attend the funeral of his wife, who had died while he was incarcerated during the trial, and that sentence would leave his three small children without anyone to care for them. These pleas were vigorously opposed by La Condamine, who proved to be an erudite and quite effective advocate, and later by Joseph de Jussieu as well. Nevertheless, in 1744 Serrano was set free, his charge reduced to "not having pursued the guilty parties with sufficient fervor," and his fine reduced to 400 pesos, which he paid to Seniergues' estate. Diego de León escaped prison in 1742, Nicolás de Neyra died there, and neither of the other two assailants served out his full term.[19]

Even before the initial sentences had been delivered in Peru, the Seniergues affair created a minor scandal in Europe. When the first letters describing the attack arrived in France in June 1740, the sequence of events was quickly garbled: The scientists had been attempting to seduce Peruvian wives; their servants had been killed and their instruments and papers destroyed. In August the news was cleared up a bit by Voltaire: "The Peruvian gentlemen are not as badly off as we have heard. Only their surgeon is dead. Their instruments and their papers are safe, and it appears that it was a private quarrel of the surgeon."[20]

La Condamine would later attempt to whitewash the circumstances of Seniergues' murder. When he eventually returned to France, his first task would be to rewrite the story as a love triangle gone horribly wrong. In a 1745 account, provocatively titled *Letter to Madame X on the Popular Uprising in the City of Cuenca Against the Academy Scientists*, he placed the blame for the attack squarely on the inhabitants of Cuenca, accused the town's justices of incompetence, and cast the expedition members as heroic and selfless. He carefully erased all incriminating evidence against Seniergues—for example, calling the two hundred lashes given to the young *mestizo* a "blameless action"—and made it seem as though the surgeon's worst crime was indiscretion in his affair with Manuela de Quesada.[21]

La Condamine himself fully understood the need for discretion, and practiced it judiciously in affairs beyond the Seniergues incident. Never once mentioned in his journals, nor in the local histories, was the fact that he fathered two daughters during his sojourn in Cuenca. The women, who became prostitutes like their mother, were still living in Cuenca when the German explorer Alexander von Humboldt passed through the city in 1802.[22]

As for Manuela de Quesada, the affair did not indelibly taint her, nor did she mourn her French surgeon for very long. Less than a year after his death, she married Lucas de Ullauri y Ortiz de Zuñiga and, according to Humboldt, lived to a ripe old age.[23]

Seniergues' tomb, like those of many of Cuenca's colonial inhabitants, has been lost to history. He was not interred in the Iglesia Matriz cathedral, as he had wished, but rather across the plaza in the Jesuit church, which has long since been demolished.[24] Seniergues' greatest wish, the promise of fortune that had brought him to Peru, may have been unwittingly granted; for he ended his days in Cuenca, the city said to once have been El Dorado.

VIII

The War of Jenkins's Ear

In September 1739, with Bouguer still recovering from his wound and the rest of the team still reeling from the attacks of a few weeks before, the expedition turned back to its mission to measure the Earth. So far the astronomers had successfully determined the exact length of the chain of triangles they had plotted along the Andes. Now they would need to fix the celestial position of each end of the chain, determining its precise latitudinal location so that they could use this information to compute the length of a single degree of latitude at the equator. Then, and only then, would the men be able to leave Peru with the answer they sought.

The scientists would fix the celestial position of their chain of triangles by carefully observing one of the brightest and most identifiable stars in the night sky. Epsilon Orion, a giant blue-white star located 1,300 light-years from Earth, appears almost directly overhead at the equator and is also visible throughout the Northern and Southern hemispheres.

Instantly recognizable as the middle star in Orion's belt, it was believed by the pre-Inca Chimu Indians of Peru to be a thief, held captive by the two adjacent stars and ready to be sacrificed to the other stars in the constellation, which the Chimu believed to be circling vultures.

The unique position and brightness of Epsilon Orion that gave it prominence in Chimu culture also made it the obvious candidate for the final stages of the French astronomers' plan to measure a degree of latitude. Years before, while still in the planning stage, they had established a procedure to fix the latitudinal endpoints of their chain of triangles by specifically observing the star. The scientists knew that they could use their zenith sector to measure the precise angle of the star from the local vertical and then compare the angles taken at the northern and southern ends of the triangle. The difference of those two angles would give the arc length of the chain (the difference in latitude between the two ends, measured in the number of degrees, minutes, and seconds from north to south). Dividing the arc length by the distance between the two endpoints would give the length of a single degree of latitude at the equator. The culmination of the team members' many years of labor, hardship, and bloodshed would be this single number, which would provide them with the true figure of the Earth.

As during the surveying stage, the expedition would work in two separate parties to take the star sightings at the southern end of the chain first and then the northern end. For both groups to make observations concurrently, they would need a second zenith sector, so Godin instructed the expedition's instrument maker, Théodore Hugo, to build a new sector with an eighteen-foot radius. This would be the first time Hugo had ever constructed an astronomical instrument from scratch and involved the meticulous crafting of the lenses and telescopic mountings, along with the fabrication of precision micrometer screws for adjusting the instrument. All of these processes would have required tools and materials that were undoubtedly scarce in colonial Quito, so Hugo would have had to improvise. Godin also instructed Hugo to repair several flaws they had previously noticed in the smaller twelve-foot sector he had obtained from the British artisan George Graham. Bouguer would use this sector for his observations.[1]

With the second zenith sector completed, the two parties (Godin's and Bouguer's) could now begin measuring the angle to Epsilon Orion from the southern end of the chain of triangles. Since they had already measured different endpoints for the chain of triangles, each team also picked different sites from which to measure this southern angle, and each would do the same on the northern end of the triangle. Effectively, the two parties were now establishing the arc lengths of two separate chains of triangles (though with most of the triangles in common). When they finished their separate observations and calculations, they would arrive, if all measurements were accurate, at the same figure for the length of a degree of latitude.

Since Godin had measured a baseline at Baños, a site somewhat closer to Cuenca than Tarqui, he and his team—now comprised of Jorge Juan and Ulloa, who had separated from Bouguer to aid the shorthanded Godin—would take their star sightings from within Cuenca itself. They set up the new eighteen-foot sector in a house in the middle of town, turning their temporary domicile into an observatory. Cuenca was far from an ideal location; emotions still ran high in the city, and the continued animosity toward the foreigners meant they could only venture outside at night. One evening, as the men were attempting to fix the position of their observatory by pacing off its distance from the Iglesia Matriz cathedral (whose bell tower had served as one of the final triangulation points for the Baños baseline), they were recognized by several local women who, believing the men were "plotting some treachery against the city," raised the alarm and chased them back to their home with sticks and rocks.[2]

While Godin's team was besieged in its observatory in Cuenca, Bouguer and La Condamine, assisted by Verguin, were making their own observations in somewhat more placid environs. Their observatory was an unfinished chapel in the hacienda Mama-Tarqui, which was owned by Pedro José de Sempértegui y Gómez del Pozo, a local politician who had notarized several testimonies during the Seniergues trial. By the end of September the second team had established the exact position of the Mama-Tarqui chapel in relation to the Tarqui baseline; Bouguer and Verguin then went to work building the roof of the observatory, while La

Condamine returned to Cuenca, where the Seniergues trial was still in full swing.

From October to December 1739, the two parties spent long nights adjusting and readjusting their sectors, waiting for the clear weather they needed to take the critical observations of Epsilon Orion. By the end of 1739, Godin's party had finished, packed up their equipment, and departed north for Quito. Bouguer's party did not finish its observations until mid-January 1740. The men carefully wrote out their findings and had the document, written in French and filled with astronomical symbols and calculations, certified by a presumably bemused notary.[3]

The two teams had been carrying out these observations under a cloud after Seniergues' death, but around Christmastime Bouguer's group witnessed something that managed to raise their spirits. A local Indian festival was being held in Tarqui, and amid the equestrian races and dressage events, some young *mestizos*, who had carefully watched the astronomers during their observations, came on stage with paper and cardboard quadrants and began mimicking their actions. As La Condamine reported, "This was done in such a comical manner that I must admit to not having seen anything quite as pleasant during the ten years of our trip. During this skit I was taken by such a strong desire to laugh that for a few moments I was able to forget even the most serious of matters preoccupying me."[4] It was precisely the release they all needed from the gloom of the recent tragedy.

On January 16, 1740, Bouguer, La Condamine, and Verguin finally departed for Quito, to begin the second set of observations at the northern end of the chain. They had carefully packed their zenith sector and arranged for it to be hand-carried by porters; it would take weeks to arrive in Quito. While their equipment inched toward the capital, the scientists took the opportunity to spend several days with their friends Pedro Vicente Maldonado and José Dávalos at the hacienda San Andrés, near Riobamba. There the Frenchmen attended the wedding of a niece of one of their friends, an event that La Condamine later described as "the most magnificent and brilliant celebration which I had ever witnessed during my entire stay in Peru."[5]

In February, Bouguer's zenith sector arrived in Quito after its long overland journey. Hugo reinforced its iron framework and transported it to the country home of Emanuel Frayré, located in Cotchesqui (Cochasquí), northeast of the city on the slopes of Mount Mojanda, a site that had excellent visibility to the Yaruquí baseline. Bouguer and La Condamine spent two months observing Epsilon Orion from Frayré's home, after which they surveyed an additional triangle in order to fix the position of the observatory as the northern extremity of the chain.

Having measured the angles to Epsilon Orion from both ends of the chain of triangles, Bouguer's team set about calculating the arc length from north to south—the crowning moment, in their minds, of the scientists' years-long enterprise. In Cotchesqui, the astronomers had determined that Epsilon Orion was 1° 26' 38" south of the local zenith; the angle measured at Tarqui had been 1° 40' 37" north. Adding the results gave them a figure of 3° 6' 43" for the length of the arc. Bouguer and La Condamine still had to perform some additional calculations and make some final observations before they could determine the length of a degree of latitude at the equator, but for all intents and purposes, they felt the job was done. As Bouguer recorded in a second notarized document, no doubt with some relief,

> we felt obliged to end our operations in Peru, and *to consider the object of our mission as fully accomplished*. Done at Quito, May 6, 1740.
>
> Signed, Bouguer [and countersigned by La Condamine and Verguin][6]

Bouguer and La Condamine may have been ready to pack their bags and head for France, but as far as Godin was concerned, there was still more work to be done. Godin had noticed, as had Bouguer's party, that the angle from each observation site to Epsilon Orion—a star that should have been stationary in the sky—varied on a daily basis, suggesting that the team's measurement methods were faulty. Bouguer believed that the variation was due to the expansion and contraction of the buildings in which the sectors were mounted, and did not appear to give it further thought. Godin suspected the inaccuracies were caused by Hugo's improvised, eighteen-foot zenith sector. On his return to Quito, Godin immediately instructed Hugo to build a third zenith sector, this one with a twenty-foot radius. Godin also thought he had detected a pattern in the variation and, to hedge his bets, ordered Verguin to make additional observations of the star.[7]

As Godin tried to smooth out the data from his star sightings, Bouguer tackled a mathematical vestige of the surveying stage of the mission: establishing the absolute height of the chain of triangles. Only by bringing the observation points onto the same horizontal plane, and then reducing the length of the entire chain to its value at sea level, could the scientists tailor their measurements to describe the precise figure of the Earth. Although both parties had made good use of the barometer to estimate altitude at various stations, Bouguer had long before decided that a direct measurement of their height was necessary to achieve the exacting level of accuracy the expedition had set for itself—and so he once again prepared to depart Quito in the service of the Geodesic Mission.

In order to precisely determine the altitude above sea level of the chain of triangles, Bouguer would need to first travel to the coast and take bearings on a mountain that was also visible to the survey stations. Once he knew the altitude of that mountain, he would return to the cordillera and take sightings of it to determine its height relative to the lowest point in the chain, which was the northern end of the baseline at Caraburo. Subtracting the height of the mountain above Caraburo from the mountain's altitude above sea level would give the exact altitude of the chain of triangles. This would then enable the scientists to reduce all the measurements to sea level.

Rather than make the arduous trek back to the coast at Guayaquil, Bouguer elected to follow a new northern road along the Esmeraldas River, which promised an easier journey to the Pacific shore. La Condamine had followed that route back in 1736, when he had hacked his way through the dense jungle from the Esmeraldas coast all the way to Quito. In the intervening years, the route had been radically transformed. Pedro Vicente Maldonado, as governor of the Esmeraldas province, had successfully petitioned the *audiencia* to approve his project for a new road between Quito and Esmeraldas, and under the supervision of a Spanish engineer—and with the labor of many, many Indian workers under the *mita* system—the road was now nearing completion. This new road, wide enough to allow mule trains to pass each other, would make the journey far easier for Bouguer, who had hoped to accompany Maldonado on the governor's first trip to oversee the final construction of the road. But Maldonado had just lost his wife, Josefa, and was now in mourning, keeping him close to home.[8] Bouguer would have to make the journey without official company.

Bouguer struck out in mid-June 1740, presumably accompanied by his servant Grangier. The scientist passed through the village of Niguas (about thirty miles from Quito), where he took sightings on Illiniza, a unique double-peaked volcano halfway between Pichincha and Corazón. Continuing north along the deserted road—which took him through villages, all strangely empty—Bouguer finally arrived at the river port of Esmeraldas. There he found that the occupants of the villages along the route had all gathered at the port to celebrate the arrival of Pedro Maldonado. Since the governor was not coming and no festivities were to be had, Bouguer had his pick of assistants. He selected three Indian guides to take him farther down the river, where he hoped to obtain a better vantage point to sight the Andes.

At the delta of the Esmeraldas River, Bouguer and his entourage set up camp on a small island called Linga, where the scientist attempted to observe the mountains of the cordillera for six weeks, without much success. On July 18 Bouguer sent a note to La Condamine, complaining that "Pichincha only showed itself once, but only for three minutes and I did not have time to examine the different points" of the mountain. He and his men had been forced to live off fruits and fish since, as he recounted, "a part of my provisions were spoiled, and the other part taken by tigers" (jaguars, actually).

Bouguer finally took advantage of a break in the weather to take a bearing and elevation on Illiniza that determined its altitude above sea level. On August 10, he and his guides broke camp and began marching back along the Esmeraldas road toward Quito. He was exhausted but no doubt relieved that this major task was finished, leaving only a relatively straightforward measurement of the height of Illiniza above the baseline, in order to determine the relative height of the chain of triangles. The additional calculations to reduce this chain to sea level could then be done at leisure. Bouguer had some small, nagging doubts about the variation in star positions that Godin had detected, but he saw nothing that would prevent the expedition's swift return home.[9]

Bouguer arrived in Quito on the evening of August 27, 1740, to find the entire city on a war footing. The militia had been mobilized, though many of the soldiers were in fact prisoners who had been taken from the city's jails and pressed into service. In the midst of militarization, Quito was also filled with hundreds of mules and muleteers, and Bouguer soon learned why.

Several weeks earlier, an enormous pack train had arrived in the city from Guayaquil, carrying six hundred chests of gold and silver. The riches had come from Lima and had originally been destined for Panama's booming Portobelo trade fair, which had been scheduled for 1739, but the fair had been canceled and the booty stranded en route. President Araujo had personally traveled to Guayaquil in May to oversee the transfer of the treasure to his capital, where, far from the ocean and under guard in the city's treasury, it could be kept safe from marauding pirates and hostile navies.[10]

The Portobelo trade fair had been called off because Portobelo itself was now in British hands. War had broken out between Britain and Spain the previous year, 1739. The two nations had been in sporadic conflict ever since the end of the War of the Spanish Succession in 1714. A continuing source of tension between the two nations was smuggling and pirate operations of British ships in the Spanish-held Caribbean. British privateers often waylaid Spanish treasure ships, while the Spanish coast guard routinely stopped and boarded British vessels suspected of smuggling, and confiscated their cargoes. One such British vessel, the merchant brig *Rebecca*, commanded by Robert Jenkins, had been boarded outside Havana in 1731. As the story went, Jenkins insulted a Spanish officer, who responded by cutting off the British captain's ear and threatening to do the same to King George II if he, too, ever broke the law by smuggling goods into Spanish territories.

At first, the story caused little excitement back in Britain, but several years later, disputes over Caribbean trade ignited British public opinion against the Spanish, the flames of anger fanned by the retelling of the mutilation of Robert Jenkins. The British government, under pressure from popular anti-Spanish sentiment, finally declared war against Spain in October 1739, a conflict later dubbed "The War of Jenkins's Ear." A British squadron under Vice Admiral Edward Vernon was already in the Caribbean when the declaration arrived there, and it had profited from the element of surprise by bombarding and capturing the ill-prepared Portobelo in a single twenty-four-hour period, in November 1739.

The capture of Portobelo had initially been devastating to Araujo, but he soon came to see in it a business opportunity. With the only transshipment point for Peru's riches having fallen into British hands, Lima's merchants had made a tactical decision to reroute their Portobelo-bound shipments of gold or silver and send it instead far inland, to Quito, where it would be safe from capture. Unfortunately, with no trade fair, the merchants could not turn a profit, and so the treasure languished in Quito. Araujo was also despondent, having counted on the contraband trade from the fair to return his 26,000 peso "investment" in the presidency. Accordingly, he began demanding bribes from the merchants for moving the coin inland and for other services.[11] Araujo, it seems, was unworried by the possibility that the government would look askance at the bribes. His *audiencia* had just been transferred from Peru to the Viceroyalty of New Granada, reestablished by royal Spanish decree in 1739. The new viceroy, Sebastián de Eslava y Lazaga, had only just arrived in Cartagena de Indias the past April and was likely too busy shoring up the Caribbean defenses against another British assault to take notice of a few minor transgressions far inland.

Araujo may have been content to use the havoc of war to his advantage, but he could not avoid his duties for long. Even though Quito was now under the rule of New Granada, the Peruvian viceroy, Marqués de Villagarcía, had no compunction ordering Araujo to bolster the defenses of Guayaquil, the main port of entry to the *audiencia* and its most vulnerable point of attack from the British. Indeed, the Spanish had already received word that that the assault on Portobelo was only one part of a larger British plan to plunder Spanish possessions in the New World, raid the treasure fleets, and, with luck, overthrow the viceroyalties in Peru and Mexico.

The next phase of the British attack on Spain's New World holdings had been set into motion in December 1739, only a month after the raid on Portobelo. In December, Commodore George Anson had been given command of a squadron with orders to enter the Pacific by way of Cape Horn, raid Peru and Panama, destroy the Spanish Pacific fleet, and capture the immensely rich Manila galleon. News of the British plot had quickly reached the ears of Spain's rulers, however, thanks to Spain's efficient intelligence network and the assistance of the French, who nevertheless had remained neutral thus far in the conflict.

By early January 1740, details of Britain's bold plan—including the name of Commodore Anson and those of the ships under his command—were known to the Spanish prime minister. He immediately sent dispatches to his viceroyalties in the Americas. When the warnings arrived in late February, Villagarcía convened his war council and developed defensive plans, including refitting four heavy warships to patrol the seas, constructing lighter war galleys for coastal defense, and reinforcing the port cities of Guayaquil and Callao in the north and Concepción in Chile (at the time an *audiencia* in the Viceroyalty of Peru) in the south. A larger fleet under Admiral José Alfonso Pizarro was being fitted out in Santander in northern Spain, but it would not be ready to sail until October and could not possibly arrive until the next year, at which point it might be too late.[12] The Spanish colonists had to prepare for the eventuality that they would have to fend off a British assault on their own.

Throughout the colonies' preparations for war, the scientists of the Geodesic Mission were conducting their usual business. In November 1740, Bouguer went south to the small mountain Papa-Urco, across the valley from the Illiniza volcano he had observed from the coast, so that he could measure the mountain's relative height to the survey stations. The expedition had already established the relative height of Papa-Urco above Caraburo, so adding the relative height of Illiniza above Papa-Urco gave the total vertical distance from the baseline to the peak. Subtracting that from Illiniza's altitude gave Bouguer the altitude of the Caraburo end of the baseline as 1,226 *toises* (7,834 feet) above sea level, a figure he would later employ to reduce the entire chain of triangles to sea level.

While his friend continued his fieldwork, La Condamine, still in Quito, was preparing a lengthy court statement for the Seniergues trial—now drawing to a close—in which he refuted, point by point, the various claims of the defendants. Hugo was in Quito as well, where he had finished the new twenty-foot zenith sector before having it transported to Cuenca, where Godin, Jorge Juan, and Ulloa were taking new sightings of Epsilon Orion in order to resolve the discrepancies they had noticed in their earlier sightings. Joseph de Jussieu was also in Quito, treating patients during the epidemics of yellow fever and diphtheria that continued to race around the globe. In December 1740, Jussieu himself would be laid so low by malaria that, anticipating his own death, "he had to put his affairs and conscience in order" before finally managing to cure himself—no doubt with cinchona bark.[13]

In the middle of this industrious time for the expedition, the war finally reached the scientists. On September 24, 1740, Jorge Juan and Ulloa received an urgent dispatch from Viceroy Villagarcía, ordering them to Lima "with a haste that admitted no delay." The summons was not unexpected, given that the entire countryside was girding for war; the two men were Spanish naval officers first and foremost, regardless of their charge to assist the French astronomers. Still, it would have been a blow to Godin, who was now left without any willing assistants for his next, crucial set of astronomical observations at the northern end of the chain. He had already lost two men to disease and murder, and now with the two Spanish officers going off to war, it was not at all clear how or even whether he would be able to finish his work. Once again, it was not the known scientific problems, but unanticipated, random, and perverse catastrophes that threatened to waylay the Geodesic Mission.[14]

Instead of going directly from Cuenca to the coast per the viceroy's orders, Jorge Juan and Ulloa accompanied Godin and his sector back to Quito, where they settled their affairs and each collected their $20,000 in back salary, which would cover the costs of the trip. The officers finally departed on October 21, arriving in Lima in December 1740.

The two Spanish officers had been well trained in the Academy of Navy Guards not only in navigation, tactics, and seamanship but also in civil and naval engineering, and once in Lima they were put to work on shoring up the colonies' defenses. During their eight-month stay in the capital, Jorge Juan was given the responsibility of constructing two light war galleys, while Ulloa was tasked with drafting plans of Callao (Lima's principal port), rebuilding its fortifications, and repairing cannon. As the year 1741 dragged on with no news of Anson's fleet in the Pacific, Jorge Juan and Ulloa became anxious to return to their astronomy. They petitioned their old shipmate, Viceroy Villagarcía, to release them, and he reluctantly agreed, signing orders that, although they allowed the officers to return to the expedition, expressly warned that he might recall them at any moment. In early August they departed Callao on a merchant vessel, heading back to join their French comrades.

When Jorge Juan and Ulloa arrived in Quito on September 5, 1741, ready to assist Louis Godin with his astronomical observations, another expedition member was also returning from the front lines of battle. In October 1740, Jean-Baptiste Godin des Odonais had decided to travel to Cartagena de Indias to trade in textiles, as Seniergues had done so successfully several years before. Godin des Odonais had been a "chain-bearer" for his older cousin, carrying and placing the survey markers during the triangulation, but those tasks had been completed, and Jean had little to show for his four years in Peru. Determined to make his fortune, he set off on the nine-hundred-mile trek overland, carrying among his supplies a trunk from La Condamine filled with "natural curiosities" to ship back to France.[15]

Godin de Odonais had reached Cartagena des Indias in March 1741, five months after he set out, but he had had the bad luck of arriving on the eve of a massive British assault. Emboldened by his easy capture of Portobelo, Admiral Edward Vernon had turned his attention to Cartagena. After several months of intermittent bombardment, Vernon had assembled a force of 186 ships carrying 31,000 men, including a marine regiment from North America containing Lawrence Washington, the elder half-brother of George Washington. Assaulting a Spanish contingent of fewer than 4,000 soldiers arrayed behind the massive stone fortifications that Viceroy Eslava had reinforced, Vernon's force would attempt the largest amphibious operation in history, until the D-Day invasion of World War II.

But Cartagena had had one advantage: Blas de Lezo y Olavarrieta, one of Spain's greatest naval strategists. A generation before Horatio Nelson was born, Blas de Lezo, also sporting one eye and limbs amputated in combat, was the most feared fighting admiral on the Atlantic. He understood siege warfare from both the defensive and the offensive perspectives, having led the 1732 invasion of Oran, during which he had besieged and then captured the Ottoman city. Now, charged with guarding Cartagena, Blas de Lezo's plan was to wage a defensive battle against the overwhelming British force, gambling that the rainy season beginning in April would force the attackers to retreat.

The battle had begun on March 15, 1741. The British fleet had stormed into the harbor, setting fire to ships there (including the one containing La Condamine's trunk of curiosities), but the assailants soon found the tide turned against them. Skillfully using the city's landscape and defensive structures to attack, retreat, and regroup, Lezo's forces had staved off the British, costing them ten men for every Spanish casualty. By mid-May Vernon, with 18,000 casualties, had retreated to Jamaica. Despite the defeat, Lawrence Washington was so impressed with his commander that he went on to name his Virginia plantation Mount Vernon, a name George Washington would keep when he inherited the property. As for Blas de Lezo, his triumph was short-lived; the admiral himself died from disease a few months later.

Effectively interned during the two-month siege of Cartagena, Jean Godin des Odonias had befriended Blas de Lezo and had even managed to conduct some business amid the terror.[16] In late 1741, after the siege was lifted, Godin des Odonais made his way back to Quito with one of the great mule trains carrying gold and silver from Cartagena for safekeeping.[17] His arrival was impatiently anticipated by his fiancée, María Isabel de Jesus Grameśón, just fourteen years old (exactly half Jean's age) but already a fetching young woman with a quick intellect who had learned both French and Quechua in addition to her native Spanish. Since the age of six she had been living in a convent—hardly a place for Jean to come across her by chance. It is likely that their engagement was arranged by Isabel's uncle José Pardo de Figueroa, the Spanish official who had initially suggested that two Navy Guard cadets accompany the

French expedition to Peru. Figueroa had since gained the title of Marqués de Valleumbroso and continued to be a strong political supporter of the Geodesic Mission.

On December 29, 1741, Jean and Isabel were married by Friar Domingo Terol in the Dominican College of San Fernando in Quito. It was a spectacular wedding by all accounts, with dancing, a great feast, and ice brought down from Pichincha to chill the drinks. Quito's elite were present, as were several of the expedition members.[18] Two were noticeably absent: the Spanish officers whose presence on the mission was owed to the bride's uncle.

On December 16, just three months after their return to Quito, Jorge Juan and Ulloa had been called to war a second time. Commodore Anson's fleet, long delayed in the voyage around Cape Horn, had suddenly appeared in the Pacific and sacked the city of Paita on the Peruvian coast. Anson's trip around Cape Horn had taken so long because of navigators' continued inability to precisely determine longitude: the other great problem of navigation in the early 1700s, besides the vagaries of latitude. While the French Academy of Sciences was busy comparing degrees of latitude to establish the figure of the Earth, its counterpart, the Royal Society of London, was helping oversee the now-famous Longitude Prize, worth £20,000 ($4 million today), to anyone with a method for determining longitude at sea to an accuracy of half a degree.

Among the many contenders for the Longitude Prize was a Yorkshire clockmaker named John Harrison, who had a particular idea for a marine chronometer driven by springs instead of pendulums, with various mechanisms to compensate for temperature and motion at sea. In 1730 he first showed his idea to the instrument maker George Graham, the same man who had furnished the Geodesic Mission with its first zenith sector. Graham was impressed and gave Harrison both money and encouragement to carry on his work. In 1736, just as the French academicians were setting foot in Peru, Harrison brought his newly completed H-1 clock aboard the sixty-gun warship *Centurion* for a trial run to Lisbon. Though the clock proved quite successful, Harrison removed it from the ship and brought it before the Royal Society commissioners with a request to make a few further improvements, which they approved (in fact, these improvements would continue until Harrison was finally granted the full award in 1773).[19]

Commodore Anson regretted that Harrison's chronometer had not remained on *Centurion*, for the vessel was now the flagship of his eight-ship flotilla, which had sorely needed the instrument as it made its way to the battlefield of the Pacific. Anson's flotilla had departed Britain in late 1740, and by March 1741 it had begun its attempt to round Cape Horn against fierce gales and strong currents. For weeks Anson navigated by dead reckoning (a seat-of-the pants calculation based on compass heading and an estimated speed of advance) until he guessed that the ships were three hundred miles into the Pacific. Turning north into what he expected were calmer waters, Anson was shocked to see Tierra del Fuego dead ahead. Without an adequate means of determining longitude, he had badly miscalculated their position and did not realize until almost too late that the wind and currents had driven them back to their starting point.

After several more weeks the British force had finally clawed north into the Pacific, the ships now scattered and a few having been lost in the process. One ship, *Wager*, went aground on the Chilean coast, where several of its crew were taken prisoner. *Centurion* finally arrived at the latitude of the uninhabited Juan Fernández Island off of Chile, a landmark then famous as the exile of the real-life Robinson Crusoe, Alexander Selkirk. The island had been the agreed-to rendezvous point for the British fleet, but Anson, unfortunately, had once again been unsure of his longitude and had spent several more weeks tracing and retracing his course until he arrived at the island on June 9. By the time the fleet was reassembled, watered, and repaired at Juan Fernández in September 1741, Anson was down to three ships and just 330 of his original 1,900 men, most of the others having died from scurvy.

Peru's own squadron of four warships had been out looking for Anson while he was groping around in the Pacific, but narrowly missed him. Having patrolled the ocean since January 1741 and touched at Juan Fernández in May, the ships were in bad shape and in need of further repair. On June 7 they left Chile to return to Callao, just two days before Anson arrived at Juan Fernández. The Peruvian authorities, after learning that the Spanish fleet chasing Anson could not round Cape Horn and was now stuck in Buenos Aires, had been convinced that the British commodore had met the same fate. It had therefore come as a complete shock when Anson appeared off Paita in November, looting and burning the city before disappearing into the offing. The same authorities, who did not know about the weakened state of the British squadron, assumed Anson would press on to attack Callao, Guayaquil, and Panama; they could not know that he was now sailing north, out of Peruvian waters to Acapulco in Mexico, where he hoped to capture the Manila galleon then sailing to the Philippines, loaded down with treasure.

When Jorge Juan and Ulloa had received their second call to action, they had been somewhat reluctant to obey. Apart from the officers' skepticism regarding military intelligence (the previous information on Anson having been badly in error), the summons was also less dignified than they would have liked; the call came not from the viceroy but from the *corregidor* (magistrate) of Guayaquil, who offered the two men the second-in-command of the city's militia, a position that the highly trained naval officers refused to take. The orders were hurriedly changed to put them in command of the militia, and on Christmas Eve 1741—just five days before the wedding of Jean Godin des Odonais and Isabel Gramesón—Jorge Juan and Ulloa arrived in Guayaquil at the head of three hundred troops, whom they had pulled together from various cities en route from Quito. For several weeks the officers reinforced land batteries and supervised the construction of oared galleys, but they finally became convinced that the danger had passed. Anxious to finish the astronomical observations, Ulloa returned to Quito, while Jorge Juan remained in Guayaquil against any eventuality of further attack.

Although they had brushed off their summons as a false alarm, the Spanish officers' travails were not over yet. Arriving in Quito on January 19, 1742, Ulloa immediately visited President Araujo, who showed him a letter from Viceroy Villagarcía commanding that the two officers make for Lima "with all possible speed." Ulloa, though exhausted from his journey, which had been made almost impossible by the incessant rains, turned around three days later and returned to Guayaquil. He picked up Jorge Juan there on February 1, 1742, and they departed by ship, stopping at Paita to obtain a firsthand account of Anson's raid and finally arriving in Lima at the end of February.[20]

Jorge Juan and Ulloa reached Lima several weeks too late to board a squadron sent to Panama to look for Anson, who was actually off Acapulco at the time. But the departure of the fleet had created a new problem for Villagarcía: The entire Pacific coast of Spain's South American territory was unguarded. Admiral José Pizarro's Spanish fleet, which had been attempting to reinforce the colonial warships in the Pacific, had made a second failed attempt to round the Horn and was in any event reduced from five ships to just one operable vessel, the fifty-gun *Esperanza*. With a sizable British force still believed to be prowling the Pacific, the lack of a defensive fleet was causing understandable consternation among the Peruvian authorities.

Although all of Peru's operable warships were off hunting for Commodore Anson, the colonial navy was not out of options. In the port of Callao were two large merchant vessels, *Nuestra Señora de Belén* and *Rosa del Comercio*. It was not unusual in that era to convert merchantmen into men-of-war and vice versa, since the two were often similar in design and construction, so the two lieutenants were now instructed to repair the ships, rearm them as frigates, and take command of them to safeguard the Peruvian coast.[21]

During the year 1742, the Spanish officers brought the two merchantmen into military service, arming each of them with thirty cannon and readying the six hundred fifty men who would sail and fight aboard each one. By the end of the year they received word that Pizarro had ordered *Esperanza* around the Horn for a third time, while the Spanish fleet's naval infantry was marching overland to Chile, where the two forces would reunite. Villagarcía's plan called for Jorge Juan, commanding *Belén*, and Ulloa, commanding *Rosa*, to join forces with Pizarro, forming a three-ship squadron to patrol the Southern Pacific.

Casting off from Callao in early December 1742, by February 1743 the Spanish officers were in the Chilean port of Valparaiso, where they rendezvoused with Pizarro, who had successfully rounded Cape Horn. From February until June, the three-ship squadron patrolled a roughly triangular patch of ocean between Valdivia, Valparaiso, and Juan Fernández

Island, without sighting a single enemy sail. In April, Ulloa ran into a terrific storm that almost submerged *Rosa*, forcing him into Concepción for repairs. By June, Anson had crossed the Pacific to the Philippines and captured the Manila galleon, which would net $60 million in prize money when he completed his around-the-world voyage the following spring. By July 1743, the Peru squadron was back in Callao, providing the viceroy and the residents of the capitol with a sense of security against any further British threat.

Though the officers were anxious to return to Quito, Villagarcía's instructions compelled them to remain in Lima. There Ulloa and Jorge Juan trained other officers, supervised coastal defenses, and completed the construction of the war galleys they had begun upon first arriving in Callao at the end of 1741. Though now shore-based, Jorge Juan and Ulloa remained commanders of their vessels, still demanding a payment of 75 pesos per month on top of their regular salaries.[22]

Finally, by the end of 1743, Villagarcía had deemed that the two officers' military service was sufficient and permitted them to return to Quito. They arrived there separately in January 1744, having been away from their astronomical work for just over three years. They had only managed to return for a single three-month stretch, back in late 1741, to attempt their primary mission of measuring the Earth. Now that they were reunited with the expedition, they learned that it was no longer the Earth but rather the stars that were refusing to be measured.

IX

The Dance of the Stars

"I am as constant as the northern star," wrote Shakespeare, a line Voltaire may well have recounted to his friend La Condamine in 1735 as he reworked *Julius Caesar* for the French stage. Of course, La Condamine knew that the stars were anything but constant; they moved around the sky in small but detectable patterns as a result of the Earth's orbit around the sun, a fact that had only recently become apparent to eighteenth-century astronomers. When Picard and Cassini had conducted their surveys in France decades earlier, their zenith sectors were not accurate enough to detect these tiny variations in stellar motion, so to them the stars had appeared fixed in their places. The new sectors used by the Geodesic Mission were far more accurate, and it was this precision that, ironically, led to the errors the scientists were now trying desperately to correct.

If the quadrant was the reliable workhorse of the Geodesic Mission, the zenith sector was its thoroughbred—finely tuned but temperamental,

Figure 9.1 Use of the zenith sector. From Charles-Marie de La Condamine, *Mesure des trois premiers degrés du méridien* (1751). Courtesy of Robert Whitaker.

and prone to complications. The instrument's purpose was to determine the latitude of its current position, and its operation was simple in principle: A long telescope mounted to a movable frame, it hung from a rafter and was gimbaled at the top so that it swung through a small arc on a precise north-south axis along the meridian, aimed at a small patch of sky almost directly overhead. The astronomer would recline uncomfortably on the floor, craning his neck up to the eyepiece. He was looking for a particular star—in this case, Epsilon Orion—to reach its maximum altitude as the Earth slowly rotated underneath. As the star entered his field of view from left to right—inverted because of the convex lenses—the astronomer would minutely adjust the telescope's position so that the horizontal crosshair aligned with the star's path. As he called out the star's crossing of the vertical (meridian) crosshair, his assistant would note the time on the wall-mounted pendulum clock. The star's exact angle north or south from the vertical, which was determined by a plumb bob, would be read from the precisely graduated limb, thus enabling the scientists to calculate the latitude of their position. Under ideal conditions, the crossing time and angle should be the same on any given night, but in order to average out any errors in the telescope or mounting, the astronomer would make the same observations over many successive nights.

The twelve-foot zenith sector that Godin had bought from George Graham was the technological marvel of its day, and a far cry from the ones used by Picard and Cassini. Those older sectors were equipped with only small, engraved scales to read off the measurements, and thus they had relatively poor angular accuracies for astronomical studies, perhaps within 20 seconds of arc, or about six inches seen at a mile's distance. Graham, widely acknowledged to be the finest instrument maker in Europe, had introduced several enhancements to the device, including a precision micrometer screw at the lower end of the tube that allowed the astronomer to finely adjust the telescope's movement, giving it an overall accuracy of about 7 seconds of arc (about two inches seen at a mile's distance). The zenith sector was by far the costliest device used on the mission; a smaller sector used by Maupertuis in Lapland set him back £130, the equivalent of $26,000 today.[1]

The precision of the Graham sector allowed its user to pick up tiny movements of the stars against the background of the night sky. Graham had originally built his sector to look for one such movement, stellar parallax, or the apparent shift in position of stars as the Earth revolves from one side of its orbit to the other, akin to the shift in position of distant objects when we close our left or right eyes, and a discrepancy that was originally believed to have allowed eighteenth-century astronomers to determine the distance to the stars. In actuality, the parallax to even the nearest star is less than one second of arc, much finer than Graham's sector could resolve (and indeed, stellar distances would not be accurately determined using parallax until a century later). In 1727 the British astronomer James Bradley was unsuccessfully attempting to measure the parallax when he discovered a new phenomenon called stellar aberration.

Stellar aberration would ultimately prove to be the true source of the errors that plagued the Geodesic Mission. The phenomenon is often compared to the apparent path of raindrops in a moving car's headlights; even when the drops are falling absolutely vertically, they appear to be coming *toward* a car as the vehicle moves forward, making the raindrops' starting point appear to be somewhere in front of the headlights instead of directly above them. Similarly, in astronomy the finite speed of light falling toward Earth causes a star's apparent position to shift back and forth as the Earth moves in its orbit toward and away from the star. This change in apparent position is about 20 seconds of arc annually—a variation well within the limits of the Graham sector's resolution, but almost invisible to the more primitive instruments of Picard and Cassini.

Though stellar aberration had been reported on at the time of the Geodesic Mission's departure from France, its particulars had not been well understood. Bradley had only reported on stellar aberration in 1729, and though it had been discussed on both sides of the English Channel, nothing substantial about the phenomenon had been resolved when the French scientists departed for the equator just a few years later. The fact that the variation in star position occurs gradually over the course of a year, moreover, meant that stellar aberration initially went unnoticed amid all the other sources of error that the astronomers were aware of and actually looking for during their two-month periods of observations in 1739 and 1740.

The astronomers had to account for several variables when taking their star sightings. A considerable source of error was the well-known phenomenon of atmospheric refraction, which Bouguer had sought to correct with his observations taken on the shores of the Pacific back in 1736. Another source of error was the astronomers themselves; none of them was particularly experienced in field astronomy except Godin, and even he had spent much of his time in the Academy of Sciences attending to bureaucratic duties rather than making astronomical observations. The academicians had to learn the practical details of observing the stars by trial and error, often placing themselves in extremely tiring positions for long periods of time, conditions that could affect the accuracy of their measurements. On two occasions, La Condamine fell down in a dead faint after straining his neck to make observations.[2]

Despite the problems with the stars and with the astronomers themselves, the most important source of known error was the zenith sector itself. Because of its extreme precision, any minor variation in the sector's assembly, mounting, or siting could translate to major discrepancies in determining the position of the stars. The expedition's original twelve-foot sector, crafted by Graham, had taken quite a beating thus far in the mission. It had been transported through extreme conditions over thousands of miles, had frequently suffered damage from mishandling, and had been rebuilt by Hugo on several occasions. Hugo had also built several more sectors of varying lengths based on Graham's design, but the materials were of uneven quality, and the structural stability of the telescopes themselves were often suspect. Although the sectors were attached to heavy ceiling beams, the natural expansion and contraction of the building from heat and humidity, coupled with the regular earthquakes, affected the astronomers' readings. Finally, the sectors had to be lined up exactly on the north-south meridian, a task made more difficult by the fact that they were sited in enclosed houses with only a small hole in the roof for observations.[3]

As a result of all of the potential errors involved in the use of zenith sectors, the scientists' astronomical observations, which at first appeared quite straightforward and easily accomplished, would actually prolong the Geodesic Mission another three years. Because the zenith sectors were so precisely made and finely calibrated, any error, no matter how minor, would instantly creep into the observational data. It took the astronomers a great deal of time to account for and eliminate each error, constantly rebuilding, recalibrating, and repositioning the telescopes every time a discrepancy was found. In correcting an error, the astronomers' first step would be to calibrate the angles on the sector's micrometer by laying the sector horizontally and observing the angles between targets several miles away, whose distance from each other (and therefore exact angle between them) was known. Next, they would position the sector on a north-south meridian and initially checked with a sundial (since the sun's shadow pointed due north at local noon). The alignment was then verified over successive nights by reversing the telescope on its gimbals through 180 degrees and timing the passage of stars, whose exact crossing time would need to be identical regardless of which way the sector was mounted. Each of these preparatory steps could occupy days or weeks, which—coupled with the frequently overcast skies—meant that it could take the scientists two months to achieve a single set of useful observations.

The problems with the expedition's zenith sectors had first emerged in mid-1740. While Bouguer and La Condamine had just finished their observations with the twelve-foot sector and had convinced themselves that they had successfully measured a degree of latitude, Godin had not shared their optimism. He doubted the reliability of his own eighteen- and twenty-foot sectors, and had ordered Verguin to carefully chart the progression of stellar aberration to verify whether that could have been the source of the errors he had noted. Meanwhile, he had decided to redo his own observations in Cuenca. By September 1740, he had claimed to have discovered a previously unknown but sizable variation in the position of Epsilon Orion, a difference of almost 50 seconds of arc over a period of nine days, far greater than the annual 16 arc-seconds of stellar aberration observed by Verguin. Such an enormous variation should have been easily visible with the telescopes of the day, but it had never been reported by dozens of reputable astronomers before the expedition. Bouguer, who learned of Godin's observations from Verguin and who still attributed the variations to the buildings in which the observations were being made, originally dismissed the findings. "I had to rub my eyes," he wrote, "as I saw nothing of the sort."[4]

Although Godin had insisted on making his observations alone and had refused Bouguer's request to communicate his findings, Verguin's revelations evoked a growing disquiet in Bouguer. After he had returned from his trip to the Esmeraldas coast (in order to measure the altitude of the chain of triangles above sea level), Bouguer reexamined his own astronomical calculations and came away with his own proof of the unreliability of the teams' observations. His findings suggested that the expedition would need to prolong its already interminable stay in Peru to make further measurements—a development that would have been almost unthinkable. Not wanting to alarm the others, Bouguer secretly obtained a house on the outskirts of Quito to use as an observatory and mounted the repaired twelve-foot sector there, working with his servant Grangier to confirm his observations.[5]

Six weeks of staring through the eyepiece of his sector did not improve Bouguer's assessment. If Bouguer had ever heard the Chimu legend that Epsilon Orion was a thief held captive by the other stars, he would have grimaced at the irony: This star was certainly *not* confined to one place, but rather seemed to move about the sky quite freely. Methodically,

Bouguer tested and dismissed various theories to account for Epsilon Orion's movements. Perhaps, he thought, the star was wobbling back and forth in response to the tug of its own planets—an astonishingly prescient observation, given that planetary wobble was not described until the mid-nineteenth century and was only successfully used in the late-twentieth century to detect extrasolar planets. Or perhaps stellar parallax really was to blame. Bouguer even tried to correct his observations for precession, or the gradual shift in the angle of the Earth's axis of rotation, similar to the wobbling of a top as it spins. None of these factors seemed to account for the dance of the stars against the darkling sky.[6]

Bouguer had reached an impasse, which frayed the astronomer's already raw nerves. He had now spent six arduous years in Peru and could see no end in sight. Even his close friendship with La Condamine had become strained; the two were now bickering like an old married couple.

The immediate cause of the friction between Bouguer and La Condamine was an argument over the use of Bouguer's new observatory. Bouguer had not yet completed his astronomical observations when, in early November 1740, he had set off to Mount Papa-Urco in order to finish the altitude measurements that he had begun in Esmeraldas. Before his departure, Bouguer had brought La Condamine to his heretofore secret observatory and, explaining his findings to date on stellar variation, had asked his friend to continue observing Epsilon Orion during his absence. La Condamine, flattered by Bouguer's confidence in him, gladly accepted the keys to the house.

La Condamine had worked for several weeks, but he, too, had been unable to obtain any favorable results that would explain the variation in the teams' star sightings. Upon Bouguer's return in late November, La Condamine had asked for more time in the observatory. Bouguer had flatly refused and demanded the keys back, insisting that he wasn't being secretive like Godin but simply wanted to be certain of his results before he communicated them. La Condamine complained bitterly at this treatment, so Bouguer reluctantly agreed to the extension, allowing La Condamine to employ his servant Grangier in order to speed up his work. La Condamine again failed to make any progress, and he complained about having to work "in a country where sometimes a fortnight will pass without being able to see the zenith." Bouguer tartly noted that during the same period, Verguin had made thirteen or fourteen successful observations. On the last day of December 1740, Bouguer had firmly demanded the keys from his embarrassed and resentful friend.[7]

By the end of 1740, it had become obvious to everyone in the expedition that the teams would have to stay in Peru in order to redo their observations, instead of heading home, as many of the members had hoped. Just at this low point, Godin came out of his isolation, no doubt because he had lost the other members of his team, Jorge Juan and Ulloa having departed for Lima just a few weeks earlier to counter the threat from Anson's fleet. On January 11, 1741, Godin wrote to Bouguer and La Condamine, stating that he, too, was mystified by the variations and suggesting a cooperative plan to resolve the problem.

Godin's plan to determine the cause of the stellar variations was simple but effective. Two of the astronomers would return to the northern and southern ends of the chain of triangles, while the third would remain in Quito. The three would then attempt to observe Epsilon Orion on the same night. By taking simultaneous measurements, they could at one stroke remove the daily variations caused by stellar aberration, as well as Godin's phantom variation.

Godin's proposal for mutual cooperation was completely out of character for him, but Bouguer and La Condamine embraced it heartily. After so many years of Godin's infighting and stonewalling, they were no doubt relieved to be able to finally collaborate. The scientists agreed that Bouguer and Morainville would go south to Cuenca with the repaired twelve-foot sector and reoccupy the observatory at Mama-Tarqui. Godin, along with Verguin and Hugo, would take the new twenty-foot sector to the northern end of the chain; instead of the observatory originally used by Bouguer, he would work out of the hacienda Pueblo Viejo, owned by Ignacio Páez, farther north near the town of Mira. La Condamine would stay in Quito to make his observations with a fixed fifteen-foot telescope, permanently trained to detect the crossing of an individual star.[8]

By April 1741, all the astronomers were in place. After the tedious process of recalibrating their instruments, they began their observations. During the months of May, June, July, and August, the astronomers nightly watched the skies and took measurements when possible, in the hopes that, on one of those nights, all three would observe the same star so they could compare the results. As usual, unexpected obstacles stood in the way of this simple plan. Tremors frequently interrupted their work and caused enough damage to the observatory structures that they had to spend several days fixing the mounts and realigning the telescopes; the earthquake of June 14 near Quito was especially destructive. Although it was no longer the rainy season, clouds frequently obscured the night sky, so that when Mira and Quito were clear, Cuenca was overcast, and vice versa.

At the end of August 1741, Godin ceased his observations and returned to Quito, having received news of the imminent return of Jorge Juan and Ulloa. The two officers had finally been released from Lima by the viceroy, Anson's fleet having failed to appear in the Pacific, and Jorge Juan and Ulloa originally intended to go back to Quito to continue making astronomical observations with Godin. Further problems with the twenty-foot zenith sector kept them from this task, however, and they were recalled to duty just three months later. Although Godin's withdrawal had made simultaneous observations impossible, Bouguer kept at his work through the end of 1741, trying to reconcile his old observations with the newer ones.[9]

By mid-December, Bouguer had reached an inescapable conclusion: Not only were the observations made during the previous six months utterly useless, but the astronomers would have to throw out all their observations from the *past two years*. The extensively repaired twelve-foot zenith sector, which the scientists had by now used at both ends of the chain of triangles, was simply too faulty to trust; they would have to build a new sector and start over, once again attempting to make simultaneous observations of Epsilon Orion from each end of the chain. Bouguer was despondent. "We are withering away here in utter exile," he wrote a colleague, "without doing anything useful on my part, and absolutely destitute of everything—books, instruments, even clothing."[10]

When La Condamine learned that the latest round of observations had failed, the revelation would have been especially bitter. A package had recently arrived in Quito from Paris, containing old newspapers that claimed the expedition had already left Quito and would arrive home on a date that was now eighteen months in the past. He lamented in his diary, "At that time I was flattering myself that all of the obstacles that had been holding us back for so long were going to be removed and that I could finally set off en route back to France." Now La Condamine would have to start his observations all over again, "so as not to bring back doubtful or uncertain results instead of the clarifications we had traveled so far to obtain."[11]

Two weeks later, when they assembled in Quito for the wedding of Jean and Isabel Godin des Odonais, the members of the Geodesic Mission would undoubtedly have discussed the implications of the astronomical setbacks. Amid the joy and festivities that they had hoped would crown their great achievement, they nursed the certain knowledge that for at least six more months—and perhaps even another year—their "utter exile" would continue.

The previous two years of 1740 and 1741 may have been fruitless with respect to astronomy, but they had not been idle ones for La Condamine. He was enthusiastically engaged in a number of activities that had nothing to do with accomplishing the goal of measuring the Earth, and indeed he seemed blithely oblivious to the fact that his projects were delaying the expedition's astronomical work and burning through its funds at an alarming rate.

With four separate trials demanding his time, effort, and money, La Condamine was becoming more adept at playing attorney than at being an astronomer, even through he protested that "the trial of the stars (*le procès avec les étoiles*) was closest to my heart." The Frenchman had spent much of the year 1740 preparing a voluminous testimony for the trial of Seniergues' murderers, and even after the sentences were handed down in August of that year, he had vigorously countered the appeals of the ex-mayor of Cuenca, Sebastián Serrano. Although La Condamine may have felt obliged to protect "the honor of the nation [France] and that of the Academy," he also went on the offensive, bringing another three lawsuits against various individuals involved with the case, including Juan Crespo, the vicar of Cuenca, in the process antagonizing the very authorities who allowed the Frenchman to be on Spanish soil.[12]

While crusading within the Peruvian legal system, La Condamine had also continued his private mission—begun at the promontory of Palmar on the equator—of creating permanent memorials to the presence of French Academy scientists within Spanish Peru. Again, as with the equatorial marker at Palmar, these memorials had no appreciable relation to the objectives of the mission, and in fact they cost the expedition precious time, effort, and money.

In 1741, La Condamine began the next set of monuments to commemorate the Geodesic Mission. He ordered a "great slab" of pure white marble to be dug from a quarry near the southern baseline of Tarqui and transported to Quito; there he had the slab cut into three tablets and then dressed and engraved to commemorate the terminal points of the triangulation: the baseline at Yaruquí and the northern and southern ends of the chain at Cotchesqui and Tarqui, respectively. Each of the tablets described, in Latin, the geodesic measurements the expedition members had made during their observations, pointedly naming the French Academy of Sciences as the sole authority over the findings. To complete the tablets, La Condamine kept an Indian stoneworker named Alvarez under lock and key for six weeks, much like the Indians the expedition's members had seen at the Yaruquí wool mill. When the memorials were finished, La Condamine distributed them throughout the region. He presented the Yaruquí tablet to the Jesuit fathers who had lodged him when he first arrived in Quito; they, in turn, mounted it on the courtyard wall of their seminary. He had the Tarqui tablet fixed to the wall of the Sempértegui hacienda the expedition had used as its southern observatory, while the Cotchesqui tablet was eventually placed in the nearby church at Quinche, where the parish priest was the astronomers' friend José Antonio Maldonado, brother of Ramón and Pedro Vicente Maldonado.[13]

La Condamine's most egregious expenditure was not the memorial tablets but rather a far more visible monument to the French Academy's presence in the *audiencia:* a pair of thirteen-foot-high pyramidal markers. Originally intended to simply denote the endpoints of the baseline at Yaruquí, these markers became an overt political statement regarding the hierarchy of royal authority, which unnecessarily set the French members against their Spanish counterparts and earned the scientists the enmity of the authorities under whose jurisdiction they were allowed in Peru in the first place. In many ways, the story of the pyramids of Quito mirrored the problems that had plagued the whole expedition: born of careful planning, lofty purpose, and backbreaking labor, the pyramids reflected the stunning arrogance of the expedition's members—an arrogance that, once again, created self-inflicted delays and disputes that would almost derail the entire Geodesic Mission.[14]

The story of the pyramids began back in France in early 1735, during the preparations for the mission. La Condamine had noted to his colleagues at the Academy of Sciences that no vestiges remained of the first survey of France, completed in 1670 by Jean Picard, and that this lack of markers had caused problems for the Cassinis when they tried to remeasure the territory thirty years later. La Condamine proposed to avoid this problem by marking the endpoints of the Peru baseline with some form of monument—as a former Egyptian explorer, he preferred the time-tested pyramid—bearing a simple Latin inscription to commemorate the expedition. The idea was warmly received, so he requested an appropriate inscription from the French Academy of Inscriptions and Literature, which composed all official inscriptions and was conveniently located just downstairs from the Academy of Sciences in the Louvre.

The approved Latin text for La Condamine's monument was straightforward and unpretentious. It identified the French king Louis XV as the sponsor of the mission, acknowledged the support of the Spanish king Felipe V, gave the names of the French academicians, made a passing reference to the Spanish officers who accompanied them, and indicated the measurements of the baseline and the date of its completion. As with all aspects of the mission, which was then in its planning stages, the inscription had quickly received widespread attention in France, including some last-minute objections from the French press, which argued that the inscription should also be written in Spanish and "Peruvian," so as not to exclude the native population from its message.

La Condamine made the pyramid project his own special task, and at first it had not interfered with his other work. Once the baseline at Yaruquí had been measured in late 1736, he obtained two millstones from a local windmill and buried them at the exact ends of the baseline, taking the added precaution of cutting a slice into each stone to render it useless and therefore less susceptible to being stolen. Jorge Juan and Ulloa, understanding the rationale behind permanently marking the position of the baseline, had witnessed all of this without the slightest objection.

A lack of funds and the continuous work of triangulation had tabled the pyramid project for several years until, in April 1740, the surveying work was essentially complete, and the construction of the pyramids could begin. La Condamine returned to the Yaruquí plateau to take some final geodesic measurements for the monuments; as before, his work did not seem to grate on Spanish observers. José Antonio Maldonado, whose parish at Quinche was near the baseline, even agreed to provide the materials and labor needed for the pyramid project. The expedition's draftsman, Jean-Louis de Morainville, had previously been hired by Maldonado to supervise the construction of his church tower in Quinche. As Morainville was now living in the parish, he was able to supervise the worksite of the pyramids while La Condamine returned to Cuenca to deal with the Seniergues trial.

Construction on the monuments began in late 1740. The squared-off pyramids were built directly atop the two Yaruquí millstones, which were left in the place where they had first been set. The ground beneath the monuments required different considerations; the southern pyramid at Oyambaro had firm foundations in the rocky hillock, but the northern pyramid at Caraburo was built on a sandy ledge and required deep wooden pilings to be sunk into the ground beneath it. The two monuments stood two *toises* (thirteen feet) on a side and equally high, faced with brick and filled with rubble. The stones had to be hauled four hundred feet out of the Guayllabamba ravine, and a six-mile long channel had to be dug to bring water from a distant pond to mix the mortar. La Condamine instructed that the brick molds be made an unusual size so the bricks would not be stolen, then demanded a hollow space be left inside each pyramid to house a small silver tablet engraved with the geodesic measurements. Finally, he ordered that each pyramid be topped with a stone fleur-de-lys, the symbol of France.

The pyramids had originally been intended as "simple structures," but all of the additional measures that went into their construction required

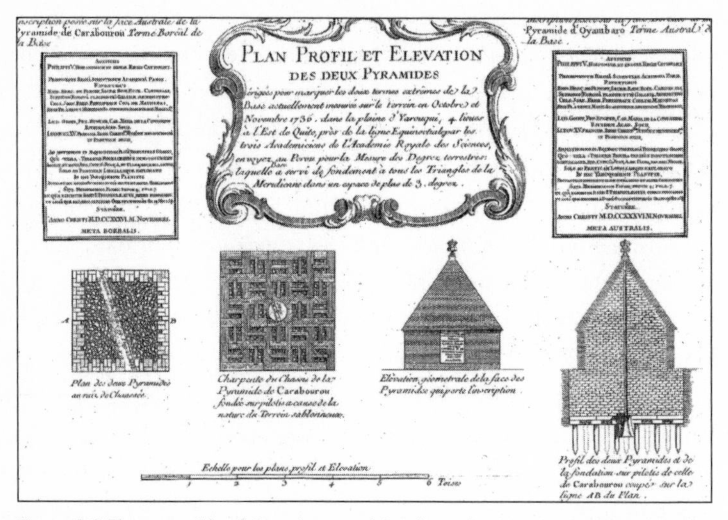

Figure 9.2 The pyramids of Oyambaro and Caraburo, by Jean-Louis de Morainville. From Charles-Marie de La Condamine, *Journal du voyage fait à l'Équateur* (1751). Courtesy of Robert Whitaker.

more time and money than anyone had expected back in France. It took over a year to build them, and the costs ran to a staggering 15,000 livres ($135,000)—more than all of the expedition's astronomical instruments and survey tools combined.[15]

The brick-and-mortar monuments were not the only memorials whose construction had spiraled out of control. The engraved stone tablets, one mounted on each pyramid, had grown from small, eight-line plaques to a pair of great monoliths, each of which devoted thirty lines to the text that the Academy of Inscriptions and Literature had provided the mission. La Condamine insisted the tablets be cut from the hardest stone so as to last through the ages—a stipulation that resulted in another six months of quarrying, dressing, and engraving.

The pyramids' tablets had been more expensive than anticipated, but the inscription itself was costlier still. The first inkling of disaster had come early in the project, when in August 1740 La Condamine showed a copy of the proposed inscription to Antonio de Ulloa and Jorge Juan. They had been unhappy with the Latin term used to describe their participation, *auxiliantibus* (assisted by), but they had little time to protest, since a few weeks later they received orders to depart for Lima to buttress the city's defenses against the threat from Anson's fleet. Undeterred by their teammates' protests, La Condamine and Morainville continued building the pyramids through the Spanish officers' absence, with the inscription unchanged.

In September 1741 Juan and Ulloa returned to Quito, intending to continue their astronomical observations with Godin. They were instead sidetracked by a closer examination of the pyramids at Yaruquí, whose inscription was just as La Condamine had showed them—a development that infuriated even the normally serene Jorge Juan. Rather than speaking directly to La Condamine, the officers quickly filed a lawsuit (*petición*) against La Condamine alone. None of the other academicians, after all, were as closely associated with the project as he was.

The fight over the pyramids had become personal. It was not the monuments the Spanish officers objected to, but rather La Condamine's insistence on using the patronizing and overtly political inscription, which Ulloa and Jorge Juan claimed insulted the Spanish crown. Not only did the term *auxiliantibus* devalue their role as envoys of the king, they argued, but the French fleur-de-lys that topped the pyramids was an obvious ploy to upstage the king himself. Jorge Juan and Ulloa demanded that the tablets be replaced with new ones and that a Spanish crown be placed at the summits of both pyramids.

La Condamine was determined to plant the French flag on Spanish soil, and the loyal Spanish officers were equally determined to stop him. At this point, the former French soldier's vanity and patriotic zeal got the better of him. Instead of acceding to the officers' requests, which would have left the monuments intact and satisfied their original purpose of marking the endpoints of the baseline, La Condamine decided to play lawyer for a fifth trial, engaging in a protracted legal battle with the officers and the leaders of the *audiencia* itself. In his legal briefs, he stubbornly argued each point in increasingly arcane terms, offering up *currentibus*, *cooperantibus*, *assistentibus*, all Latin terms denoted varying degrees of assistance, rather than any equality of labor. The Spanish officers countered with their own text. Louis Godin, whose zenith sector was still under repair, now entered the fray and proposed his own inscription, further muddying the debate. La Condamine tried a new tack, arguing that the fleur-de-lys could equally apply to the Spanish king, since he was a grandson of Louis XIV. All of the Frenchmen's claims were rejected. By now, the only astronomer actually doing any astronomy was Bouguer, who was two hundred miles away in Cuenca, still trying to comprehend the chaotic stellar data from the past two years.

The debate over the pyramids had taken on a life of its own. The arguments were interrupted in December 1741, three months after they had begun, when Jorge Juan and Ulloa were recalled to Guayaquil at the sudden reappearance of Anson in the Pacific. But by now the colonial government was backing the officers' claims, and the *audiencia* continued to prosecute the case in their absence. In April 1742, while the Spaniards were refitting the two merchantmen in Callao for military duty, the *audiencia* handed down its verdict: the *petición* of the Spanish officers was legitimate, so La Condamine was ordered to remove the fleur-de-lys and change the inscription to describe the Spanish officers as *concurrentibus*, reflecting the equal role that Jorge Juan and Ulloa had played in carrying out the mission.

After months of resistance, La Condamine would be forced to change his cherished monument—but the ruling could hardly have come as a surprise. The crown attorney for the case had been none other than Juan de Valparda, the host and confidant of Juan and Ulloa ever since their arrival in Quito six years earlier. La Condamine had no choice but to give in to the first part of the judgment, grudgingly agreeing to place a bronze Spanish crown atop the fleur-de-lys, thereby acknowledging the supremacy of the Spanish king above his own. Nevertheless, he stubbornly persisted in denying an appropriate inscription for the Spanish officers, dragging out the legal arguments for another six months. In fact, whether out of stubbornness or deference to the royal office that had decreed the inscription, La Condamine would never change the wording on the tablets, leaving it to the Spanish government to decide whether his defiance warranted the destruction of the monuments. By then, La Condamine, Jorge Juan, and Ulloa would be long gone.

On January 3, 1742, Bouguer returned to Quito, so that he and La Condamine could decide how to repeat their astronomical observations. Although they were still uncertain about the precise effects of stellar aberration on their measurements, the two men continued to believe that the simultaneous observation of Epsilon Orion at the two ends of the triangulation was the best way to circumvent the phenomenon. In their next round of sightings, they would at least have the advantage of more reliable instruments and the hard-won lessons of the previous two years.

The astronomers planned to conduct their next round of sightings using one refurbished zenith sector and one entirely new one. They would abandon their old twelve-foot sector and have Hugo construct a new one with an eight-foot radius, for Bouguer to use at his Tarqui observatory. For the northern measurements, La Condamine asked Godin to move the twenty-foot sector (which Hugo had finally repaired) from Godin's observatory at Mira to La Condamine's old one at Cotchesqui. Godin readily agreed to this, since he was unable to make any observations without Jorge Juan and Ulloa, who had left for Guayaquil the previous month.

On January 19, one of the Spanish officers returned from the war for just long enough to hamstring the French scientists. Ulloa, who had missed the recent wedding of Odonais and Isabel after he and Jorge Juan had been called away on duty, suddenly reappeared after a three-week slog through the rain from Guayaquil. But instead of assisting Godin, he fired a shot across the Frenchman's bows that immediately halted the scientists' plans.[16]

The recent affair of the pyramids had had a chilling effect on the previously good relations between the Spanish officers and the French scientists, and now it appeared that the ill will would derail the Geodesic Mission as well. Jorge Juan and Ulloa, once the moderating force in the expedition, no longer trusted their French teammates—least of all La Condamine, whom they particularly resented for his intransigence. Now, upon learning of Godin's plan to give La Condamine the twenty-foot sector, Ulloa petitioned his old friend Valparda to immobilize the instrument at Mira, citing the recent furor over the inscription as one of the reasons for the impoundment. Since Ulloa had been called back to the coast as soon as he had arrived in Quito, he further demanded that "for no reason or occasion should [the sector] be given to anyone," even in situ, until he and Jorge Juan returned. Valparda, like Ulloa offended by La Condamine's recent actions and eager for retribution, readily agreed.[17]

The Spaniards' vengeance was now complete. Without a second sector, the French scientists could not make the simultaneous observations necessary to accurately measure the length of a degree of latitude, leaving the entire Geodesic Mission dead in the water. The scientists' only hope was that Hugo could repair the first, faulty zenith sector while also constructing a new one.

Hugo had already built one zenith sector for Godin, but now his skills would be tested as never before. He would have to construct the new eight-foot sector from scratch, using Bouguer's design; in the best of circumstances this would take many weeks, but Hugo was also faced with a shortage of critical materials, especially brass for the micrometer.[18] He would also have to take apart and repair, yet again, the dilapidated twelve-foot sector to take the place of the impounded twenty-foot instrument.

The instruments were not the only things fraying at the edges; Bouguer and La Condamine were now sniping at each other almost continuously. Bouguer was particularly incensed by La Condamine's extracurricular activities over the past year; the pyramids, and now *five* separate lawsuits, had, in Bouguer's estimation, delayed the expedition even more than the astronomical problems had. He demanded to know "the necessity to hide the real intention for your voyages to Cuenca," perhaps insinuating that he was actually visiting the prostitute who fathered his two daughters. La Condamine, in turn, accused Bouguer of reneging on their agreement to share information and of "concealing his results." Godin, meanwhile, withdrew once again; he stated that he had "no obligation to fulfill" his role in the joint effort he had originally proposed. He refused to communicate the results he had already attained except in cipher, sending off a letter that gave his initial estimate for the length of a degree of latitude beginning with the code "aabcccdeefff. . . ." Godin's actions meant that Bouguer and La Condamine would have to work out the problems of the unexplained stellar motions on their own.[19]

With no astronomy possible for at least the four or five months it would take for Hugo to finish the zenith sectors, the expedition members once again found themselves with little to do but wait. This unintended work stoppage actually came at an opportune time, for the rainy season of 1741–1742 was one of the worst on record, with a constant cloud cover that permitted almost no observations. This was likely due to a strong El Niño season, when warm currents enter the eastern Pacific Ocean, causing torrential downpours in the Andean regions and a resultant plummet in the harvests. "The famine caused the price of wheat, corn and all other grains to increase eightfold," explained Bouguer, and the spike in prices further diminished the scientists' already meager cash reserves.[20]

With money short and food at a premium, Louis Godin attempted to fill his near-empty coffers by hunting for a treasure that had been lost the year before. When his cousin Jean Godin des Odonais had accompanied the great mule trains carrying gold and silver from the war-torn city of Cartagena to Quito for safekeeping, some of the mules had lost their footing while crossing a bridge over the Pisque River, about thirty miles northeast of Quito, and had pitched into the river, losing 80,000 pesos ($3.6 million) of treasure. When Godin heard of this, he assembled a team of Indians and attempted to divert the river to find the treasure. Despite a considerable effort, three attempts over the course of many months yielded nothing.[21]

By May 1742, as the rainy season wound down, the zenith sectors were still not finished, and the three academicians were still estranged. The Jesuits of the University of Santo Tomás in Quito provided an opportunity for the scientists to come together on neutral ground, inviting them to the presentation of a thesis dedicated to the French Academy of Sciences. Louis Godin acted as the oral examiner for the young scholar, and the Jesuits presented the scientists, in commemoration of the event, with an engraved silver plaque, which was eventually brought back to the French Academy of Sciences.[22] Although they did not know it at the time, the presentation at the university would be the last time that the three academicians would find themselves together in Peru.

The effort of the Jesuits seems to have lessened the bickering between La Condamine and Bouguer, and now La Condamine extended an olive branch to his old friend. In June, under clear skies that must have maddened the scientists, as their sectors still weren't ready, the two men ascended the active volcano of Guagua Pichincha, known as the Vesuvius of Quito. They had previously set up a triangulation station on the rocky peak of the dormant Rucu Pichincha, about three miles away from

Guagua Pichincha, but they had never crossed the lunar landscape to the actual caldera. As they set out for the climb, a local priest spoke with them "with a great air of mystery" about a gold mine reputed to be hidden in the mountains, which he assumed—as did many others—was the real objective for their expedition to Peru.

The two scientists set out for Guagua Pichincha, Bouguer marching ahead while La Condamine waited for his guides to show up. Snowfall from the evening before had obliterated the trail, forcing La Condamine—still without his guides, and now utterly lost—to spend the bitterly cold and foggy night out in the open. He finally got his bearings and caught up with Bouguer a day later, after which the two spent several days plowing through the deep powder to find a route to the crater.

La Condamine and Bouguer reached the western lip of the volcano on June 17, 1742. Together the two men peered cautiously over the edge. Pichincha, one of the most active volcanoes in the Andes, was quiescent at the time, and the scientists were able to look down at the "blackened and charred sides," as La Condamine noted, the result of the last explosion eighty years before, when the volcano had showered Quito with almost a foot of ash. "Everything I could see," he later wrote, "seemed to be the debris of the crumbled summit of the mountain when it caught fire: a confused mass of enormous rocks, shattered and haphazardly piled on top of each other, which offered to my eyes the very image of the chaos of the Poets."[23] If there were an underworld, it would likely look much like this.

As the two men were descending Pichincha on June 19, they happened to notice smoke rising from the summit of Cotopaxi some thirty miles away. Even from that distance, they could hear the rumbling of what they soon learned was a sudden, unanticipated eruption. It would turn out to be the beginning of a two-year sequence of violent explosions that covered the countryside with ash and melted hundreds of years of accumulated snowfall, which quickly inundated the valleys and ravaged the nearby towns of Latacunga and Napo. "Cataracts of fire pried open new routes down the mountainside," La Condamine recorded, "where avalanches of half-melted snow barreled down the mountain into the plains below. In just a few minutes a sea of boiling water covered the land for several leagues around, with fiery blocks of lava, blocks of ice and fragments of rocks tumbling pell-mell into the mass."[24] The two scientists would have shuddered when they remembered that, just a few years earlier, they had camped for two weeks near the summit of the now-raging volcano.

Returning to Quito, La Condamine and Bouguer were greeted by Hugo, who informed them that the new eight-foot zenith sector was completed and the original twelve-foot sector had been rebuilt. They now had two sectors with which to complete their simultaneous observations of Epsilon Orion, from both the northern and the southern ends of the chain of triangles.

Bouguer and La Condamine realized that, if all went according to plan, the next series of simultaneous observations would be their final ones in Peru. Once these sightings were completed and the calculations finished, the scientists would have the single number they had spent almost a decade trying to determine—the length of a degree of latitude at the equator. Armed with that single number, they would at last be able to go home.

With the end of the mission in sight, Bouguer and La Condamine prepared to go their separate ways. For some time now, the two scientists had known that the Geodesic Mission would not be returning to Europe as a single body. Most of the mission's surviving members had already dispersed, as numerous problems—Godin's toxic leadership not least among them—had led to irreparable ruptures among the men. The Spanish officers seemed to have broken from the Frenchmen entirely, while Verguin, whom La Condamine had charged with fixing several mistakes that Morainville had made on the pyramids, complained that Morainville was "a strange man in more ways than one."[25] Even Bouguer and La Condamine, as close as they had become, had entirely different goals for their return.

Bouguer and La Condamine would be traveling back to France by different paths, and so their final observation posts would reflect their chosen routes home. Bouguer was anxious to get back to his beloved Academy of Sciences to present the findings of the expedition. He therefore resolved to take the eight-foot sector to the northern observatory at

Cotchesqui, since that location would allow him to return to Europe by the shortest possible route: traveling north along the Andes to Cartagena de Indias and from there by some Spanish vessel to Saint Domingue, where he would find a French ship to take him home. "La Condamine the inquisitive," as he was later dubbed, settled on a more adventurous route rejected by Bouguer: a journey down the largely unexplored Amazon to the Atlantic and from there to Europe.[26] In light of his plans, La Condamine would go south with the twelve-foot sector to the observatory at Tarqui, since it was en route to the Amazon port of Lagunas. Fortunately, there was enough money left over from La Condamine's letters of credit, as well as from Seniergues' estate (which La Condamine and Jussieu jointly administered), to cover the travel expenses for the two scientists.

Unlike La Condamine and Bouguer, Godin did not have immediate plans to return to Europe after the expedition was complete. He had already indicated that he would stay in Peru for some time and perhaps even go to Chile to make other observations. It was clear that he could not return to France after his disastrous leadership of the Geodesic Mission had besmirched his name back at the Academy of Sciences. The remaining members of the French party—Jussieu, Verguin, Morainville, and Hugo—would have to fend for themselves. No money had arrived from France since 1738, and none seemed to be forthcoming. Maurepas had coldly cast the Peru expedition adrift, leaving them to find their own way home.[27]

The months of July and August were bittersweet ones for Bouguer and La Condamine. They were leaving Quito, their home for the past six years, for the last time, and although they were anxious to return to France, they were also leaving behind many friends and colleagues. Letters from across the viceroyalty, from colleagues living nearby in Quito to those as far away as Arequipa in southern Peru, attest to the esteem in which the Frenchmen had been held.[28]

Bouguer, traveling light, quickly set his affairs in order and made preparations for the long journey home. Aware of the dangers awaiting him, he had made a copy of his magnum opus, the treatise of naval architecture he had written during the long periods of bad weather on the Andean mountaintops, and left the copy "in the hands of a reputable person" in case something happened to him during his return. By July 17 he had packed up his bed and luggage and left for the Cotchesqui observatory with Hugo and Grangier to assist him.[29]

La Condamine, as usual, took rather longer to get moving. He had to settle his affairs and wrap up his various court trials; since he would no longer be making triangulation surveys in the field, moreover, he sold his tent to a local official, his quadrant to a priest, and his Graham pendulum clock to Domingo Terol, the friar who had presided at the wedding of Isabel and Jean Godin des Odonais. The scientist crated his partially repaired sector and sent "the skillful Morainville" ahead with a dozen Indian porters to carry it to Tarqui. Books, instruments and clothes were loaded onto mules. In late August he made one last journey before his departure, first traveling to Yaruquí to inspect his pyramids and then stopping at the home of José Antonio Maldonado in Quinche for a final conference with Bouguer, who came the short distance from Cotchesqui.[30]

Over the course of several days, La Condamine and Bouguer planned their operation like a military campaign. This time they would capture that thief Epsilon Orion and determine its position once and forever. The extended Maldonado family would play a vital role in this operation, for in order for the two scientists to ensure that they were in fact achieving simultaneous observations of the same star on the same night, they would have to confirm the sightings by quickly sending messages from one end of the baseline to the other. The homes of José Antonio Maldonado in Quinche and his cousin José Dávalos near Riobamba would serve as relay stations. La Condamine would send a "reliable express messenger" to Riobamba, halfway between the two observatories, and from there Dávalos would send a pair of runners to Quinche; Maldonado would then bring the message the final ten miles to Cotchesqui. Bouguer's response would take the reverse route. In this way, a message and its reply could be sent and received in just two weeks.[31]

On August 27, 1742, the two friends parted company, after reminiscing over the adventures and troubles they had shared. They would not see each other again until they met in the halls of the Academy of Sciences, where the mission had first been conceived. Any sadness at seeing Bouguer go, however, was quickly displaced by the theft of some of La

Condamine's possessions. As he wrote of the incident in his journal, his ambivalence over his years in Peru was palpable: "One can well imagine that this event, after all of the disagreeable matters that I had to deal with in Quito over the past two years, was very helpful in moderating any regrets I had about leaving a place with such a wonderfully agreeable climate and in which I can flatter myself by stating that I left behind some friends."[32]

Although the final stage of the mission was now at hand, La Condamine was in no hurry to get to work. Stopping in Ambato, just north of Riobamba, he visited his friend Pedro Vicente Maldonado, who had recently completed his work on the Esmeraldas road—as Bouguer had seen firsthand—but who was still in mourning for his wife. Now, with La Condamine's time in Peru drawing to a close, the two tentatively agreed to a plan they had been considering for several months: Maldonado would accompany La Condamine down the Amazon and then sail to Europe, where he could introduce the Spanish king and the Council of the Indies to his road project and request their support for further development in Esmeraldas. With his wife dead, there was little left to hold Maldonado in Peru, but he still vacillated under his family's influence. Finally, the two friends agreed that, if Maldonado decided to go, they would meet at the port town of Lagunas the following year.[33] Finally, La Condamine made a quick visit to the Elén hacienda of José Dávalos, where, apart from confirming the messaging system that he would use to communicate with Bouguer, he had one last reunion with the eldest daughter, María, whom he wistfully termed "the French muse of the province of Quito." Peru and its denizens, it seems, had not entirely lost their hold on the Frenchman.

La Condamine finally arrived in Cuenca a full two months after Bouguer had begun his work up north. There he met Morainville, who had accompanied the zenith sector that La Condamine was to use in his observations. The two men soon discovered that Pedro de Sempértegui, the owner of the hacienda in nearby Tarqui where they made their observatory, had locked the chapel because the team members had neglected to pay him back rent, so in order to continue their work the Frenchmen had to break into the building. Fortunately, La Condamine encountered Sempértegui at his daughter's wedding (La Condamine was a friend of the groom's family), and soon charmed himself back into the owner's good graces, no doubt aided by his gift of a 16-*doblón* gold coin (about $3,000 today) to the newlyweds.[34]

Once set up in his observation post in Tarqui, La Condamine set to work. He rebuilt the twelve-foot zenith sector, which had been damaged in transit, strengthening it in the process and installing new pinnules (sights) at the eyepiece to eliminate problems of parallax. With that completed, another delay awaited; yet again, the weather conspired against both La Condamine and Bouguer. During November 1742 it rained continuously, and frequent earthquakes made regulating the instruments almost impossible. As the year drew to a close, the two astronomers became testy and wrote long letters filled with real and imagined grievances, accusing each other of withholding vital information about their observations.

Despite their frustrations, both La Condamine and Bouguer persevered. They continued watching the heavens, and though neither knew it at the time, they made their first simultaneous observation of Epsilon Orion the night of November 29, 1742. They repeated it again the following night, then several more times during the months of December and January. Throughout this period, letters containing their observational data were relayed back and forth via the Maldonado network.

Having simultaneously observed and measured Epsilon Orion from both observatories, they could now begin the final calculations needed to complete their mission. The apparent path of the star across the zenith was actually in between the two observatories (i.e., it crossed the sky to the south of Cotchesqui and to the north of Tarqui). It was therefore simply a matter of adding the zenith angles to find the arc of latitude between them. Taking the average of several simultaneous observations, Bouguer measured the angle to Epsilon Orion as 1° 25' 48.3" south of his local zenith; La Condamine measured it as 1° 41' 10.7" north. Adding their angles together and adjusting the sum for refraction, the two astronomers agreed that the arc of latitude between them was 3° 7' 1". Using Bouguer's initial north-south distance between the baselines of 162,965 *toises*, and adjusting the distance to account for the positions of the two observatories, gave a new meridian length of 176,940 *toises* (La Condamine's adjusted distance was almost identical). Now they had only to use long division and several more adjustments to reduce the baseline to sea level, in order to determine the length of the first degree of latitude at the equator.

Bouguer had received La Condamine's results first, so by late January 1742, he had the answer they were seeking. The culmination of years of work, countless setbacks, and innumerable tragedies had come down to a single number: 56,753 *toises* for a single degree of latitude at the equator. Remarkably, the figure that the Geodesic Mission had arrived at is within fifty yards of the modern accepted value; even by today's standards, the accuracy of the scientists' measurements and calculations was breathtaking.[35] The true figure of the Earth was contained in their single measurement, the equivalent of 362,899 feet, or 68.7 miles. Compared with the measurements taken at the Arctic Circle (57,437 *toises*) and Paris (57,060 *toises*), that figure proved that the length of a degree of latitude shortened considerably toward the equator, as a result of its bulging out from the axis. Bouguer and La Condamine had confirmed that the planet was indeed oblate and that Newton was right.

La Condamine received Bouguer's news of the results in mid-February. Elated, he wrote Bouguer a long letter on March 9, 1743, recapitulating not just their recent work but that of the previous years. He intended to have the letter become part of the permanent record of the expedition, but the missive never reached Bouguer, for on February 20 he had packed his gear and his papers and set out north for Cartagena in the company of his servant Grangier and a slave. La Condamine did not learn of this until April, by which time he'd received another letter: Pedro Vicente Maldonado would indeed join him at Lagunas to accompany him down the Amazon and was in fact on the point of leaving for the rendezvous. It was certainly past the time for La Condamine to leave, if he were to join his friend on the journey of a lifetime. On May 11, 1743, he and his slave departed the plain at Tarqui and struck out east toward the Amazon. The Geodesic Mission to the Equator had achieved its end, though the expedition was far from over.

X

The Impossible Return

The diaspora of the members of the Geodesic Mission began in the spring of 1743, almost as soon as Bouguer and La Condamine had completed their observations. Within a few weeks the two men, who had worked almost as one for the past seven years, were embarked on separate paths home. It was the inevitable end to an expedition that had been disintegrating from the moment it left French soil. The other members would slowly drift their separate ways, some going back to Europe, others remaining behind in Peru with little hope of return.

Nevertheless, the two scientists themselves had no guarantee that they would ever get home. The Atlantic Ocean was now far more dangerous than during the expedition's first crossing almost a decade earlier. After Vernon's aborted attack on Cartagena de Indias, Britain had sharply reduced its maritime aggressions against Spain, but a naval flare-up could happen at any moment. Indeed, by 1742 the War of Jenkins's Ear—which saw several British assaults on the coasts of Spanish America—had merged into the much wider War of the Austrian Succession. That war began as a land grab by Prussia and Bavaria for a piece of the Austrian empire, but in short order the fighting spread across Europe, polarizing all the major powers; Britain supported Austria, while France and Spain were part of the Germanic alliance. France, despite this alliance and its Bourbon Family Compact with Spain, had no desire to engage in a ruinous war with Britain, so for now it remained officially neutral in the Spanish-British side of the conflict. But this neutrality was precarious and could end at any moment, so the scientists ran the risk of being captured or killed, regardless of whether they took passage aboard a French, Spanish, or even neutral vessel. With news taking up to a year to reach the shores of Peru, any ship sailing from the Americas could unknowingly run into a firestorm.

The scientists' perils would not be confined to the high seas. Even the most direct, well-traveled route home—the one upon which Bouguer, his servant Grangier, and a slave had set off in February 1743—required reaching the port city of Cartagena de Indias by a seven-month trek through nine hundred miles of mountains, rivers, and jungle. Bouguer's tiny band would be easy prey for jaguars and brigands—even the lethal infections that could develop from a simple scratch. After striking north from Cotchesqui on February 20, Bouguer and his companions took the same route as Seniergues and Godin des Odonais had when they traveled to Cartagena to seek their fortunes years before. Following the valley between the two chains of the cordillera, by July the party arrived at Popayán, called the White City for the color of its beautiful homes. From there they crossed the eastern cordillera to the headwaters of the Magdalena River, which they would follow all the way to the Caribbean.

The path over the eastern cordillera of the Andes was as dramatic as it was perilous. The route was punctuated by large chasms, which were spanned by immense rope bridges whose origins preceded the Inca. Though constructed solely of woven reeds and grasses, these bridges could support the weight of fully laden pack animals like llamas and mules or even a horse and its rider. Bouguer, always fascinated by great feats of engineering, wrote, "The most extraordinary bridge is that of La Plata [where the river is a hundred twenty feet wide]: one cannot construct as quickly so solid a bridge with such fragile materials." Bouguer described how the locals would twist reeds into thick ropes that would span the river in a large, inverted arch. He was impressed with the attention to detail, noting that the bridges were secured to the riverbank to prevent the winds from overturning them.[1]

After passing over the eastern ridge of the Andes, Bouguer and his company followed the banks of the Magdalena River until they reached the port town of Honda, a short distance from Bogotá, at which point the river became navigable by boat. Boarding a riverboat there, the men spent a leisurely two weeks floating downstream. With the work of the Geodesic Mission now complete, Bouguer could simply watch the scenery glide by, punctuating his reverie with the occasional compass reading and star sighting to verify the group's longitude and latitude. By September 30, he and his attendants had reached Cartagena de Indias, where they would look for a ship to take them to Saint Domingue.

No records remain of how or when Bouguer made the voyage to Saint Domingue, but letters from the colony indicate that by January 1744 he was in the port town of Léogane, out of funds (he drew 2,000 livres from the intendant), and recopying his memoirs while looking for a ship to take him across the Atlantic. While in Saint Domingue, Bouguer also freed his servant Grangier, releasing him in the same colony where the scientist had first bought his services. Grangier's vast training and experience under the Geodesic Mission quickly gained him a position as a royal surveyor for Saint Domingue.[2]

Bouguer presumably sold his slave while in the colony, for while slavery was permitted in France's colonies, it was banned in the country itself; Bouguer would have had to leave his slave in the New World. To all appearances, the colony's plantation economy was little changed from Bouguer's first visit a decade earlier—but unknown to either him or Grangier, the seeds of that economy's destruction had already been sown. The year before, 1743, a slave named François-Dominique Toussaint-Louverture was born on a plantation in the northern part of the colony. Toussaint-Louverture would become the leader of the Haitian Revolution of 1791–1804, which would overthrow French colonial rule and lead to the world's first black republic. The subsequent loss of sugar revenue would force Napoleon to unload Louisiana onto the nascent United States, and the repercussions of the revolt would help break the back of the European slave trade.

Though the end of slavery in Saint Domingue was already in the offing when Bouguer arrived in the colony, it was the slave trade that would ultimately bring him home. Indeed, Bouguer was intimately familiar with that trade, and not just because he had recently been a slave owner himself. The Bouguer family—father, Jean, and sons, Pierre and Jan—had trained hundreds of masters and pilots from Nantes, the leading slaving port in France, which was located not far from the family's school of hydrography in Le Croisic. They were so well respected that the Montaudouin family, owners of a leading merchant shipping firm, named one of its slaving ships *Le Bouguer*. As he walked along the wharves of Léogane, Bouguer undoubtedly saw many familiar faces amid the whirl of activities.[3] One of them, he surely hoped, would grant him a trip back to Europe.

Soon enough, Bouguer found a slave ship to ferry him across the Atlantic. The two-masted *Triton* had arrived in November 1743, crammed with four hundred slaves from the Ivory Coast. Like many Nantaise slavers, *Triton*'s owners were Irish, as was its master, Jack Shaughnessy, who was to stay behind in Saint Domingue to drum up additional financing. He thus left the ship's second-in-command, Pierre Fouré, one of a large, well-known family of mariners, to pilot *Triton* back to Europe. The voyage would be the third leg of the so-called triangle trade, in which ships carried manufactured goods from Europe to Africa, then ferried their human cargo across the Atlantic to the slaveholding colonies before bringing colonial products back to Europe, where the process began anew.

Before *Triton* could make its return to France, the slave ship had to first be wiped clean of all traces of its human cargo and converted to carry its new cargo of clayed sugar, cotton, and six passengers (including Bouguer) back to France. It was first scrubbed down with buckets of seawater, washing out the detritus of the enslaved Africans who had been shackled there; the ship was then disinfected with vinegar and smoke in a vain attempt to remove the stench of the thirty-two deaths that occurred during their horrific voyage. The wooden barricades that had separated the male and female slaves were taken down, as were the platforms that had divided the already-cramped decks into two levels, so as to wedge in more slaves. Rigging was retuned, planking recaulked, and stores replenished for the long voyage home.[4]

On March 18, 1744, *Triton* cleared the harbor of Léogane and sailed around the island of Saint Domingue to embark on an easterly course toward France. Fouré was short-handed, as eleven of his thirty-six-man crew had drowned on the coast of Africa or died at sea. Bouguer assisted with navigating *Triton* and gave Fouré additional instruction in long-distance piloting; meanwhile, he profited from his sea voyage through observing the behavior of the ship in order to verify the naval architecture treatise he had written in the mountains of Peru.

The ship made the crossing to Europe in just over two months without any interference from the British, even though the War of the Austrian Succession had expanded back in January 1744, with France and Britain now officially at war with each other. The passage was not entirely carefree, as two more of *Triton*'s crew members died at sea before the vessel moored on May 28, 1744, in Paimboeuf, just downriver from Nantes at the mouth of the Loire. After a journey that spanned nine years and over 14,000 miles, Bouguer was the first member of the Geodesic Mission to step foot again in France.

Once back in his native land, Bouguer did not immediately head to the Academy to deliver his findings. He first spent a week in Nantes staying with friends and likely visiting his brother in nearby Le Croisic. He then struck out on the final leg of his journey, taking the coach to Paris, where the other members of the French Academy awaited.[5] The ride along the rain-rutted roads would have been uncomfortable in more ways than one; Bouguer had had no news from Paris for several years now, and he simply would not have known if he would enter the Academy of Sciences wreathed in glory for his work at the equator or be relegated to a follow-on act for Maupertuis' victory at the Arctic Circle.

If Bouguer was nervous about how he would be received in Paris, La Condamine had felt no such compunctions as he embarked on his return trip to Europe. His voyage down the Amazon was certain to be the adventure of a lifetime. Only a handful of Europeans had ever navigated the great river: the first, Francisco de Orellana, had propagated the myth of the women warriors that gave the Amazon its name; Pedro Texeira, a Portuguese explorer, attempted to mark the boundary between Spanish and Portuguese territories on the river; and Samuel Fritz, a Jesuit, published the first map of the Amazon in 1707.[6] Now La Condamine undoubtedly hoped to place his name along theirs in the annals of history.

His plan to traverse the Amazon had been long in the making and had involved considerable planning on his part, even though he was already occupied with numerous extramural affairs. Although the Amazon ran through Portuguese territory, Portugal was at peace with France and tied to Spain by royal marriage, so any French visitors coming from Spanish Peru would be seen as at least nonbelligerent. Louis Godin had originally suggested that the entire expedition return by way of the great river, but when it became obvious that Godin would be staying in Peru and that Bouguer wanted to return by the shortest route, La Condamine realized that he would have to make the trip alone. He wrote to the French authorities requesting permission and passports to travel by himself through the Portuguese lands. When those documents arrived in 1742, La Condamine informed his friend José Pardo de Figueroa—Isabel Godin's uncle, now titled the Marqués de Valleumbroso—of his upcoming voyage and asked for information about the Amazon.

Pardo de Figueroa, a great-nephew of Texeira's geographer, excitedly provided La Condamine with information about the Amazon but also charged him with a colossal task. Pardo de Figueroa gave La Condamine a copy of Samuel Fritz's map, as well as more up-to-date observations by other Jesuit missionaries living along the river, but cautioned La Condamine to take the information with a large grain of salt and urged him to make a comprehensive scientific survey of the Amazon's geography. Although it would be an astonishing feat for one man to accurately map the world's largest river in one rapid passage, La Condamine readily agreed to Pardo de Figueroa's suggestion. As always in the Frenchman's varied career, his unique combination of curiosity, bravery, foolhardiness, and sheer good fortune would see him through this latest adventure.[7]

Luck was certainly with La Condamine as he began his journey in May 1743, accompanied by one of the two slaves who had formerly belonged to Seniergues (it is not clear which of the two accompanied him or what happened to the other). They departed from Tarqui and headed south to collect some cinchona saplings from Loja and to see the gold mines of Zaruma, but his choice of stopovers might have cost him his life. One of the ferrymen he encountered along the way discovered that La Condamine was the man responsible for the imprisonment of Cuenca mayor Sebastián Serrano in the Seniergues trial, and he informed the Frenchman that Serrano's henchmen had been lying in wait for him along another, more commonly used road. La Condamine was, the man said, "very fortunate to have taken this route instead of the other."[8]

Though La Condamine had managed to avoid Serrano's brigands, the jungle route was still treacherous. Crossing steep mountain valleys along the same type of rope bridges that Bouguer was at the same time encountering on his journey to Cartagena, La Condamine and his slave arrived in late June at the port of Jaen, on a small river that fed the Marañón River, one of the primary tributaries of the Amazon itself. La Condamine waited a week while his balsa-and-vine raft was constructed, copying his notes and sending them to a Jesuit missionary "in case I should die en route."[9] On July 4, 1744, he and his slave boarded the raft and departed downstream, without guides, toward the Amazon. During the journey, La Condamine carefully noted the size and direction of the river with his watch, compass, and quadrant and took barometric readings and sun and star sightings whenever they beached.

By July 12, La Condamine had arrived at the upper reaches of the Amazonian basin: the notorious Pongo de Manseriche, a six-mile stretch of whitewater rapids where the Marañón gushes through a tall, narrow gorge before leaving the last ridgeline of the Andes and spreading out onto the vast Amazon plain below. The fragile-looking raft, loaded with hide-covered baskets carrying his journals and books (already soaked from the journey), was dashed repeatedly against the rocks as it shot through the rapids. Nevertheless, its elastic construction flexed under the impact, and in just under an hour La Condamine and his slave found themselves "in a new world, far from any human commerce, upon an inland sea in the midst of a labyrinth of lakes, rivers and canals which disappeared everywhere into the immense forest."[10] After years of preparation, La Condamine had finally reached the fabled Amazon jungle.

Figure 10.1 La Condamine's Map of the Amazon. From Charles-Marie de La Condamine, *Relation abrégée d'un Voyage fait dans l'intérieur de l'Amérique Méridionale* (1778). Courtesy of Robert Whitaker.

Although abundant with many varied Indian cultures, the Amazon was not—as La Condamine soon found—entirely devoid of Western civilization. Just below the Pongo de Manseriche rapids, at the town of Borja, La Condamine met a scientifically minded Jesuit, Jean Magnin, who gladly shared his encyclopedic knowledge of the Amazon with the traveler. A week later La Condamine was in the port city of Lagunas, where his friend Pedro Vicente Maldonado had been waiting over a month. Maldonado, recently widowed, had struggled with the decision to leave his nation and his family but ultimately had decided to accompany his friend to Europe. Once there, he would request support from the Spanish court for further development of his Esmeraldas road and the elevation of Riobamba to city status. From the start, however, Maldonado knew that his journey was likely a one-way voyage; two months before he left Riobamba, he had hastily remarried so that his only daughter would have a mother to raise her.[11]

Together, La Condamine and Maldonado hired a pair of dugout canoes over forty feet long, each equipped with eight rowers, and began the downstream journey at a rapid clip. The canoes traveled day and night so they could arrive at the mouth of the Napo River, the accepted demarcation between Spanish and Portuguese territories, in time to take a critical sighting for La Condamine's map of the Amazon. On the night of July 31, one of the moons of Jupiter would move behind the planet; La

Condamine wished to observe this with his telescope in order to determine the exact time of occultation. Once his observation was compared with those made in Europe, he would establish the longitude of the Napo and thus provide a clear delineation of the two territories. As it happened, the conditions for astronomy were poor on the night of July 31, and La Condamine's only observation of Jupiter's moon gave an incorrect position for the Napo; his findings would erroneously move the boundary between the two territories two hundred miles in Spain's favor when the borders were later established by the 1750 Treaty of Madrid.[12]

Having raced to the border, they now slowed their pace considerably. Leaving Napo and entering the Portuguese territory of Brazil, La Condamine, Maldonado, and their entourage drifted and paddled down the Amazon for two months at the pace of a slow walk, leaving plenty of time for geographical surveys and observing the natural history of the region.

La Condamine was astonished with some aspects of the Amazon region and dismissive of others. The colossal river earned his enduring amazement; at times the travelers could not see from one bank to the other, and La Condamine's sounding lead sometimes did not touch bottom at three hundred feet. Stopping at missions and villages along the way, the men inquired after such legends as the tribe of Amazon women and the city of El Dorado. They found the real human populations of the Amazon far less impressive.

To La Condamine, the Indians who occupied the Amazon were not even humans but "forest animals," eking out their existence by hunting and gathering like the other primates of the jungle.[13] He and Maldonado observed with disdain the Indians camped along the shores of the river in tiny, widely separated settlements, which were further estranged by their disparate languages and cultures. La Condamine could not have understood that the biological wealth of the Amazon region was in fact the product of these peoples: thousands of years of Indian civilizations had transformed the rain forest into a giant farmland with hundreds of domesticated varieties of trees and crops. The Indians had carefully cultivated these plants using advanced agricultural techniques that continually enriched the nutrient-poor soil and gave rise to the miraculous flora that La Condamine witnessed on his journey.

Just as La Condamine did not realize that the Amazon owed its abundance to the Indians who once peopled it, he was also unaware that their current, relatively primitive state of existence was less than two hundred years old. The first European explorers had written of massive cities dozens of miles across, each housing tens of thousands of people, but the European public soon scoffed at such notions, for later travelers saw only small tribes fleeing into dense, unkempt jungle. In the interval between those two groups of explorers, European diseases had surged through the Amazon, wiping out perhaps 90 percent of the Indian population in less than a generation and, in the process, plunging some of the most advanced cultures in the world back to the Stone Age. The jungle did the rest, swallowing up the ruins of the once-great cities. Now, floating by in their canoes, La Condamine and Maldonado were oblivious to the wreckage of the once-thriving civilizations, older than the Egyptian pyramids, that had long existed on the banks of the Amazon.[14]

On September 19, 1744, four months after La Condamine left Tarqui, he and Maldonado arrived in the town of Belém do Pará, the Atlantic gateway to the Amazon. It was the end of one journey and the beginning of another, for now the two men would have to make their separate ways to different destinations in Europe. In December, Maldonado obtained passage with a Portuguese fleet bound for Lisbon, from where he would travel to Madrid to make his petitions to develop his Esmeraldas road and to grant Riobamba city status.

Meanwhile, La Condamine planned to return directly to France, but events outside of his control would greatly complicate his crossing. Wanting to sail directly to France via the nearby French colonial city of Cayenne, he stayed in Belém until he could secure coastwise passage in a canoe. During his stay there, his inquisitive gaze fell on the practice by Carmelite missionaries of successfully inoculating Indians against smallpox, a procedure only recently introduced in Europe, and filed that fact away for future use. Finally, in February 1744, he arrived in Cayenne, where he repeated Jean Richer's pendulum experiments while awaiting the annual French warship to take him home.

Eventually La Condamine learned of the declaration of war between France and Britain and decided that, rather than sailing on a French warship, which might be prone to attack if the hostilities spread, his wisest course of action would be to secure passage on a neutral Dutch vessel from the neighboring colony of Dutch Guyana. The French governor provided a royal canoe for La Condamine's short coastal trip to the small Dutch territory, and the Frenchman set off once again.

By September 1744, La Condamine was on a Dutch merchant ship carrying a load of coffee back to the Netherlands. After a harrowing two-month journey in which the vessel narrowly avoided pirate ships, at dusk on November 30 he "disembarked in Amsterdam, and placing my feet on dry land, forgot all else."[15] Six months after Bouguer had arrived in France, the expedition's second member was safely back in Europe. For the expedition members they had left behind in Peru, the voyage would take much longer.

In January 1744, while both La Condamine and Bouguer had been on the shores of the Atlantic awaiting separate passages home, Jorge Juan y Santacilia and Antonio de Ulloa had just been returning to Quito from the shores of the Pacific after a futile three-year hunt for Anson's fleet. Back in the capital, they found Louis Godin preparing for his own departure, though not back to France; he had just been offered the chair of mathematics at the University of San Marcos in Lima, on the death of its previous occupant, Pedro de Peralta y Barnuevo, the expedition's long-time friend and correspondent.

The offer of a university position in South America would certainly have come as a relief to the expedition's erstwhile leader. Godin had long before decided that he had no future in Paris, and ever since the brief return and second departure of Jorge Juan and Ulloa in late 1741, he had abandoned all his work on geodesy. This appointment offered him the ideal pretext to remain in South America. Moreover, it would give him the steady income that he had been lacking for many years. Godin was so deeply indebted that he could not repay Pedro Vicente Maldonado the 3,400 pesos he'd borrowed back in 1737 at 5 percent annual interest; instead, he had requested that Maldonado demand the equivalent sum from Godin's wife, Rose Angélique, when the *criollo* official got to Paris, as he planned to do after petitioning the Spanish king in Madrid.[16]

Godin was ready to depart for Lima at once, but on arriving in Quito, Jorge Juan and Ulloa prevailed on their one-time leader to delay his departure. The two men would be happy to accompany Godin to Lima—they were planning to embark with a small fleet scheduled to depart Lima for Europe later in the year—but they could not abandon the geodesic work they had come so far to achieve. No records indicate whether they knew that Bouguer and La Condamine had already completed their observations, but the two officers would have seen the completion of their own measurements as necessary to fulfill their orders to the Spanish crown, as well as to provide an independent measure of a degree of latitude. They convinced Godin to quickly finish the task before their departure.

Together Godin, Jorge Juan, and Ulloa would complete their measurements, although using a somewhat less accurate approach than that taken by Bouguer and La Condamine. Godin's party raced to extend the chain of triangles to their new northern observatory at hacienda Pueblo Viejo near Mira. There they would take their star sighting with the twenty-foot zenith sector, which Ulloa had previously ordered impounded so that the French scientists could not make use of it. Since Godin's party now had only one sector, they could not make simultaneous observations of Epsilon Orion at both ends of the chain of triangles, as Bouguer and La Condamine had done to eliminate errors caused by stellar aberration. Instead, Godin, Jorge Juan, and Ulloa resolved to simply add the northern angles they hoped to obtain in Mira with the southern ones they had observed four years earlier in Cuenca. By late May they were done, and after making several corrections they determined the length of a degree of latitude to be 56,767.8 *toises*, just thirty yards different from the determination by Bouguer and La Condamine.[17]

With their measurements complete, the three men could finally begin their exodus from Quito. Godin undoubtedly said his good-byes to his cousin Jean Godin des Odonais, now dabbling as a textile merchant in Quito, and to his newly pregnant wife, Isabel. Jorge Juan and Ulloa made a final visit to the treasurer of Quito, their erstwhile adversary Fernando García Aguado, who coughed up the officers' back salary and expenses one last time. By his careful reckoning, the Geodesic Mission had drained 13,125 pesos, or half a million dollars, from the Quito treasury. In late

June the three men were on their way to Lima, but while passing through Guayaquil, Jorge Juan was handed a letter sent from the viceroy, asking him to carry out one last mission. Villagarcía had been so impressed with the young officer's shipbuilding skills during the Pacific campaign that he asked him to prepare plans and estimates for constructing a new sixty-gun warship. Jorge Juan's design would later be built as the *San José el Peruano*. The delay was fairly short; by September 1744, the three men were all in Lima, where Godin took up his post at the University of San Marcos, while the two officers prepared for their voyage home.[18]

The Spanish officers soon encountered the ships that would take them back to Spain. Anchored in the nearby port of Callao were three armed French merchantmen, which had originally been chartered several years earlier in Cadiz to carry Spanish cargo to Peru under a neutral flag, in order to avoid entanglement with British warships, since Spain and Britain were at war. Jorge Juan and Ulloa had taken note of the ships while in Chile during their unsuccessful search for Anson's fleet. Since the vessels had at the time been en route to Callao, where they would spend many months awaiting new cargo for their return voyage to Europe, the officers had resolved to sail home aboard them. The news of the declaration of war earlier that year between France and Britain had not yet reached Lima, so they still believed that they would be sailing safely under a neutral flag. On October 22, 1744, Jorge Juan boarded *Lys*, while Ulloa embarked on the smaller *Notre Dame de la Délivrance*, separating—as during their first voyage—to ensure that at least one man (and his duplicated reports) made it safely back to Spain. The voyage round Cape Horn was a hazardous one, even in the summer, when westerly winds astern would push them around the Horn.

Jorge Juan made a relatively speedy return to Europe. Soon after clearing port, *Lys* separated from the other vessels and made for Valparaiso in Chile, where the crew learned of the outbreak of war between France and Britain. Jorge Juan soon received another surprise, when *Lys* embarked an additional complement of passengers: the captured crew from the wrecked frigate *Wager*, one of Anson's ships that Jorge Juan had been chasing around the Pacific. Prisoners or not, adversaries or not, they were now shipmates, and *Wager*'s captain John Byron got along famously with Jorge Juan, whom Byron later described as "a man of very superior abilities."[19] In July 1745, after pausing for a month to repair a sudden, massive hull breach, *Lys* arrived in the Caribbean, where it joined a French convoy bound for Brest. Jorge Juan arrived in the Brittany seaport in October 1744, becoming the expedition's third member to reach Europe. On his way back to Spain, he paid a visit to Paris, where he met with Bouguer and finally reconciled with La Condamine, who named him a corresponding member of the Academy of Sciences. By early 1746 he was back in Madrid.

Antonio de Ulloa's return voyage in *Délivrance* was decidedly more harrowing. Its captain and crew having learned of the war during a port call in Concepción, Chile, *Délivrance* rounded the Cape in January 1745 and sailed north, avoiding all Atlantic ports for fear of being captured by British forces. This danger was confirmed when the vessel narrowly escaped a pair of British corsairs after watering at an archipelago off the tip of Brazil. The merchant ship was undermanned and too lightly armed to fend off a serious attack by a heavily gunned warship, so its captain decided to make for Louisbourg, the fortified port in French Canada. Unknown to him, British forces from the American colonies had already captured the town, and on August 6, *Délivrance* sailed right into their waiting arms. Ulloa desperately threw most of his papers overboard, except the ones related directly to the geodesic measurements, before he was captured by the British forces.

As an enemy officer, Ulloa was made a prisoner of war of the British, but his membership in the Geodesic Mission would prove to be his salvation. After his capture in Louisbourg, Ulloa's remaining papers were confiscated, and he was taken aboard the frigate *Sunderland*, which was headed to Britain. Arriving just before Christmas, Ulloa spent several months in prison near Portsmouth before being granted preferential treatment at the behest of the Royal Society and allowed to travel to London. Martin Folkes, the president of the Royal Society, was well aware of the Geodesic Mission to the Equator and overjoyed to have one of its members in his midst. Folkes quickly read through and summarized Ulloa's confiscated papers on the Geodesic Mission, after which he returned them to the Spanish officer and arranged for his release. Folkes presented an account of Ulloa's recent geodesic exploits before the Royal

Society, invited him to attend several meetings, and ensured that the Society conferred membership on the officer for his scientific work.[20] Now that the British scientists had been thoroughly debriefed, the British Admiralty secured Ulloa's passports and travel to Lisbon, a neutral territory from which he would be able to make his way to Spain.

On July 25, 1746, Ulloa arrived in Madrid to find the city in deep mourning for Felipe V, who had died suddenly two weeks earlier. Both Ulloa and Jorge Juan had been out of sight for a decade, and with the king who had authorized their departure now dead, any possibility of royal recognition or advancement in their naval careers seemed unlikely. The Spanish officers could only hope that their decade-long service to the crown had not been completely forgotten.

If Jorge Juan and Ulloa were worried about their futures in Spain, there could have been no doubt that Louis Godin's career back in France was already dead. In December 1745 Godin had been unceremoniously removed from his senior post as pensioner astronomer and banished from the Academy of Sciences; Jacques Cassini's son César-François, known as Cassini de Thury, now took his place. The official justification for Godin's removal was his breaking the Academy's rules by accepting a post at a foreign institution (the University of San Marcos), but this was a convenient façade; the real reason, as Maupertuis put it, was that Godin had been "dishonored by the embezzlement attributed to him; for having prolonged by ten years what it took us six months to do; for having sown hate and discord among his colleagues; and to top it off, for cravenly and dishonestly taking refuge in Peru."[21]

Like his career in France, Godin's marriage appeared dead as well. By now his wife Rose Angélique had heard quite enough about his overseas liaisons and misadventures, and his disbarment from the Academy had left her penniless; so when Pedro Vicente Maldonado arrived at her doorstep demanding settlement of Godin's 3,400 peso loan plus interest, it was the last straw; she spat back that her husband could come to France and pay all his debts himself.[22]

Although disgraced in Europe, Godin made himself quite comfortable in South America. Once he was established in Lima following his arrival in September 1744, his career came alive. His post as mathematics chair at the prestigious University of San Marcos, the oldest university in the New World, carried with it the title of "royal astronomer" and publisher of the nautical almanac, netting him the equivalent of $200,000 per year, a generous salary that Godin apparently squandered, as he had done with the expedition's funds beforehand. His position placed him at the right hand of the new viceroy, José Manso de Velasco, who had replaced the outgoing Villagarcía in 1745. Godin served as cartographer, surveyor, and architect for the capital and its surroundings, and he examined pilots for their navigation skills.[23] For two years, everything seemed to be going exceedingly well, though it was merely the calm before the cataclysm.

At 10:30 PM on October 28, 1746, some families in Lima were sitting down to a late dinner, while others were taking their evening *paseos* through the streets of the city, lit by an extraordinarily bright full moon that illuminated even the darkest corners. The evening seemed like any other in the City of Kings.

At first, the low rumbling did not cause alarm; the entire coast of Peru was constantly subject to *terremotos* (earthquakes) that shook the region but generally lasted only a few seconds and did relatively little damage. Then came the first shocks, which threw people off their feet. A few were able to recover, run into the central patios of their buildings, and hide in their earthquake refuges—shelters specially constructed of flexible cane and reeds to withstand the vibrations of earthquakes while also blocking any falling debris that might cascade down into the courtyard. There they waited for the quake to end, but it kept on; for four minutes the entire earth shuddered "like a great beast shaking the dust off itself," as one eyewitness recounted. The cathedral bells pealed and clanged frantically until the towers collapsed into the nave. The buildings shook with unbelievable speed, casting their occupants about like sailors in storm-tossed ships before the adobe and stone cracked and collapsed, burying scores of people in the rubble. Nine miles away from Lima, in the port of Callao, the sandy soil had simply liquefied, collapsing most of the buildings into the quicksand.[24]

Then, as suddenly as it began, the shaking stopped. The noise did not; the rumbles and screeches of the ravaged earth were immediately replaced by the groans of crumbling masonry and the screams of terror and pain throughout the city. And the destruction was far from over.

Those brave enough to ascend Callao's still-intact city walls shortly after the last tremor took in a heart-stopping sight; the ocean was rapidly receding, the harbinger of an incoming *maremoto*—a "seaquake" or tsunami. At 11 PM they could see, glinting in the moonlight, the first great wave coming in from the northwest: a cliff of ocean fifty feet high, trailed by a giant plateau of water. Ships in the harbor had their anchor ropes snapped like threads and were carried inland on the tide, which overtopped the walls and submerged the city. One vessel, the warship *San Fermín*, hove up a mile inland; another, a merchant ship providently named *Socorro* (Aid) and carrying a full cargo of grain, was gently set down on the beach. As the ocean rushed over Callao's narrow peninsula, the San Agustin church was ripped whole off its foundations and carried out to sea, eventually landing on San Lorenzo Island, some three miles offshore.

The combined damage of the earthquake and tsunami was nothing short of apocalyptic. The earthquake, later calculated to be a staggering 8.5 on the 10-point Richter scale, had been caused by a sudden, massive vertical displacement of a thrust fault in the Nazca Plate, about sixty miles offshore. As the dust settled, the scale of damage became clear. Every home and building in Lima was destroyed or damaged, most beyond repair. Callao had simply ceased to exist. Thousands had died, and the total damage was estimated to be the equivalent of $30 billion. Most of the navy was also wiped out, leaving the entire coast defenseless.

Viceroy Manso de Velasco was quick to survey the damage and begin relief efforts in the wake of the calamity. Velasco had seen this sort of natural disaster before, when, as governor of Chile, he had rebuilt Valdivia after its destruction by an earthquake in 1737. After touring the rubble-strewn streets of Lima on horseback, he realized that the first priority was to secure the city's food supply and shelter its residents. He immediately ordered the bread-ovens to be rebuilt, sent messengers to the provinces to obtain additional grain, set up tents, and posted guards around the city to prevent looting. The grain warehouses in Callao had been washed away along with the rest of the city, but the grain in the merchant ship *Socorro* would provide some relief until more could be brought in. Manso de Velasco also quickly took charge of the rebuilding of his cities and their defenses (he was later granted the royal title Conde de Superunda, Count Conquer-the-Waves, for his actions). It was a massive undertaking for which he turned to his right-hand man: Louis Godin.

Godin's surveying experience during the Geodesic Mission proved invaluable in rebuilding the demolished cities of Callao and Lima, as well as their defenses. After considering and quickly rejecting the idea of moving the capital thirty miles east, near modern-day Chosica, Godin quickly measured up the existing sites and mapped novel layouts that would breathe new life into the cities. Godin's plans called for the cities to be reconstructed with much lower buildings to control future earthquake damage, and with wider streets and plazas to provide refuge and allow passage even if buildings fell. The viceroy quickly adopted the idea of a "humbler" capital and instructed Godin to design and oversee its rebirth.

Under Godin's guidance, Lima was transformed into one of the most gracious cities in South America. The new City of Kings sported ample boulevards, open, friendly plazas, and a well-developed sense of proportion that had been lacking in its previous incarnation. The city's architecture also became more modern and secular, with churches and monasteries giving way to civic buildings like theatres, and with prayer rooms in private homes being converted to French-style salons. Godin had been given the additional task of developing a new, stronger fortress to protect the strategically vital port of Callao, and his design for the massive, pentagonal Real Felipe fortress featured the latest concepts in siege warfare derived from his famous countryman, the great military engineer Vauban.

In many ways, the reconstruction of Lima and Callao became Godin's redemption, but although it may have made up for his previous misdeeds, the process showed that Godin's arrogance was little changed. In his efforts to improve urban safety, the Frenchman imperiously ordered that all homes be rebuilt without the upper stories and ornate facades favored by the wealthy, but whose collapse during the earthquake killed or wounded thousands. When the upper-class residents rebelled against the changes that would reduce their living space, Godin belittled them, labeling their concerns vain and ostentatious, and attempted to ram through his designs over their objections. In the end, although some of Godin's recommendations were rejected, his overall vision of a modern city took hold and even today can be seen in the colonial center of Lima.[25]

North of Lima, in Quito, Joseph de Jussieu was struggling with a natural disaster of his own. In March 1745, the doctor had been preparing to travel down the Amazon in the wake of La Condamine when a smallpox epidemic struck the city. The president of the *audiencia* ordered Jussieu to remain in Quito until more doctors could be brought in. The incessant parade of epidemics kept the Frenchman busy and well-paid, though his immobility apparently plunged him into an increasingly deeper state of depression.

In September 1747, belated news from France helped lift Jussieu's spirits; he had been elected to the Academy of Sciences and was now an associate botanist. This promotion was quickly followed by Jussieu's first official order as a member of the Academy: Maurepas commanded Jussieu to locate Louis Godin, no longer an Academician, and relieve him of the Academy's geodesic instruments. Maurepas had sent no letters of credit to pay for Jussieu's passage back to France or even a pension for his new post, yet the assignment did come with a substantial benefit: It gave Jussieu the justification he needed to have the president's restraining order voided and, more importantly, to plan his voyage home to France.[26]

Released from his service in Quito, the new associate botanist planned one expedition of his own before setting out to find Godin. Jussieu wished to visit the fabled Canelos region, whose name was derived from the Spanish word for cinnamon. It was a lushly forested area noted for its botanic treasures as well as being another of the many mythical locations of El Dorado. In December 1747 he left Quito in the company of José Antonio Maldonado, the priest who had helped La Condamine with his pyramid project (and whose brother Pedro Vicente had successfully descended the Amazon and was by then traveling around Europe). Maldonado's family estate near the town of Baños gave the pair access to the Pastaza River, a direct route into the heart of Canelos. For many weeks the two men botanized the region, taking copious notes on the varieties of cinnamon and bamboo they found and collecting seeds that

Jussieu subsequently shipped back to his brothers in Paris. It was a happy, almost carefree time for the oft-tormented doctor. Resting for several weeks at Elén, the Dávalos hacienda near Riobamba, he wrote a torrent of letters revealing that his mood had considerably lightened and that he was eagerly anticipating the journey home.

Refreshed after his journey through Canelos, Jussieu finally struck southward toward Godin, following the coastal roads and arriving in Lima in mid-August 1748. He had doubtlessly heard about the earthquake, which had been felt eight hundred miles away in Quito, but he would not have been prepared for the devastation still evident, nor for the rapid pace of reconstruction he saw. Jussieu would have also been astonished at the transformation in Godin. In Quito, the scientist had been withdrawn and practically inert, but here in Lima he was a whirlwind: surveying building sites, making calculations, writing bills of material, and directing construction of the Real Felipe fortress, whose cornerstone had been laid the previous year.

Though Godin's behavior was surely remarkable, the scientist had another surprise for Jussieu: He was on the eve of his departure for Spain. Had Jussieu arrived a week later, Godin would have been gone. Jorge Juan and Antonio de Ulloa had recommended the ostracized Frenchman for two newly created positions within Spain's scientific community.

The careers of the two naval officers had not suffered for long after they had arrived in Spain. The new king, Fernando VI, and his navy minister, Zenón de Somodevilla, the Marqués de la Ensenada, were bent on remolding the navy to the new age of Enlightenment. Jorge Juan and Ulloa were just the sort of "scientific officers" the nation's new leaders were looking for. Among the reforms that Ulloa and Jorge Juan proposed—and Ensenada accepted—were an overhaul of the Academy of Navy Guards, which was to be given a more scientific curriculum, and the establishment of a royal observatory in Cadiz along the lines of those in Greenwich and Paris. The officers recommended Louis Godin as head of both institutions, and Ensenada accepted this proposal, too.

In April 1748 Godin, still in Lima, received Ensenada's offer to come to Cadiz. Godin, claiming penury despite his generous salary, accepted on the condition that his 3,400 peso debt to Maldonado be forgiven. (It was, in a sense; the viceroy had his treasurer repay it.) Godin had literally been packing his bags when Jussieu arrived in Lima to seize his geodesic instruments in the name of the French Academy.

With Godin about to depart for Spain, Jussieu found himself in a difficult position. *He* was now the Academician and thus had authority over his one-time superior, and yet, when he demanded that Godin hand over all the Academy's instruments, Godin refused to do so. In reality, there was almost nothing of value left. La Condamine had already sold his pendulum clock and quadrant (which the Jesuit priest Jean Magnin, who helped La Condamine navigate a short stretch of the Amazon, eventually bought).[27] The expensive twelve-foot zenith sector from London had been rebuilt too many times to salvage. Godin's only instrument of any value was the six-foot iron *toise*, precisely fabricated by Claude Langlois to validate the geodesic measures. Godin would not relinquish the *toise* to Jussieu, preferring to take it back to Europe himself in order to present it to the French Academy of Sciences and perhaps return himself to their good graces.

In accordance with Maurepas' orders to make certain the *toise* arrived safely, Jussieu took the only dutiful option available to him: He resolved to travel with Godin back to Europe, to ensure that Godin brought the instrument to Paris. The men were uncertain of when another ship would depart Callao to make the difficult passage around Cape Horn, so they planned to travel by foot to Buenos Aires, the capital of the *audiencia* of La Plata and Peru's main port in the South Atlantic. There they were certain to find passage back to Europe.[28]

Godin and Jussieu left Lima on August 27, just days after the doctor's arrival, and headed for Buenos Aires. They first took the road that led across the western cordillera to Cuzco, the ancient seat of the Inca Empire. It was an arduous journey, but the Frenchmen were not in any hurry, especially not Jussieu. He busied himself botanizing en route, taking copious notes and samples along the way, and also showed a keen interest in the environment through which he and Godin were passing. Arriving in Cuzco at the beginning of 1749, Jussieu marveled at the massive Inca structures that surrounded the city and the fortress of Ollantaytambo near the site of Machu Picchu, which at the time lay undiscovered just a few miles away. Once back on the trail, the two men rounded Lake Titicaca and traversed the cold, bone-dry altiplano, arriving at La Paz in May 1749.

At La Paz, Jussieu decided to take a detour to do some long-delayed botanizing in the Andes but promised to rejoin Godin in Buenos Aires. After so many years of strife between them, they had now reached a rapprochement and parted ways amicably. Jussieu even seemed committed to restoring Godin's standing in France; he sent a note to his brothers in a parcel containing the seeds he'd collected, pleading for their help in reinstating Godin in the Academy.[29]

Having promised to wait for Jussieu in Buenos Aires as long as possible before setting sail, Godin continued east, finally arriving at in the port city in December 1749. Shipping traffic had not yet recovered from the recently ended War of the Austrian Succession (concluded in 1748), so Godin had to wait six months until a Portuguese vessel appeared. In July 1750, after sending a letter to Jussieu forewarning him of his voyage, Godin embarked for Lisbon.

Although it should have only taken him two months to cross the Atlantic, Godin's journey back to Europe continued in fits and starts for another year. As it happened, the Portuguese vessel upon which he had left Buenos Aires went only so far as Rio de Janeiro, where it inexplicably languished in port. No other ships arrived in the harbor until January 1751, when a French armed merchantman came into Rio. Aboard, surely much to Godin's surprise, was his old friend and fellow astronomer Nicholas Louis de La Caille, on a mission to survey and map the southern skies. The two men had much to discuss during La Caille's month-long stopover, but as he was remaining in the South Atlantic, his ship would be unable to take Godin back to Europe anytime soon.

Finally, a few months after La Caille's departure—and after almost a year of waiting—the Portuguese vessel on which Godin was traveling cleared Rio. The Frenchman arrived in Lisbon in July 1751, after a voyage that had taken him nearly three years. The hardest part of the journey was yet to come: Now Godin had to return to Paris to face the Academy—and his wife.[30]

Godin had betrayed not only his wife and the Academy of Sciences. Having almost bankrupted the mission, alienated political authorities at every turn, and estranged his colleagues, he had abandoned his subordinates to their fates, leaving them to fend for themselves. But Bouguer and La Condamine, now safely back in France, had also betrayed their colleagues; they never fully assumed the mantle of leadership that was so desperately lacking. Except for when they required their services, the two gave little thought to the welfare of their surviving countrymen. And when it finally came time to depart for France, the scientists inexplicably left their assistants, Verguin, Morainville, and Hugo, behind in Peru. For two of those men, the voyage back to France would prove to be the impossible return.

XI

A World Revealed

When Bouguer arrived from Nantes at the walls of Paris in June 1744, he would have seen the great city girding for war. The War of the Austrian Succession had by then expanded to include most of the great European powers. France had allied with Spain and various German states against the Austrian-led coalition, but initially it stayed neutral with respect to the British part of that alliance for fear of a potentially ruinous conflict with its adversary across the Channel. This cautious truce did not last long. French troops, supporting the Germanic alliance, had in 1743 skirmished with British troops supporting Austria. French ports had also provided refuge to several Spanish naval squadrons that had confronted British warships in the Mediterranean. By January 1744, France had declared war on Britain. In March the French army ratcheted up its plan to invade the Austrian-held Netherlands, as British troops stationed there presented an existential threat directly on France's northern borders. Just as Bouguer was entering the capital in mid-June 1744, the whole population was turning out to cheer King Louis XV as he personally led his troops into battle.

Now that he had arrived in Paris, Bouguer's own battle was about to begin. If he were to receive the recognition he so craved, he would have to fight for it—for his colleagues at the French Academy had already begun to write him out of the debate over the figure of the Earth. When Maupertuis' expedition returned with evidence that the Earth was flattened at the poles, skeptics had continued to defend the idea of an elongated Earth, until 1740, when Cassini de Thury remeasured his father's survey of France and announced that the new results confirmed the flattening. With Bouguer still in the Andes, Maupertuis, now director of the Academy, and his companion Alexis Clairaut began rewriting the history of the Newtonian-Cartesian polemic to include the latest results. Clairaut's work pointedly omitted Bouguer's previous contributions to determining the shape of a spinning Earth, despite the fact that it formed the basis for his own physical explanation of how much the Earth was flattened at the poles.[1] Bouguer was now determined to thrust himself back into the Academy's limelight.

While the Newtonian-Cartesian debate had been raging nine years earlier, Bouguer's stance at the Academy had been that of a neutral skeptic. In the intervening years, after Maupertuis' expedition had settled the debate, Newtonianism had become the talk of Paris, replacing Cartesian vortices as the subject of afternoon salons. Dog-eared copies of Fontenelle's *Conversations* were replaced on bookshelves by Francesco Algarotti's *Newtonianism for the Ladies* and later by Emilie du Châtelet's translation of Newton's *Principia*. Bouguer returned to France with findings that were sure to ingratiate him with the Newtonian crowd. He came to the Academy armed not only with the results of his geodesic observations, but also with those of his gravity experiments on Chimborazo, the first direct confirmation of the theory of universal attraction.[2] Bouguer would out-Newton even Maupertuis—if only he could get attention for his work.

Bouguer set foot in the Academy of Sciences on June 27, 1744, and soon learned that his geodesic observations were eagerly sought after, and not just as a simple rehash of those by Maupertuis. The Arctic Circle expedition, he would have learned, had effectively settled the debate over the Earth's shape, but it had not sufficiently settled the question of the true dimensions of the planet. It was this measure—the actual degree of flattening of the Earth—that the French Navy saw as vital to long-distance ocean navigation, and it was the reason Maurepas funded the expeditions in the first place. When Maupertuis' data was analyzed and compared with the earlier survey of France, it showed that the Earth's diameter through the poles was smaller than its equatorial diameter (the polar flattening) by 1 part in 178 (1/178). This was much flatter than what Newton had originally predicted (1/230) or what Clairaut's updated equations showed (between 1/573 and 1/230). Suspicion was immediately cast on the veracity of Maupertuis' observations, with critics attacking the short length of his chain of triangles and his lack of a verifying baseline, as well as questioning whether he had used his "English" instruments properly.[3] It also called into question, among diehard Cartesians, the whole premise of a flattened Earth. Bouguer's observations, the Academy hoped, would settle the matter of the true figure of the Earth.

On July 29, just over a month after he had returned to the French Academy, Bouguer began reading his account of the Geodesic Mission before a rapt audience of Academy members. He no longer had reason to fear that his work would go unrecognized; with Maupertuis' dubious results having created a strong demand for new data about the Earth's shape, Bouguer's report was certain to eclipse that of his erstwhile foe, at the same time vaulting Bouguer into the upper reaches of the Academy. Maupertuis' account had been like that of an adventure story in which a quest reveals the answer to a long-held mystery. Bouguer's tale would be that of a Greek drama: The conclusion, that the Earth was flattened at the poles, was already known, but what held the audience's attention was the voyage through adversity to achieve that understanding.[4]

That audience pointedly included Maurepas, whose presence at the Academy was all the more significant as it came while he was preparing for a potentially existential war with Britain. By turns magnanimous and cold-gutted, Maurepas had sponsored both the Geodesic Mission and Bouguer's rapid rise in the Academy, and his appearance at the Academy gave particular standing to Bouguer. Yet this same minister had cut off all funding when the Arctic Circle expedition returned with its results, leaving many of Bouguer's companions still shipwrecked in Peru.

Bouguer read his account of the expedition at sporadic intervals for the next seven months, both to the Academy of Sciences and to the wider public. He began his narrative by recounting the difficulties and dangers the expedition faced simply to arrive in Peru, followed by a description of the country that was their home for the next seven years. He interspersed his tale with historical and cultural observations of Peru and its peoples, providing European audiences their first scientific look at the previously closed-off kingdom at the middle of the world.[5]

The second half of Bouguer's narrative detailed the exacting procedures they took to measure the baseline and establish the chain of triangles, and then reduce it to sea level. He explained the difficulties they endured in correcting for the various astronomical errors, finally arriving at the angular measure they had sought, thereby establishing the exact length of a degree of latitude at the equator. He finished his account with the matter-of-fact statement that with three data points on the globe—the lengths of a degree of latitude at the Arctic Circle, in France, and at the equator—he could mathematically define the precise curve of the Earth's shape and confidently report that it was flattened by 1 part in 179 (1/179). This confirmed Maupertuis' data and lifted the pall of doubt that had been cast over the entire project to measure the Earth.

The Geodesic Mission to the Equator, in combination with the Geodesic Mission to the Arctic Circle, marked a watershed in the progress of the Enlightenment. Together they expanded the framework of science from the national to the global level, deliberately coordinating and analyzing data from around the world to provide a comprehensive measurement of the planet. The science of navigation no longer had to be expanded in a piecemeal, localized manner; the two Geodesic Missions had shown nations how to cooperate over vast distances in order to achieve the high degree of navigational knowledge and accuracy needed in the increasingly interconnected world.

Bouguer's memoir would become the official account of the Geodesic Mission to the Equator. It was a heroic, if also self-serving, narrative of arduous toil under grueling conditions in the noble cause of science. In a dig against his erstwhile adversary Maupertuis, he noted that the Peru expedition was far more difficult than the one to the Arctic Circle, measuring a distance three times longer. Bouguer's status was further enhanced by Maupertuis' recent departure from France. Still smarting from the criticisms of his geodesic work and insulted by Maurepas' derisory pension of 1,200 livres, Maupertuis had subsequently accepted an invitation from Friedrich II ("the Great") of Prussia and was now on his way to preside over the Berlin Academy of Sciences, the newest rival to the French Academy.[6]

In his memoir Bouguer took pains to acknowledge the contributions of his colleagues, but his use of the royal *nous* (we) did not fool anyone: as summer and fall gave way to winter, no one in the Louvre doubted who had hoisted the Geodesic Mission from certain defeat to resounding victory. In November the newspaper *Mercure de France* published a glowing account of Bouguer's public speech, followed with a commemorative poem by his childhood friend Paul Desforges-Maillard. The poet spoke of the sacrifices his friend had made and the hardships he had endured to carry out the mission: "You leave behind your days pure and serene / to satisfy the wishes of the king / willing to endure, in defiance of the Fates / the highest mountains where never-ending winter awaits." For Bouguer, who had only reluctantly agreed to join the mission in the first place, these hardships had at last borne fruit. He had indeed returned to the Academy in triumph, as the poet noted, with "eternal honor and laurels upon his brow" bestowed by his king, his nation, and his peers.[7]

Bouguer's scientific achievements quickly made him one of the most esteemed scientists in France. Thanks to Maurepas, who topped off Bouguer's salary with an annual pension of 1,000 ecus (3,000 livres, almost triple that given to Maupertuis), he moved from Le Havre to Paris, living in a home on the rue des Postes, a prestigious neighborhood for astronomers, where Godin had once lived. He quickly grew to be the favorite scientist of Maurepas, who continued to pull him up the political ladder of the Academy. Bouguer became the Academy's director after Maupertuis decamped for Berlin, a doubly sweet victory. He was soon its most prolific author, with a stream of works on astronomy, physics, mathematics, navigation, and ship theory, including the fruits of his idle moments while in the mountains of Peru: *Treatise of the Ship*, the very first book of naval architecture.[8]

Bouguer went on to publish his geodesic findings from Peru in *The Figure of the Earth* and later in the best-selling *Treatise of Navigation*. Bouguer's books gave ships' pilots their first comprehensive set of tools on how to account for the earth's out-of-roundness while calculating longitude and latitude at sea. The actual effect of the Earth's flattening "fortunately is not very great," as Bouguer admitted (a difference of perhaps twenty miles over a transatlantic voyage), but such accuracy was increasingly warranted by recent improvements in navigational instruments such as the sextant and the chronometer. Bouguer's methodology became the standard across Europe and in the Americas. It was highlighted in the most widely read navigational text of the day, Robertson's *Elements of Navigation*, which was used well into the nineteenth century.[9]

Maurepas' investments in Bouguer and the Geodesic Mission had paid off handsomely, not merely scientifically but also in the wider realm of international politics. Britain and France both had undertaken to solve the great navigational problems of the age—Britain's quest was longitude; France's was the Earth's shape. The means by which each nation solved its problems were poles apart, as different as the Royal Society was from the French Academy. The British Longitude Prize, favoring individual entrepreneurship, was still up for grabs—thirty years had passed, and John Harrison's marine chronometer was still not accepted (and would not be for another twenty years). France, meanwhile, had brought the full force of government-backed science to bear on its problem and solved it in just ten years at a total cost of $2 million, half that of the Longitude Prize. Besides showcasing the institutional might of French science, Maurepas had shown potential allies that France could carry out large-scale missions under the banner of international cooperation. In the coming years the French navy would deploy scientists on long-range cruises, in collaboration with the Spanish and Portuguese authorities, to accurately chart coastlines and the heavens around the globe. His British adversaries, Maurepas would have gloated, were still dispatching lone scientists on derisory quests. France's overall investment in science would also bring substantial returns in colonial trade and international geopolitics. With improvement in navigation made possible, in part, by a more accurate knowledge of the figure of the Earth, increased trade in the Americas and in far-flung regions such as the Pacific and Indian Oceans brought millions of livres into French coffers, and French naval presence in those regions provided an effective counterweight to British influence.[10]

Bouguer's seven-month discourse had had to compete with a series of public spectacles that was transfixing Paris. The public mood was wrenched from deep mourning at the news that Louis XV was gravely ill at the battlefront and had received last rites, to elation at his sudden recovery, to mourning again at the death of his favorite mistress, and finally to glee at the betrothal of the king's son Louis, the Dauphin of France. On February 27, 1745, Bouguer stood before the Academy and triumphantly finished his account of the Geodesic Mission to the Equator. As he walked out onto the streets of Paris in the frigid twilight, the entire city was awash in candlelight as crowds flocked to the masked balls at the Hôtel de Ville held in honor of the Dauphin's marriage several days earlier. Bouguer, basking in the Academy's admiration and the glow of the candles, might have followed the throngs, whom he could imagine were also celebrating his own crowning achievement.[11]

One of the attendees at Bouguer's final pronouncement was hardly in a celebratory mood. Charles-Marie de La Condamine had arrived in Paris just two days earlier, having spent the previous two months holed up in Amsterdam awaiting passports and reading accounts of Bouguer's triumphal exposition. Denied his share of the scientific glory, La Condamine took a different approach when his turn before a public session of the Academy came on April 28, 1745. He was reduced to mouthing a travelogue about his adventures on the Amazon, encounters with wildlife and Indians, and threats of piracy on the high seas, but where Bouguer's parchment-dry officialese was quickly filed away and forgotten by the public, La Condamine's chatty, almost poetic narrative quickly caught Europe's attention. Though he never attained a reputation as a top-flight scientist, his descriptions of the New World launched his career as one of the great popularizers of science and natural history during the Enlightenment.[12]

While Bouguer was burnishing his credentials in the scientific community with dense mathematical memoirs, La Condamine pursued a public path to fame by breathing new life into the European vision of South America. European audiences were increasingly aware that their conception of the continent was derived from distorted information. Spain had long kept South America off-limits to foreigners, and the Inquisition had placed tight controls on what Spanish authors were allowed to publish. By the beginning of the Enlightenment, scholars were calling into question the reliability and accuracy of the chronicles left by the Spanish conquistadores and missionaries, who drew on biblical references to support their vision of a continent and its peoples in a primitive state, crying out to be civilized.[13] The few foreign travelers who had managed to penetrate the veil surrounding Peru, such as the French scientists Feuillée and Frézier, had shown how to write critical, scientific accounts of the New World.

La Condamine capitalized on the hunger for information about South America by embellishing his scientific narrative with a storyteller's gift for the ripping yarn and the novelist's eye for the telling detail. He turned his Academy of Sciences memoir of his voyage down the Amazon into a best-selling book, *Abridged Relation of a Voyage Made in the Interior of South America*, which—besides boasting of the mathematical precision of his navigation—featured, among other thrilling details, a breathless travelogue replete with the dangers of rafting through whitewater canyons and the tale of the courageous monk who first introduced inoculation against smallpox at his mission. La Condamine accompanied his book with the first detailed map of the Amazon River basin, filling in many of the blanks from previous expeditions. He also appended to the memoir his page-turning account of the assassination of Seniergues, titled *Letter to Mrs. X, on the Popular Uprising Raised in the City of Cuenca*.[14]

By publishing less scientific, more popular accounts of the Geodesic Mission's exploits, La Condamine was also pulling back the curtain on the previously veiled kingdom of Peru. At the same time that he was transforming the expedition's travels into a thrilling narrative of adventure, he devoted considerable effort to producing a comprehensive atlas of the *audiencia* of Quito. His friend Pedro Vicente Maldonado had begun this project while still in Peru, but his untimely death from measles in London in November 1748 left La Condamine to complete the work himself. The finished atlas was released to public acclaim, coming at a time when European society was transfixed by all things Peruvian. Plays, operas, ballets, and novels featuring Incan and Peruvian themes began sprouting up. Françoise de Graffigny, the powerful opinion maker (and close friend of La Condamine) who ran Paris's most prestigious salon, penned a satirical novel titled *Letters of a Peruvian Woman*, which—like Voltaire's *Alzire*—featured a Peruvian princess. Graffigny used her royal character quite differently from Voltaire, by depicting the princess making the journey to France and providing bitter social commentary on the status of women there. *Letters of a Peruvian Woman* became one of the most widely read books of the eighteenth century.[15]

Although La Condamine had succeeded in capturing his own share of the limelight, Bouguer was clearly in the ascendancy, and his friend knew it. Bouguer had garnered full credit for the success of the Geodesic Mission, leaving La Condamine, accustomed to being at the center of Parisian political influence, now standing powerless on the sidelines. "I give you my compliments on the new position you have obtained in the Observatory," La Condamine wrote in fawning terms to Bouguer, at the same time asking that he "request of the King the money that I had advanced" to the expedition while in Peru. Through Bouguer's intervention, La Condamine was eventually reimbursed in full.[16]

The men remained close friends for as long as their separate visions of Peru did not conflict, but with both of their accounts in high demand, competition between the two egocentric scientists soon erupted into open warfare. The rupture came in November 1748, when both men were at the home of Maupertuis. The former director of the French Academy had returned to Paris during a vacation from his new post in Berlin. Since he and Bouguer were at separate institutions and thus no longer mutual threats, they had struck up a friendship of sorts. At that time, Bouguer was preparing to publish in book form the memoir of the mission he'd already presented to the Academy of Sciences. Suddenly La Condamine announced to his colleagues that he was planning to publish *his* account of the geodesic results. Bouguer was aghast. It was perfectly fine for La Condamine to spin exciting yarns about his adventures on the Amazon, but not to print a potentially conflicting version of the science that

Bouguer, as the expedition's de facto leader, believed was his responsibility alone to describe.[17]

Tensions between the two scientists escalated the following April, when Bouguer's mentor Maurepas was abruptly dismissed from Versailles after losing his political sea-sense and writing one too many satires about Madame de Pompadour, Louis XV's new mistress. Bouguer's wagon, once firmly hitched to Maurepas' star, was now veering into a ditch. Coming at this critical moment, La Condamine's book was no mere annoyance to Bouguer, but rather a threat to his standing in the Academy. The two colleagues now threatened to stop publication of the other's book, each demanding that various Academy members take one side or the other.

In 1749, over La Condamine's objections, Bouger published *Figure of the Earth*, his narrative account of the mission and its scientific results. La Condamine's version of the scientific results, *Measure of the First Three Degrees of the Meridian*, had been scheduled to be printed at the same time as Bouguer's, but had subsequently been delayed two years by the latter's protests. With Bouguer having beaten him to the punch over the scientific findings, La Condamine used the forced interruption to polish his day-by-day narrative of the mission, *Journal of the Voyage to the Equator*, which was published simultaneously with *Measure of the First Three Degrees* in 1751.

The two friends' dispute soon left the confines of the Louvre and went public, with each printing brochures that attacked the other. The press closely covered the dispute, with newspapers like *Mercure de France* providing updates and editorials. A handful took one side or the other, but most provided reasonably objective reports while deploring the baseness of the whole affair.[18]

The newspapers had good reason to criticize both parties; in reality, the principal points in the dispute between La Condamine and Bouguer were petty and not worthy of either of them. One of the arguments concerned the choice of operations, measuring either longitude or latitude, which the scientists had quarreled over early in the mission. Bouguer accused La Condamine of siding with Godin in proposing the longitude approach, which would have derailed the whole expedition; La Condamine denied it and pointed a finger at Bouguer, claiming he had initially resisted redoing the simultaneous stellar observations that were key to the mission's success. Each man also took issue with minor editorial details in the other's account of the expedition.[19]

The entire quarrel was not only unnecessary but bewildering to those who witnessed it. Bouguer had given La Condamine sufficient credit in his initial work, and La Condamine's own books did not dispute any of Bouguer's conclusions. Many reasons for the row were advanced by subsequent historians, but ultimately it remained unfathomable, as the Marquis de Condorcet noted in La Condamine's eulogy: "One wonders what were the objectives of the dispute which arose between Bouguer and La Condamine, between two men who, during several years had slept in the same room, under the same tent, often on the ground, wrapped in the same blanket, all the time holding each other with reciprocal esteem."[20]

By 1754 the dispute had burned itself out, but much had been lost in the fire. La Condamine, far better connected socially and with a wittier pen than Bouguer, had won out in the public arena over his erstwhile friend, who continued naïvely to believe that scientific reasoning should trump his opponent's facile, silver-tongued arguments. In the eyes of the public, "La Condamine was assumed to have done everything," an astronomer later observed, "and Bouguer was there simply to help out." Bouguer's reputation within the Academy of Sciences had also been damaged by the affair, and with no supporter as powerful as Maurepas waiting in the wings, he never quite attained the standing of other scientists that he had seemed due to receive.[21]

While Bouguer found his position in the Academy weakened after his dispute with La Condamine, the fall from grace did not prevent him from reaching great heights within other arenas in France. His works on naval architecture and navigation kept him closely connected to—and highly regarded within—the French navy. Bouguer became an instructor at the first school for ship designers in Paris as well as a founding member of the country's new Academy of Navy, which brought scientifically educated officers in close contact with mathematicians and politicians. While assisting the French navy, he did not neglect his scientific duties. He edited an academic periodical and wrote so many books and memoirs that eight of his works had to be published posthumously. His range of interests was remarkable; in later years he would be recognized as the founding father of naval architecture, photometry (the study of light), and gravimetry (the study of gravitational anomalies).

Bouguer never quite found the depth of comradeship with others that, through shared adversity and purpose of mission, he had developed with La Condamine. During his life, Bouguer maintained good relations with many colleagues, but few were close to him apart from his brother and his childhood friend Paul Desforges-Maillard. He never married, and there is no indication that he ever had a love interest. He spent his last years entirely in Paris, working at his beloved science.

In the end, though, Paris killed him. After surviving almost a decade in the harsh mountains of Peru, Bouguer had found himself living in the largest city in Europe with all the amenities of civilization but with all of its hazards as well. Like most cities of the day, Paris was inadequately supplied with fresh water; its stores were continually polluted with sewage, making them a breeding ground for *entamoeba histolytica*, a lethal species of amoeba that causes dysentery and liver infection. This pathogen attacked Bouguer in the spring of 1758, weakening him over several months. He managed to complete his seminal work on photometry from his bed before succumbing on August 15, 1758. He was buried in the city, but when the cemeteries were emptied in the 1780s, his bones were removed to the now-famous Catacombs. Bouguer's mortal remains are now dispersed somewhere below the streets of Paris.[22]

Ultimately, La Condamine enjoyed greater recognition than had his onetime friend. His *Journal of the Voyage to the Equator* had been an instant success when it was finally released. An engaging, insightful, and often witty account of his ten years spent in Peru and his voyage home, it quickly became the most recognized story of the Geodesic Mission. It was widely reviewed and excerpted in the popular press, and it became the primary source for information on South America in two eighteenth-century best sellers: Antoine Prévost's *General History of Voyages* and Diderot and D'Alembert's monumental *Encyclopedia*.[23]

La Condamine's writings changed many European misconceptions about South America—but not all of them. Within a decade of La Condamine's return, the popular image of Peru had been transformed from a place of material riches and unspeakable cruelty to a world revealed as a fountain of botanical and natural treasures. La Condamine's *Journal* nevertheless reinforced the European view of Indian inferiority, which in turn influenced authors such as the naturalist Comte de Buffon (now the intendant of the Royal Garden of Paris) to claim that New World plants, animals, and peoples were smaller, weaker, and altogether inferior to those in Europe as a result of the hotter climate in which they grew. Buffon often lifted his phrases straight from La Condamine and Ulloa, for example accusing Indian men of "resting stupidly on their thighs or lying down all day" in lieu of working.[24] It would take several more generations of explorers, culminating in Humboldt's grand voyages, before most of those pernicious ideas about South America and its native peoples were laid to rest.

Although La Condamine was a great popularizer of science, he never accomplished much through direct research into any one discipline. He was among the first to propose creating a standard length of measure for all nations. His concept was based upon the length of a seconds pendulum at the equator, a politically savvy plan that would have linked France with South America and would have extended French influence internationally. (The concept of a standard international measure later became the basis for the metric system.) Most famously, he became enmeshed in the controversy over inoculation against smallpox, reaching back to the successful inoculation of Indians in Belém in support of the practice. His mathematically based arguments, that the risks from contracting the disease far outweighed those from the inoculation, influenced families in high society and royalty to inoculate their own children, paving the way for wide public acceptance of the procedure.[25]

While Bouguer lived out his days in relative solitude, La Condamine surrounded himself with friends and family. In a similar fashion to his good friend Voltaire, La Condamine fell in love with his own niece, the much younger Charlotte Bouzier d'Estouilly. Unlike the playwright, he married her, traveling to Rome to receive papal dispensation for the betrothal. The couple was childless but lived a happy life at their home near the Louvre palace.

Like Bouguer, La Condamine received distinctions in his later life that were not materially related to his work on the Geodesic Mission. In

1760 La Condamine, who had by then utterly lost his sense of hearing, was elected as one of the "forty immortals" in the Académie Française. The election made him one of the few scientists to be charged with overseeing the French language and its culture. One of his colleagues, Alexis Piron, amusingly penned the epigram:

> La Condamine is today
> Received in the immortal troupe
> He is deaf: good for him
> But not mute: too bad for them.[26]

La Condamine's profound deafness was accompanied by a growing numbness in his limbs, to the point at which he could not feel whether he wore shoes. He attributed the afflictions to his exertions in Peru, but it is much more likely that he suffered a rare form of maternally inherited diabetes. That is not what killed him, however. In 1774, at age seventy-three, he called a young doctor to his home to operate on his hernia. It was a new procedure, and he remained awake throughout in order to report on it later. But two days after the surgery, on February 4, he was dead, possibly from an infection but most certainly as a result of an overdose of insatiable curiosity.[27]

❧

The weather in Cadiz, on the Spanish coast, was calm and clear on the morning of November 1, 1755—just a year after the final argument between Bouguer and La Condamine had sputtered out. Long before most of the inhabitants of that great port city, Louis Godin felt the faint rumblings that signaled a major earthquake. *Terremoto*, would have been his first thought, not *tremblement de terre*. After twenty years living away from France, he was thinking in Spanish, not his native French. Although it was a Sunday morning, he was not at home with his wife and daughter, but rather at his office in the observatory, surrounded by his scientific instruments. As the tremors began, his eyes instinctively flashed to the wall clock—exactly 9:52 AM. "I began to feel a very light, very soft movement of the Earth," he later wrote. "This movement was felt some time later by those who were not accustomed to it; but for me, who had felt this many times in Peru, I paid far greater attention to the event." A veteran of the horrific Lima earthquake nine years earlier, Godin immediately recognized the danger. The rumblings grew more intense, building into a violent shaking that lasted three minutes, followed by gradual subsidence.

Miraculously, the quake left most of the buildings in Cadiz still standing, and the city was also spared the horror that had befallen Callao in the wake of the 1746 quake. Like Callao, Cadiz was located on a spit of land and might have been flooded by the *maremoto* that hammered in an hour later, but the waves were just twenty feet high and barely overtopped the seawalls. Other cities were not so fortunate. Two hundred miles north of Cadiz, the city of Lisbon was utterly destroyed by a tidal wave in the worst natural catastrophe to strike Europe in living memory. A thrust fault located far out to sea off Gibraltar had slipped, and the resultant seismic and oceanic wave patterns meant that Lisbon was flattened, but Cadiz was saved.[28]

Godin had been at his post as head of the observatory and of the Academy of Navy Guards in Cadiz for less than a year when the earthquake struck. He had spent a year in Paris, from 1751 to 1752, successfully wooing back his wife and reuniting with his family. He was less successful petitioning the Academy of Sciences to reinstate him, even though he had reconciled with Bouguer and La Condamine and had even brought back the celebrated *toise* of Peru as an offering—Jussieu never having joined him in Paris to escort the rod back, as planned. When it became clear that Godin had no real prospects in Paris, his family decamped to Madrid, where they lost a son to smallpox, and then finally moved to Cadiz, where Godin took up his joint post, six years after Navy Minister Ensenada had offered it to him.

Godin's work in Cadiz was productive but short-lived. While there he created new mathematics courses for the Navy Guard cadets and brought into the observatory the most up-to-date astronomical instruments and techniques. He only served in his position for three years; a year after the earthquake, in 1756, he traveled to Paris, where he was finally reestablished as a veteran pensioner in the Academy of Sciences. Upon his return to Cadiz he fell seriously ill for many months. In 1760 his daughter, his only remaining child, died, a blow from which Godin did not recover. On September 11, 1760, he suffered a fatal stroke, at just fifty-six years of age.[29]

By the time of Godin's death, Jorge Juan and Ulloa—who had originally recommended the disgraced, then rehabilitated Frenchman to Ensenada—had grown from neophyte assistants to influential officials of the Spanish crown. The Geodesic Mission had launched their careers; they had proven their mettle by operating independently far from home, on both political and military fronts. The reports they delivered, publicly and in private, had made them household names across Europe and the Spanish king's favored troubleshooters. Given what the two men had originally endured on their return to Spain, the recognition would have come as a considerable relief.

The future had not looked so promising for the Spanish officers when they had first arrived in Madrid in 1746. Their original patron for the mission, José Patiño, had died soon after they had arrived in Peru, and the government reports from Lima and Quito barely noted the Spaniards' existence; so for ten years their names had not been heard in the king's court. When the officers came to Ensenada, telling their story without any written proof, they were initially dismissed as mere reward seekers. However, Admiral José Pizarro, who commanded the fleet in which Jorge Juan and Ulloa hunted for Anson, had also just returned to Madrid and vouched for their exploits. Intrigued, Ensenada ordered the two officers to write their memoirs for publication by the crown.

After gaining the credence of Navy Minister Ensenada, the two men rapidly agreed to produce three separate works. Collecting their remaining notes (much had been cast overboard when Ulloa was captured), the officers spent the next two years preparing the three texts for their intended audiences, both public and private. The first became the *Historical Relation of the Voyage to South America*, a sort of gazetteer primarily authored by Ulloa that described their travels and the sites they visited. The second became *Astronomical and Physical Observations*, Jorge Juan's detailed account of the geodesic operations over their ten-year stay. The third work, a jointly written manuscript titled *Discourse and*

Political Reflections on the Kingdoms of Peru, was a scathing criticism of political corruption that was circulated within the court but was never intended to be published.[30]

The officers' public writings, released under the auspices of the Spanish crown, helped to bring the wonders of Peru to Europeans, much as would La Condamine's later publications. *Voyage to South America*, published along with *Observations* in 1748 (well before the accounts of Bouguer and La Condamine came out), marked a dramatic shift in Spanish policy. Previously, the crown had kept Peru under wraps, fearful of revealing any military vulnerability or reinforcing what was later called "the black legend" of Spanish conquest and violent oppression of the indigenous peoples. Now a newer generation of administrators, weaned on the Enlightenment philosophy of more open government, proudly displayed the jewel in its crown. A total of 2,500 copies of the two books were printed on expensive parchment at great cost (equivalent to $1 million) and distributed around Europe. Spanish was not widely spoken in Europe, so translations in French, English, and German soon followed, which were snapped up by an audience hungry for information about the New World. *Observations* was a dense treatise filled with tables of measurements and equations for calculating the Earth's shape. The far more popular *Voyage to South America* kept to a fairly standard formula: Each section provided a short account of their activities in a particular region, followed by descriptions of the cities, inhabitants, climate, and economic activities, often accompanied by detailed maps of the area. In these public writings, the authors avoided any mention of corruption or mistreatment of the local Indians, still careful—despite the crown's pretensions of casting aside secrecy—to showcase Spain and its colonies in the best possible light.

Ulloa and Jorge Juan were not so restrained in their privately circulated *Discourse*. As inheritors of Spain's "enlightened" reformist movement (of which Antonio de Ulloa's father had been a part), they had been on the lookout for colonial misrule while traveling in South America, having made firsthand observations and collected secondhand accounts throughout their journeys. Jorge Juan and Ulloa were not the only ones sensitive to the plight of the indigenous peoples of the continent; on their return to Madrid the officers found that numerous political tracts were

Figure 11.1 The Flattened Earth. From Jorge Juan and Ulloa, *Relación histórica del Viaje á la América Meridional* (1748). Credit: Special Collections & Archives, Nimitz Library, U.S. Naval Academy.

already circulating around the court, decrying the abuses of the Spanish viceroyalties in America. Their *Discourse* synthesized these ideas into two major themes: how to reform the colonial navy and military and how to reform the colonial government to avoid the worst excesses of systems like the *mita*. The document became one of many that influenced the course of colonial reforms. But as numerous administrators had already found, carrying out such reforms in the face of established interests was an entirely different matter.[31]

Antonio de Ulloa was neither the first nor the last to try reforming colonial administrations, and like his predecessors and successors, he mostly failed. His first foray into colonial service was in 1758, when he relinquished his naval commission for the post of governor of the deteriorating Huancavelica mercury mine in Peru. Mercury was a vital part of the silver industry. It was mixed with silver ore so that it would bind to the silver, leaving the rock and earth to be washed away; when the resulting amalgamation was heated, the mercury vaporized, leaving the silver behind. The mining operations had suffered a loss in production caused by technical problems of extraction, lax safety conditions that condemned workers to injury and death, and widespread pilfering of the precious material. Ulloa's prior experience in Peru, his passion for reform, and his extensive knowledge of metallurgy (he was, for example, the first to scientifically describe platinum) made him appear to be the right candidate to fix these problems. However, he was unable to muster the political support needed to establish strong governmental authority over the entrenched union control of the mines and thus fix the many abuses of native laborers that they engendered. After six years of frustration, Ulloa was worn down so much that he asked to be transferred.

That transfer was soon in coming. In 1765 he was appointed governor of Louisiana, a French colony in North America that had just been ceded to Spain as part of the resolution to the Seven Years War. Once again, Ulloa failed to muster sufficient political and military support to keep the colony firmly under Spanish control, and in 1768, with a popular uprising threatening his life, he left under a cloud of failure. Ulloa returned to the navy and went on to command a fleet arrayed against the British during the American War of Independence, later rising to become vice admiral of the fleet.[32]

If Ulloa's efforts as a reformer ended in catastrophe, his personal life was much more fruitful. Ulloa had married a young Peruvian woman, Francisca Ramírez de Laredo, during his stay in Louisiana, and despite his advancing age they had nine children who would become his primary joy; one, Francisco Javier de Ulloa, later rose to be secretary of the navy. Ulloa himself was active in the navy and in research into natural history until his death in Cadiz on July 6, 1795, at the age of seventy-nine. He died surrounded by his family and his immense collection of scientific instruments, mineral samples, and Indian artifacts that marked the successes and failures in his remarkable life.[33]

Jorge Juan y Santacilia took a much different path from his colleague's, remaining in Europe and devoting himself entirely to the Spanish military. In 1749 Ensenada, recognizing Jorge Juan's talents as both a shipbuilder and a careful observer, sent him to Britain on a spy mission to steal plans of ships and equipment and to bribe shipbuilders to come to Spain. In one year the officer managed to entice over eighty British shipbuilders to work in Spanish shipyards, and many of these builders went on to construct the very ships that later fought the British fleet at Trafalgar. Turning a blind eye to these acts of treachery, the British made Jorge Juan a fellow of the Royal Society for his astronomical work, just as they had done for Ulloa.

In 1752, Jorge Juan was appointed as the overall head of Spanish naval construction, during which time he also worked with Godin to establish the new Royal Observatory in Cadiz. Never married, he placed all his energies into his widely varied naval career, rising in rank as the king sent him from one location to another to solve increasingly complex problems. He followed in Bouguer's footsteps by writing highly influential treatises on naval architecture and navigation. His one foray into diplomatic affairs was in 1767, when he successfully negotiated a peace treaty between Spain and Morocco.

While the two Spanish officers had been constantly united during the Geodesic Mission, Ulloa would end up outliving Jorge Juan by over twenty years. Jorge Juan's peregrinations were often interrupted by severe gastrointestinal distress, which he always sought to relieve by returning to the thermal springs near his home in Alicante. On one of those visits he apparently was infected by a rare "brain-eating" amoebic parasite that thrives in warm water, resulting in his untimely death on June 21, 1773.[34]

The legacies of Jorge Juan and Ulloa have been inextricably linked, and the two men are still much venerated by the Spanish navy they so ably served. From 1851 until the modern day, three pairs of warships have borne their names. Jorge Juan is interred in the Pantheon of Illustrious Mariners in Cadiz, near the marker of his lifelong friend Antonio de Ulloa, whose remains are in the Church of San Francisco in Cadiz. The two officers are the only two members of the entire Geodesic Mission whose burial places are known today.

Two years before Jorge Juan's untimely death, one other member of the Geodesic Mission finally returned to his native soil. Of all the expedition members who had been left in Peru following the scientists' departures some twenty years earlier, Jussieu—who had planned to meet up with Godin in Buenos Aires—ostensibly had the best chance of making it back to France. After all, he had been ordered by Maurepas to ensure that the team's *toise* was returned safely to France, an order that provided the doctor himself with a mandate to return to his mother country—although not with the money to make the journey. That he had been expected to provide for himself.

Whatever Godin might have thought about the reasons for Jussieu's delay, he surely could not have guessed the truth. For all his interest in botany, Jussieu still had the humanitarian impulses of a doctor, and when he stumbled on some of the most horrible human conditions in colonial Peru, he simply could not tear himself away. Despite Jussieu's longing to return to Paris, he would remain in the Viceroyalty of Peru for another two decades.

In 1749, Jussieu had parted ways with Godin at La Paz and botanized the Peruvian countryside for over a year, collecting trunks of samples and copious notes that would have formed an unprecedented survey of South American botany. By July 1750 he had arrived at the infamous silver mines of Potosí. The Cerro Rico ("Rich Mountain") had been the primary source of Spanish wealth for two hundred years, at one time accounting for over 90 percent of Spain's total silver production. Its output had been steadily declining for a century, however, and the Spanish had attempted to reverse the downward trend by overworking the adult miners who they conscripted to work at the site, while also forcing children into the mines.[35]

At Potosí, Jussieu witnessed ravages of the *mita* system even worse than those the expedition members had first encountered in Quito. Silver mining was a slow death sentence in colonial Peru, since silver extraction required the workers to mix mercury with the ore using their bare hands. As the amalgamation was heated to separate the silver, the workers also inhaled the toxic mercury vapor that was released. Vision and coordination were often the first to deteriorate, rendering mine workers even more likely to suffer accidents. Numbness and loss of memory followed, and the children in particular suffered horrible rashes and kidney failure.

Jussieu had intended merely to stop over at Potosí on his way to Buenos Aires, where he planned to reunite with Godin and return to Europe, but the doctor was haunted by the appalling conditions he saw at the mines and could not bring himself to walk away. He would spend the next five years tending the sick and injured as well as serving as a sort of field engineer to repair and rebuild dams, bridges, and pumps. His charitable intervention would not go unpunished. By 1755 he was in the grip of mercury poisoning, and while he was sometimes lucid, he was at other times afflicted with memory loss so profound that he was incapable of writing even in his native French. His friends took him to recover in Lima, where he spent the next fourteen years tending to the most destitute residents of the city while passing up several opportunities to return to France. Even when Antonio de Ulloa had showed up in Lima in 1758, on his way to assume the governorship of the Huancavelica mercury mine, Jussieu would not budge when offered a position in his government.[36]

While Jussieu attended to the population of Lima, his mental and physical health continued to deteriorate; he was finally forced to leave the New World altogether. In October 1770, his friends arranged for his return to France, promising to care for his botanical collections and papers, which, in his weakened state, he could not bring with him. In the company of two Frenchmen also traveling from Lima, he journeyed by way of Panama, Cuba, and Spain, arriving in Paris in July 1771 after thirty-six years' absence. But neither France nor his family could cure Jussieu of the toxic effects of mercury. He sank further into silence and torpor, from time to time hallucinating that he was back in the Caribbean, where he had first experienced the freedom of botanizing in a new world. He did not notice the death of his brothers, nor the news from Peru that most of his papers and collections, a veritable treasure trove of botanical knowledge, had been destroyed; the descendants of his now-deceased friends had not recognized their worth and pitched them into the fire. Finally, on April 10, 1779, at his family home in Paris, Joseph de Jussieu died from an infection at age seventy-five.[37]

The scientists from the French Academy of Sciences all saw France again during their lifetimes, but the secondary members of the expedition, those who were not members of the Academy, were not all so lucky. Verguin, Hugo, Morainville, even Godin's cousin Godin des Odonais—all were quickly forgotten by the scientists and politicians who had sent them to Peru in the first place and who had benefited from their work on the behalf of the expedition. Bouguer and La Condamine had left Peru with their servants and slaves, as they were considered personal property, but had not seen fit to include their countrymen in their return travels. Maurepas, who was most responsible for these men's fates, had been content to leave them stranded in Peru even as he demanded that Jussieu travel eight hundred miles to fetch a rod of iron. And yet, remarkably, two of the men did make it home—for better or for worse.

For Jean-Joseph Verguin, whose crucial astronomical observations had gone almost unremarked by Bouguer and La Condamine, the return home proved far more painful than any of the hardships he had suffered in the mountains of Peru. Delayed by a long illness, he had finally departed Quito in summer 1745, arriving in Toulon, France, in late 1746 to find his wife dead and his two children, whom he had not seen since infancy, living with their grandmother. His nomination as a corresponding member of the Academy of Sciences would have done little to assuage the pain of his loss. Nevertheless, Verguin did remarry and resumed his old position as civil engineer for the port of Toulon, constructing buildings and harbor defenses until his death on April 29, 1777.[38]

For Théodore Hugo and Jean-Louis de Morainville, there would be no return home. Both men sought in vain to collect compensation from Maurepas, though they eventually had to reconcile themselves to life in the Spanish colony. Hugo, the man who had built and maintained the expedition's instruments, was particularly bitter toward Godin and Maurepas, whom he felt had cast him adrift without any resources or even recognition by the Academy of Sciences: "I am therefore stuck here in Quito where I spend my life working off my debts by the sweat of my brow." He abandoned his former career as an instrument maker and set up a tileworks, eventually marrying a local woman. Morainville, for his part, occasionally received letters from his wife and eventually a notice of her death. He became an itinerant artist, moving around the *audiencia* to take various jobs for the church as an architect and painter, leaving behind several artworks that are visible even today.[39]

The two Frenchmen continued to try to raise money for their return home but would ultimately live out their days in Peru. In 1753 they joined with Antonio de Ormaza, the unscrupulous owner of hacienda Guachalá, where they had first stayed while surveying the plain at Cayambe, to open a silver mine on the peak of Mount Pichincha. The project ended two years later in mutual recriminations and lawsuits, but no silver.[40] Hugo passed away in Quito in 1781, leaving behind a family of eight, including a son, Luis Hugo, who would later become part of Quito's nascent independence movement. Morainville, who apparently never remarried, had meanwhile moved from Quito to Riobamba, where in 1764 he witnessed a massive Indian rebellion against the *mita*, one of many indigenous uprisings that sporadically arose across the viceroyalty. Morainville helped engineer the city's defenses against the two-month-long siege, during which time the Church of Nuestra Señora de Sicalpa was apparently damaged. Morainville died while repairing it, in a fall from the scaffolding.[41]

The last member of the Geodesic Mission to return to France was Louis Godin's cousin, Jean-Baptiste Godin des Odonais, who traveled by way of the Amazon. But it was the story of his wife, Isabel Godin's, harrowing descent of the river that caught the public's imagination, when the couple finally arrived in France—some thirty years after the first members of the mission had departed Peru.[42]

Isabel and Jean had moved from Quito to Riobamba in 1744, surrounded by Isabel's family but also increasingly immersed in debt, as Jean's business ventures failed to generate much income. They had had numerous opportunities to return to France together, but each time Isabel had been pregnant and unable to travel. The couple had originally intended to join La Condamine on his Amazon journey, but Isabel's pregnancy had dashed that hope; to make matters worse, that pregnancy, and each of two subsequent ones, ended in the death of the child. In 1749 Jean was called back to France to take over his family's estate, but Isabel was once again pregnant and could not journey. Jean landed on a bizarre plan: to reconnoiter the Amazon and then somehow return upstream to fetch his wife when he had determined it was safe to travel.

Jean descended the Amazon without trouble and made his way to the French colonial port of Cayenne. He soon encountered physical as well as bureaucratic barriers in trying to return to collect Isabel. The Portuguese authorities who controlled much of the river refused him a passport for his return voyage; he could not circumnavigate the continent to return to Peru, as no vessels sailed from Cayenne along that route; and he could neither send messages to his wife nor receive any from her. He was stuck in Cayenne, writing to the authorities back in Paris, asking for their help in obtaining a Portuguese passport to return up the Amazon. It was unthinkable that he would board one of the ships bound for France without his wife and child. This included the vessel that arrived in 1753 carrying the astronomer La Caille, still on his South Atlantic expedition and bringing notice of Jean's cousin Louis Godin.[43]

After fifteen years, the Portuguese government finally acceded to an official French request and allowed a message to be sent to Isabel, along with a boat that would wait at the headwaters of the Amazon for her. When Isabel finally received Jean's message, their only surviving child had recently died; grief-stricken and with nothing more to lose, Isabel resolved to descend the Amazon to rejoin Jean, the only remaining link to her previous life. Selling her property and gathering forty family members, servants, and porters to accompany her on the journey, she set out in 1769 to follow the various rivers that would lead to the Amazon and the boat that would take her to her husband.

Isabel's voyage quickly went to pieces. Abandoned in the middle of the jungle by their guides, the group decided to split in two; a small party would head downstream to seek rescue, while the larger group, led by Isabel, hunkered down on a sandbar to wait. After a month it was obvious that no rescue was coming, but while Jean had dithered in his plans to traverse the river, Isabel was now cold-bloodedly decisive. Gathering their belongings, the group began to travel on foot along the shoreline, downstream toward their only hope of salvation. Heat, thirst, and starvation took the members one by one, until only Isabel was left alive.

After stumbling through the jungle for several weeks, delirious and alone, Isabel saw a group of Indians canoeing down the river. At first she thought it was another hallucination, but stepping out of the undergrowth she spoke to the startled Indians in Quechua, barely able to get the words out: Would they take her downriver? They healed her wounds and brought her to the Portuguese boat, still waiting to transport her to her husband, whom she had not seen in twenty-one years. They finally met aboard a small boat at the mouth of the Amazon, the happiness of their reunion almost making them forget the grief and misfortune they had so long endured.

It was not until 1773, thirty-eight years after a freshening wind had carried *Portefaix* and the members of the Geodesic Mission into the Atlantic, that Isabel and Jean sailed from Cayenne and came to the Godins' ancestral home in Saint-Amand-Montrond. Jean received a royal pension for his work in South America and settled into a peaceful life with Isabel, bereft of children but surrounded by his family. After living apart for so many years early in their lives, they could not do so at the end; in 1792 Jean and Isabel died within six months of each other.

By the time of Isabel and Jean's death, the eighteenth century was drawing to a close, and the world had changed completely from the time the expedition members had set out to measure the Earth. Science was on the rise: Not only were new discoveries and inventions being reported from around the world, but an increasingly educated public was becoming more receptive to the effects on their daily lives. Monarchies were on the wane: An independent nation, the United States, had been carved out of British North America, and the revolution in France was just beginning to revise the political landscape in Europe. The members of the Geodesic Mission had lived long enough to see the impact their expedition had on the rise of science, but they could never have envisioned the far greater impact the mission would have on overturning colonial rule in South America.

Epilogue

THE CHILDREN OF THE EQUATOR

I had explored the mysterious sources of the Amazon,
and then I sought to ascend to the pinnacle of the Universe.
I strove bravely forward in the footsteps of La Condamine and
Humboldt, and nothing could hold me back.

—Simón Bolívar, *Mi delirio sobre el Chimborazo*
(My Vision on Chimborazo), 1822[1]

Nearly a century after the Geodesic Mission first set sail from Europe, a Venezuelan aristocrat named Simón Bolívar led an army to unshackle much of South America from Spanish Royalist forces. His mission was to create an independent South American nation, and he would soon succeed. By 1822, he had liberated the Viceroyalty of New Granada, including Quito. Known as the Liberator, Bolívar—along with the Protector, Argentine general José de San Martin—was the driving force behind South America's struggle for independence from Spain.

Bolívar's epic vision on Chimborazo, written soon after his liberation of Quito and describing a feverish dream in which the revolutionary retraced the steps of great European explorers and scientists, is notable for two reasons: how deeply entwined the Geodesic Mission had become in the idea of South America as a distinct entity, not simply a part of the Spanish Empire, and how thoroughly La Condamine had supplanted the other members of the expedition as its symbol. The two facts are inseparable. Although the accounts of Bouguer, Jorge Juan, and Ulloa were well received, it was La Condamine's *Journal* that captured the public imagination. Later travelers such as Alexander von Humboldt specifically referred to La Condamine as their inspiration for exploring South America, a pattern that would ultimately lead to Charles Darwin's famous journey to the Galapagos Islands.

The Geodesic Mission to the Equator accomplished the original goal that the French and Spanish governments had established for it, but this was only a small part of its historical legacy. Its immediate effect—redefining the shape and figure of the Earth in order to improve navigation at sea—was rather minor, amounting to a few dozen miles over a typical ocean voyage and burnishing the scientific reputation of the expedition's sponsor nations, France and Spain.

The expedition's real impact was the way it reshaped our world, in ways its founders could never have imagined. The Geodesic Mission inaugurated a spate of large-scale international scientific expeditions that rewrote our understanding of the planet, and it gave us the concept of South America as a unique place, separate from its mother country of Spain, which would eventually give birth to the new nations of Latin America.

Scientists across Europe looked on the Geodesic Mission as the model for future scientific expeditions. These would not be mere voyages of geographic discovery; the heroic age of Columbus and Magellan charting unknown seas and claiming new territory had long since passed. Nor would they be the work of solitary observers toiling in some distant corner of the globe. Science, employed in the name of empire, would now be the driving force behind large, usually government-funded enterprises that could involve many nations and last a decade or more. Just as the goal of the Geodesic Mission was never to simply map the equator but to accurately determine the figure of the Earth, these new expeditions were aimed at improving the fundamental tools of ocean navigation such as cartography, geomagnetism, and astronomy, and to prospect for new plants that could feed, clothe, and cure an ever-growing European population.[2]

The most famous of these international scientific enterprises were the cooperative voyages from 1761 to 1769 to observe the transit of Venus, so as to establish its distance from the sun to allow more precise calculations for planetary astronomy. Scientists from Britain, France, Russia, and Austria coordinated their observations around the world and corresponded with each other about the results, despite the intervening Seven Years War (1756–1763) that pitted their nations against each other.

The frequent collaboration between academicians in different nations did not diminish the role of scientific knowledge as a potent political and military weapon, and scientific voyages were more often than not undertaken by the navies themselves. The circumnavigations of sailor-scientists such as Louis Antoine de Bougainville, James Cook, and Alessandro Malaspina were carried out by, respectively, France, Britain, and Spain to expand each nation's own global cartographic knowledge. Even William Bligh's famous voyage on the naval ship *Bounty* was for the express purpose of carrying out botanical experiments with breadfruit trees.

Spain took a special interest in the natural treasures of South America, but it now turned to France for help in mounting an expedition. The books and memoirs of La Condamine and Ulloa, replete with scientific descriptions of cinchona, rubber, cinnamon, and dozens of other plants and herbs, were the driving force behind several botanical expeditions under the Spanish crown to carefully study and bring back samples of nature that could improve both science and commerce. The Royal Botanical Expedition to Chile and Peru (1777–1788), led by Hipólito Ruiz and José Pavón, included a French doctor Joseph Dombey (a disciple of the elder Jussieu brothers), whom the French court had sent in the hope that he would fulfill the botanical mission that Joseph de Jussieu had attempted but ultimately failed to complete. The expedition was a resounding success; over the course of a decade the men sent back hundreds of samples and illustrations to the botanical gardens in Madrid and Paris. The cinchona tree remained of particular interest, and Dombey, in addition to describing the trees in their natural state and analyzing the chemistry of the bark, also recovered the manuscripts on cinchona that Joseph de Jussieu had left behind in Lima and that had miraculously escaped destruction.[3]

Spain's Royal Botanical Expedition to Chile and Peru was soon followed by another botanical expedition through modern-day Colombia. This mission was conducted over a twenty-five-year period starting in 1783 by the Spanish doctor José Celestino Mutis, then living in Bogotá, who also founded the first observatory in South America. Mutis was later joined at the observatory by Francisco José de Caldas, a young lawyer inspired to become a naturalist by the accounts of La Condamine and his "companion" Bouguer. Even at that early date, it was La Condamine's name that was invoked when referring to the Geodesic Mission to the Equator.[4]

The most influential scientific expedition in South America was conducted not by a Spanish scientist but by a wayward German bureaucrat named Alexander von Humboldt. He had literally missed the boat for two different expeditions before finding himself in the company of a French doctor and botanist, Aimée Bonpland, on a five-year journey through Latin America. This voyage became the basis for groundbreaking works in which Humboldt attempted to synthesize disciplines such as meteorology, biology, and geography into a single, scientific worldview. Humboldt and Bonpland were keenly aware of the latest European discoveries and thinking on the natural sciences including biology, botany, geography, and meteorology. They brought this knowledge to bear on the still largely unknown world south of the equator.

Humboldt's voyage contained explicit echoes of the Geodesic Mission. In 1801, after making important discoveries in the Amazon region, Humboldt and Bonpland arrived in Bogotá, where they worked on botany with Mutis and Caldas for several months before traveling to Quito in the company of Caldas. Humboldt, like Caldas, was smitten with the accounts of the Geodesic Mission and constantly compared his own scientific observations with those of La Condamine. He made several detours specifically to visit the important sites of the Geodesic Mission, such as Yaruquí and Pichincha, which he considered to be "hallowed ground."[5]

More than the latest European discoveries, Humboldt brought some important new ideas and convictions to his South American colleagues. The American War of Independence and the French Revolution had spawned political unrest everywhere, and Humboldt was a font of both knowledge of and support for the liberal ideas of independence and freedom from royal tyranny. Yet it would be far too simple to state that Humboldt was the catalyst for the nascent independence movements in South America. That process had already begun, with the awakening of a sense—inspired in part by the Geodesic Mission—that South America was a distinct place on Earth with its own unique identity, not simply the El Dorado that provided Spain with its fantastic wealth.

Some of the first stirrings of South America's unique identity were the direct result of the Geodesic Mission. The French astronomers galvanized many of the locals whom they met, by examples both positive and negative. The Jesuits of Quito, who had first lodged La Condamine, emulated the French Academy of Sciences by creating the Academia Pichinchense, which encouraged scientific learning and, most famously, rebuilt the deteriorated sundial that La Condamine had constructed in their church courtyard. Maldonado's celebrated map of the *audiencia* of Quito, completed posthumously, was the first to accurately reflect the geography of the region, leading some historians to label him "Quito's first scientist." Quito's former president, Dionisio Alsedo y Herrera, was motivated by the French mission to write the first wide-ranging histories of the viceroyalty. Juan de Velasco, who as a boy witnessed firsthand La Condamine's attempts to portray Indians as subhuman, later wrote a history of the *audiencia* in an attempt to refute those very misconceptions.[6]

The currents of independence in South America were not simply notional, and it was not merely words that were being thrown about. The uprisings of 1739 in Cuenca and 1764 in Riobamba, followed by a major revolt in Quito in 1765, marked the beginnings of patriotic revolutions against Spanish domination. When Humboldt, Bonpland, and Caldas arrived in Quito at the beginning of 1802, they found a flourishing community of liberal-minded reformers, the Patriotic Society of the Friends of the Nation, which had been founded in 1791 by the doctor-turned-polemicist Eugenio Espejo. This Society, like dozens of similar ones around Spanish America, was originally intended to spur economic growth, but now it was fast becoming the center of a separatist movement, promoting emancipation of the colonies from Spain and the establishment of republican governments in each of them. Among its members was Luis Hugo, a pastor in Riobamba and son of Théodore Hugo, the Geodesic Mission's instrument maker.[7]

Humboldt arrived in Quito at a key time, when the sentiments that had given rise to numerous small rebellions were being distilled into a single, organized movement. Humboldt and his colleagues stayed in Quito for six months, in the home of Juan Pío Montúfar, a cofounder of the Society and its intellectual engine. Their conversations would have turned on the key ideas manifested in the recent American and French revolutions, and the men would have compared these concepts with the separatist movement now gaining strength in Quito. Humboldt soon departed with Juan's son Carlos Montúfar, who accompanied him on the rest of his travels, but the revolutionary ideas remained and matured in Quito. On August 10, 1809, a group of patriots met to overthrow the Spanish monarchy and organize a new government, with Juan Pío Montúfar as its president. Their declaration of independence and, later, their constitution were modeled on the founding documents of the United States of America. For three years they attempted to ignite popular revolts against the monarchy to declare Quito's independence, but each time they were crushed by royalist forces.

The homegrown separatist movement in Quito would have to wait for professional revolutionaries to throw off the shackles of imperial power. By 1813 Simón Bolívar had begun his campaign to liberate New Granada, while in the south José de San Martin successfully routed Spanish forces in San Lorenzo, Santiago, and Lima, eventually carving out the new nations of Argentina, Chile, and Peru. The fighting dragged on for a decade, until the final battles in May 1822 on the slopes of Mount Pichincha freed Quito and put an end to Spanish rule in New Granada. It was shortly after that battle and the liberation of Quito that Bolívar, on his way to Guayaquil and a strategic conference with San Martin, ascended Mount Chimborazo, "in the footsteps of La Condamine and Humboldt," on his journey to create the nation of Gran Colombia and drive the remaining Spanish armies out of Peru.

For eight years after the Battle of Pichincha, the old *audiencia* of Quito was a part of Gran Colombia, but even at the beginning there were signs of unease with this arrangement, and a sense that the old names of "Quito" and "Colombia" no longer served. A new, modern name was required, one that both located the nation geographically and distinguished it from its previous colonial masters. The term "equator" had frequently been used by La Condamine to describe not simply the line, but also the region in which the Geodesic Mission made its observations. It was this term the people of the region now employed to describe their homeland—not "Quito" but "Ecuador." By 1821, Bolívar himself was referring to the residents of Quito as "*los hijos del Ecuador*," meaning "the children of the Equator." The declaration of independence, signed in July 1822 to affirm the old *audiencia*'s integration into Gran Colombia, used the term "Ecuador" to denote the region, and the 1824 constitution referred to the "State of Ecuador in Gran Colombia."[8]

The birth of any nation is tumultuous, and that of Ecuador was no different. In 1828 the Peruvian army invaded the region but was repulsed at the Battle of Tarqui, within sight of the old observatory of the French astronomers. By that time, internal rivalries were fracturing Gran Colombia, with Venezuela splitting off in 1829. A second new nation emerged from the remains of Gran Colombia on May 30, 1830, declaring its independence under the identity of La República del Ecuador, the Republic of the Equator, its name a testimony to the Geodesic Mission.

Two years after its independence, Ecuador's president, Juan José Flores, annexed the Galapagos Islands to his country, calling them the Archipelago of Ecuador. In September 1835 a small British warship, *HMS Beagle*, toured the archipelago for a month before continuing its round-the-world scientific survey. Aboard was a young naturalist named Charles Darwin, who had developed a "burning zeal" to explore nature in its original state after reading Humboldt's narratives of his South American journeys.[9] Exactly a century after the Geodesic Mission set out to change our knowledge of the Earth, Darwin's voyage to the equator—spurred, to a degree, by the seminal influence of Bouguer, La Condamine, and their cohorts—would revolutionize our understanding of humankind.

The Geodesic Mission to the Equator allowed us to understand our planet as never before, but it was only the beginning of a body of research that continues to this day. The expedition inaugurated an effort to determine the figure of the Earth even more precisely and to pinpoint exact positions on its sphere—an ongoing scientific odyssey that now involves every nation on the globe. Beginning with the first results from Peru, Lapland, and France, geodesic measurements from other nations were continually integrated to form an ever-more-accurate portrait of the planet. Over the next half-century, geodesic observations from La Caille in southern Africa, Mason and Dixon in North America, and others showed that Bouguer's estimate of Earth's flattening (1/179) was too great and that it was actually closer to 1/298. In 1784 French and British astronomers laid out a geodesic baseline and took surveys across and on both sides of the English Channel in order to accurately link the positions of the London and Paris observatories, allowing them to directly incorporate each other's results in their own calculations.[10]

One of the most important steps in scientists' efforts to integrate their studies of the earth's shape was to create a uniform set of measures (length, weight, etc.) that could be used internationally. In 1790, the French National Constituent Assembly—newly elected after the revolution began the year before—authorized the Academy of Sciences to develop a scientifically based system of measures to supplant those of the old regime. La Condamine's original proposal based on the length of a seconds pendulum was rejected in favor of a "meter," which measured one-ten-millionth of the distance between the North Pole and the equator.

In order to facilitate this international metric system, yet another geodesic expedition would have to be organized. Two astronomers were charged with conducting a new, more accurate French survey, from Dunkirk to Barcelona, to supersede the older ones done by the Cassinis. When combined with other results from around the world (including those of Bouguer and Maupertuis), the exact curve of the Earth could be determined and a new unit of measure calculated. The survey, which began in 1792, as France was being plunged into war and chaos, ultimately took seven years to complete, just as long as the Geodesic Mission to the Equator required on considerably more difficult terrain. The meter was confirmed by an international commission in 1799, and over the next two centuries the metric system was adopted by almost every nation on the globe.[11]

Geodesic surveys also became more widespread and more integrated internationally, as the triangulations conducted in one country were slowly linked to others to build ever-more-accurate maps of the Earth. Two of the most famous surveys were conducted during the early 1800s: the Great Survey of India under George Everest (which first established the Himalayas as the highest mountains in the world) and the Struve Arc from Norway to the Black Sea, which among other feats remeasured the Arctic Circle arc with far greater accuracy than Maupertuis did.[12]

In 1864 a new scientific body, the International Geodesic Association, was formed to coordinate and standardize geodesic surveys around the globe. After repeated calls within the Association to remeasure the equatorial arc with more modern precision instruments, in 1899 the French Academy of Sciences and the government of Ecuador agreed to a new Geodesic Mission to the Equator.

The second Geodesic Mission was led by army officer Georges Perrier and included the young anthropologist Paul Rivet, who later established the Museum of Man in Paris. The team was assisted by the Ecuadorian army and equipped with the latest instruments, including the theodolite, which can simultaneously measure angles in both the vertical and the horizontal planes—unlike the quadrant used by the first mission, an instrument that could only measure horizontal angles.

Arriving in Ecuador at the beginning of the twentieth century, just as the Wright brothers were preparing for their first flight, the expedition found itself in exactly the same predicaments as its predecessors two hundred years earlier. The Ecuadorian railroads were not yet built and automobiles not yet available, so the scientists employed mules and Indian porters to carry their tons of equipment to the surveying sites. The modern theodolites were frequently rendered useless, with mountain survey stations obscured for months on end by fog, clouds, and rain. Violent winds and snow squalls buffeted the camps and ripped up the scientists' tents. Local Indians made off with their signals, certain—just as locals had been generations before them—that the expedition was really in search of gold, forcing the teams to remeasure hundreds of angles. Expedition members were afflicted with yellow fever and other diseases. Perrier's survey ultimately required seven grueling years to complete, the same as the first Geodesic Mission to the Equator, for an arc of latitude not much longer than the original laid out by Godin, Bouguer, and La Condamine.[13]

Few physical relics of the original Geodesic Mission survive until the present day. One of these is the *toise* of Peru, which was brought back to France by Godin and is now at the Paris Observatory. Another is the original 1742 watercolor by Morainville showing the pyramids and the plateau at Yaruquí, a painting currently exhibited in the Museum of the Banco Central del Ecuador in Quito; Morainville's watercolors of plants and birds are in the National Museum of Natural History in Paris. The Graham pendulum clock that La Condamine sold to Domingo Terol was later bought by Francisco José de Caldas and brought to his Bogotá observatory. A traveling French chemist, Jean-Baptiste Boussingault, found it in a state of disrepair when he visited in 1823, but the instrument has since vanished.[14]

Two of the three marble tablets that La Condamine had fabricated to commemorate the endpoints of the triangulation survive. The one for the Yaruquí baseline remained affixed for many years to the courtyard wall of the Jesuit seminary in Quito, as was noted by numerous visitors, including Humboldt. This tablet is now at the Quito observatory, mounted on the stairwell at the base of the main dome. The tablet of Tarqui was found after Caldas read an account of it in the newspaper *Mercurio Peruano* in 1804; his rage over its "miserable and barbaric treatment"—it was being used as a footbridge over a ditch—"made me decide to seize it and take it to Bogotá," where he displayed it in his observatory. The Tarqui tablet was brought back to nearby Cuenca in 1886, and though it was presumed lost for several decades, it is now on display at the city's Museo Municipal Remigio Crespo Toral. There is currently no trace of the tablet at Quinche.[15]

Just as most of the relics of the Geodesic Mission have vanished, the expedition's traces in the population and the landscape have all but disappeared. Most of the expedition members died without issue, or their lineage was cut short. Antonio de Ulloa is the only one known to have modern-day descendants, via the Spanish family titled the Marqueses de Torre-Milanos.

A few landmarks in Ecuador still evoke the mission's presence. Two separate mountains, Pambamarca near Quito (a survey station) and Puguín near the Tarqui baseline, are known by the mixed Spanish/ Quechua name Francés-Urcu, or "French mountain." Various haciendas in which the expedition members stayed and worked—San Augustín de Callo near Cotopaxi, Guachalá in Cayambe, Pueblo Viejo in Mira—remain much as they were in colonial days. Other sites have changed over time. In Quito, one of the houses occupied by the mission members (at the corner of *calles* Benalcázar and Manabí) was known as the "Observatory of Santa Barbara" until a new building was erected on the site in the 1930s.[16] Plaza San Sebastián in Cuenca, the scene of the deadly riot against Seniergues, is now a beautiful park with trees and a fountain.

The most emblematic artifacts from the Geodesic Mission, the pyramids marking the ends of the baseline at Oyambaro and Caraburo, gradually fell apart over time. Just one of the inscribed stone tablets now remains, and its contentious inscription—which almost brought an end to the Geodesic Mission—has long since been erased. In 1751, when La Condamine wrote his acid-penned version of the story of the pyramids, he led off with the epigram *Etiam perire ruinae* ("even the ruins were destroyed"), believing that his pyramids had already been torn down by small-minded Spanish authorities. This was not the case. The original 1746 order to destroy the pyramids—in response to La Condamine's defiance of the government's order regarding the inscriptions—had quickly been rescinded before it could be carried out, and it was replaced with another order to erase the offending inscriptions but to leave the pyramids intact. These orders were carried out in October 1747 by the aptly named magistrate Francisco Javier de Piedrahita ("Stone Marker"), who directed a mason named Juan Chahuagna to chisel out the inscriptions and carefully remove the fleur-de-lys atop each pyramid.[17]

Within half a century the pyramids were almost unrecognizable. When Humboldt visited the "hallowed ground" of Yaruquí in March 1802, he found the brickwork scattered over the ground, though the foundations were intact and the circular millstones marking the exact ends of the baseline were still in place. (Interestingly, he did not hesitate to desecrate the shrine by stealing a piece of the fleur-de-lys.) He found the inscribed tablets at the hacienda in Oyambaro. The northern tablet (from the Caraburo pyramid) was broken in three pieces, but the southern one (from Oyambaro) was intact though barely legible, lying in the courtyard where horses regularly trampled it. Twenty years later the chemist Boussingault saw the same stone, by then elevated and used as a mounting block for mules.[18]

The idea of rebuilding the pyramids was first mooted by Juan Pío Montúfar and was enthusiastically endorsed by Humboldt, but the subsequent independence movement derailed it. Simón Bolívar himself had considered rebuilding the pyramids in 1829, but he was too enmeshed in the revolution to give it more than a passing thought. Several years after the Republic of Ecuador was founded, its second president, Vincente Rocafuerte, decided to erect a new pair of pyramids for the centennial of the completion of the baseline. France was one of the first nations to recognize Ecuador as a new nation, and its ambassador Jean-Baptiste-Washington de Mendeville became one of the prime movers for the pyramid restoration project, placing a French naval engineer, Jean-Hippolyte Soulin, at the Ecuadorian government's disposal.

On November 25, 1836, President Rocafuerte ceremoniously laid the cornerstone for the restoration project. By then the remains of the old pyramids were virtually gone. Soulin carefully excavated around what was left of the foundations, closely following the measurements and alignments of the monuments recorded in La Condamine's journals. Eight months later he finished the project, ruefully noting that since the millstones had previously been disturbed, the pyramids could only serve as historical, not geodesic, markers. Indeed, for this reason Georges Perrier decided against using the pyramids in 1899 for his own baseline. Today, visitors can climb the small hill to see the replica of the Oyambaro pyramid, overlooking the dozens of greenhouses that make Ecuador one of the world's leading exporters of flowers. The Caraburo replica pyramid stands on the property of the new Quito International Airport, half a mile from the runway on the edge of the ravine, clearly visible to airline passengers as they take off and land.

Reconstructed Pyramid at Oyambaro (built in 1836). Author's photo.

The single remaining tablet from the two pyramids, the one that La Condamine had mounted on the pyramid at Oyambaro, remained at the nearby hacienda for years. It continued to serve as a mounting block, as the British mountaineer Edward Whymper noted in 1880, when he passed through the region. Around the turn of the twentieth century, the tablet was inexplicably moved from Oyambaro to the hacienda Guachalá, under the care of the hacienda owner Neptalí Bonifaz. In 1913 the short-lived Franco-Ecuadorian Committee moved the tablet to its present location, in the garden of the Quito Observatory in the Parque de la Alameda.[19] What text was not chiseled off the tablet is now worn almost smooth, the words "*meta australis*" or "southern marker" barely discernable at the base of the stone. It now stands prominently at the entrance to the observatory, with a copy of the original text engraved on a clear protective covering.

Today, the best-known and most visited monument to the Geodesic Mission to the Equator has nothing to do with the original expedition. The first Mitad del Mundo ("Middle of the World") monument was a thirty-foot stone tower erected in 1936 under the direction of the great Ecuadorian geographer Luis Tufiño. He placed it in the town of San Antonio de Pichincha, for the simple reason that it is where the equator

The Oyambaro Tablet at the Quito Observatory. Credit: Observatorio Astronómico de Quito.

passes closest to Quito. A careful historian, Tufiño never made any claims that the expedition members had traveled to the region to determine the location of the equator, as many of his countrymen had insisted. Instead, he inscribed the sides of the monument with dedications to the three nations—Ecuador, France, and Spain—and to the scientific achievements of the men who "in measuring the arc of the meridian at the equator, described the shape of the Earth."

In 1979 the provincial government of Pichincha undertook a project to greatly expand the Mitad del Mundo into a major tourist complex, including museums and a planetarium. The old tower was disassembled and moved four miles west to the town of Calacalí. In its place rises a massive one-hundred-foot stone tower, positioned directly on the equator, where tourists can now straddle the line between north and south. Along the main avenue of the complex stand two rows of busts that represent each of the men on the Geodesic Mission.[20] The statues are facing south, gazing fixedly along the avenue of volcanoes that formed the route of the great survey that the original members first undertook back in 1736.

❧

Each day at first light, long before the sun peered over the eastern ridge of the Andes, the men and women were at work in the open fields, adjusting their instruments to make the delicate measurements for the new project now under way. There, on the plateau of Yaruquí at the dawn of the twenty-first century, the state-of-the art Quito International Airport was being built astride the old baseline laid out by Godin, Bouguer, and La Condamine almost three hundred years earlier. Instead of measuring rods, the surveyors for the runway at Yaruquí used electronic theodolites and a global positioning system (GPS). The project was a long time coming; the original Quito airport was now hemmed in by the expanding city, and a new facility had been desperately needed for years.[21]

It was perhaps inevitable that the new airport would be built atop the Geodesic Mission baseline, for the simple reason that geodesic baselines as well as airports require long, level, flat terrain. The baseline at Yaruquí was certainly not the first to be appropriated for such a purpose; the geodesic baseline at Hounslow Heath, which was used to connect the observatories of London and Paris, was itself later covered by Heathrow Airport.

The Yaruquí baseline and the airport that has risen atop it are both manifestations of humankind's drive to better understand the planet on which we live. The GPS used by today's surveyors, as well as the navigation systems aboard the aircraft, are all based on knowing the precise orbits of a constellation of satellites. This knowledge in turn depends on highly accurate models of the figure of the Earth and its gravitational makeup. These models were begun on that very site by the Geodesic Mission, which gave the world a precise calculation of the length of a degree of latitude, as well as Bouguer's revelation of gravitational anomalies at Chimborazo.

Underneath the scientific history and technological advancements, a lingering human connection can be felt between the site at Yaruquí and the events that transpired there in the receding past. Standing on the grounds of the airport, one can see the perfect snow-capped cone of Cotopaxi rising far in the distance, just as it did when the members of the Geodesic Mission stood on that very site to lay out the baseline. Yet the mountain seems much closer now, as does the rest of the world, because of what those men did three centuries ago.

AFTERWORD

The Geodesic Mission to the Equator was, by any measure, one of the most famous and widely published tales of science, travel, and adventure during the eighteenth and early nineteenth centuries. The names of the principle characters—Pierre Bouguer, Charles Marie de La Condamine, Jorge Juan, and Antonio de Ulloa—were recognized by every educated person in Europe and the Americas. But by the twentieth century, those names and that expedition had fallen off the map. Even in France, the country of origin, or Ecuador, where it all took place, you would today be hard pressed to find an otherwise well-read scholar who knew that such an expedition ever occurred.

Why has the story of the Geodesic Mission been forgotten? The simplest explanation is that the expedition was a victim of its own success; by opening Europe's eyes to the New World, the mission paved the way for luminaries like Humboldt and Darwin to make their own voyages of scientific discovery to South America, and these scientists' accounts eclipsed those of mission, consigning it to relative obscurity.

It did not help that the Geodesic Mission had never been seriously analyzed until the late twentieth century. Most of the histories of the expedition were derived from the popular accounts of La Condamine and Ulloa—memoirs that more often than not were biased and incomplete. In the 1870s, Jules Maillard de La Gournerie, a French engineer and mathematician, began collecting original letters and memoirs from both French and Spanish archives, with the intention of writing a complete history of the expedition, but he died before completing any more of his project than making various notes. That task was undertaken in 1955 by Robert Finn Erickson, who relied on La Gournerie's unedited notes to create a magisterial doctoral thesis on the expedition—a work that remains, alas, unpublished.[1]

The modern public was reintroduced to the expedition by the 1945 book *South America Called Them* by explorer and anthropologist Victor Wolfgang von Hagen. It was a classic ripping yarn, full of adventure and fascinating portraits, but also replete with egregious errors and entire events fabricated from whole cloth, which have since been propagated in dozens of other books.[2] In 1979, Florence Trystram's riveting novel of the expedition, *Le Procès des Étoiles* (The Trial of the Stars), appeared, later translated into Spanish but not yet into English.[3]

In 1987, a well-researched and eminently readable study of the mission, *Los Caballeros de Punto Fijo* (The Gentlemen of the Fixed Point) by Antonio Lafuente and Antonio Mazuecos, was published. Patrick Drevet's haunting novel *Le Corps du Monde* (The Body of the World), about the wanderings of Joseph de Jussieu, appeared in 1997. More recently, Robert Whitaker's *The Mapmaker's Wife* (2004) dramatically tells the story of Isabel Godin's survival in the Amazon, set against the backdrop of the Geodesic Mission. Other works by James R. Smith (*From Plane to Spheroid*, 1986) and Michael Hoare (*The Quest for the True Figure of the Earth*, 2005) convincingly place the mission in the larger historical odyssey to determine the Earth's shape. Several academic conferences and museum exhibitions have also drawn international attention to the mission.[4]

I came across the Geodesic Mission almost twenty years ago, while researching *Ships and Science*, my book on the history of naval architecture, a discipline that Pierre Bouguer created single-handedly during his idle moments in the mountains of Peru. The epic saga gripped me almost immediately, and for two decades now it has refused to let me go. I found other small but dynamic groups of individuals who were also fascinated by this little-known expedition. I was fortunate enough to work with a dedicated team at the BBC that produced a television documentary on the Geodesic Mission for the series *Voyages of Discovery*, which gave me a wonderful opportunity to view the expedition as a bold adventure as well as a grand scientific achievement.[5]

My early research into the available material brought the realization that there were many gaps and inconsistencies in the various accounts; a complete history of the Geodesic Mission was still waiting to be written. In order to do justice to the story, I sought out the original letters and memoirs of the mission's members and their contemporaries, in order to understand how the members thought and felt at the time—and not merely in published retrospect—and more importantly, to view the mission through the eyes of the local populace. I also determined to set the science of the expedition firmly in the political environment of the era and to highlight the impact that the mission had both on the European view of South America and on the self-awakening of South Americans themselves.

Documents were often hard to come by. Although I explored archives and libraries in France, Spain, Ecuador, Peru, Chile, Britain, and the United States, I discovered that much of the original material has been lost. For example, Peru's National Library has suffered several disasters, including fires (some original documents were badly burned and barely legible) and the removal of books and manuscripts to Chile as spoils of war after the War of the Pacific (1879–1884). In other cases, important documents found in European estates and libraries were sold into private hands at auction, as early as 1844 and as late as 1991, placing them out of reach of researchers.[6]

To complete this work I have drawn on several previously untapped archives, notably those in Ecuador and Chile. Especially critical were those archives concerning the Seniergues affair. All previous histories of the events in Cuenca have relied upon La Condamine's skewed and ultimately self-serving account. Using actual court testimonies, I have reconstructed, for the first time, the dramatic events surrounding the surgeon's assassination.

As part of my research, I crisscrossed the sites where the expedition set foot. I have met, spoken, and corresponded with many authors and experts on the various disciplines that the mission touched on. I owe them my thanks. Sharp-eyed readers will note that I pay homage to some of them in the chapter headings: "Degree of Difficulty" is from an article by David A. Taylor,[7] "The Dance of the Stars" is from *Los Caballeros del Punto Fijo* by Lafuente and Mazuecos, and "The Impossible Return" is from Florence Trystram's *Le Procès des Étoiles*.

A NOTE ON LANGUAGE

I have consciously used several terms that improve readability but are nevertheless somewhat inaccurate. At the top of the list, the term "geodesic mission" is not quite right. The precise term today would be "geodetic mission," but the original works in French and Spanish referred to it, respectively, as *mission géodésique* and *misión geodésica*, and the phrase "geodesic mission" is far more commonly used in English-language histories to describe the expedition. The official title of the French Academy of Sciences was actually the Royal Academy of Sciences of Paris. I frequently refer to the men of the Academy as scientists, but they called themselves savants, academicians, and astronomers, since the term "scientist" only came into popular use in the late 1800s. I refer to the region where they operated as Peru; that was its name at the time, and more importantly, it was the notoriety of the Geodesic Mission that led to the nation becoming known as Ecuador. Finally, I refer to the indigenous people of South America as Indian for the same reason Charles Mann explains in his preface to *1491*: They describe themselves as Indian.[1]

UNITS OF MEASURE AND CURRENCY

Just as I have translated French and Spanish into the English language, I have also translated units of length and currency into their modern American equivalent. That simply makes it easier for the reader to intuitively grasp what was happening, instead of performing mental gymnastics to get into the mind-set of the era.

Length

Ligne = 0.09 inches
Pouce (inch) = 12 *lignes* =1.07 inches
Pied (foot) = 12 *pouces* =1.07 feet
Toise (fathom) = 6 *pieds* = 6.39 feet
Pulgada (Spanish inch) = 0.9 inches
Vara (Spanish yard) = 2.75 feet

Angular Measures

1° = one degree = 1/360 circle = 90 feet seen at a mile's distance
1' = one minute = 1/60 degree = 18 inches seen at a mile's distance
1" = one second = 1/3600 degree = 1/3 inch seen at a mile's distance

Currency

It is difficult to convert the money of three centuries ago to present values; not only were the commodities different (e.g., horses versus cars), but the proportions of salary spent on, say, housing and food are poles apart. However, economists have developed estimates of inflation that permit a rough comparison of currencies across time. The principal currencies referred to in this book are the French livre, the Spanish peso, and the British pound. Using several sources,[1] I have derived the following equivalent measures for converting those currencies in the year 1740 (the midpoint of the expedition timeline) to the American dollar of the year 2010:

1 livre = $9

1 ecu = 3 livres = $27

1 peso = $45

1 escudo = 2 pesos = $90

1 *doblón* = 4 pesos = $180

1 pound = $200

ACKNOWLEDGMENTS

First and foremost, to my agent, Michelle Tessler, who first saw the possibilities in the rough, unalloyed material; my editor Lara Heimert at Basic Books, who forged it into proper shape; and Alex Littlefield, associate editor at Basic, who brought it to a fine temper.

Special thanks to Robert Whitaker, author of *The Mapmaker's Wife*, who literally and figuratively journeyed many of the same paths as I did and helped guide me over them.

More people than I can mention assisted in my research in libraries, archives, and museums around the world. But I must personally thank a few whose help was indispensable in creating this work, listed here alphabetically and by nation.

BRITAIN: Annabel Gillings, Michael Rand Hoare, Nicola Lees, Paul Rose, Caroline Sellon, James R. Smith

ECUADOR: Diego Brito, Cristóbal Cobo, José Luis Espinoza, Luis Gallegos, Nelson Gilberto Gómez Espinosa, Lorenzo Saa Bernstein, María Antonieta Vásquez Hahn

FRANCE: Laurence Bobis, Jean-Francois Caraës, René and Ghislaine Chesnais, Suzanne Débarbat, Patrick Drevet, Danielle Fauque, Florence Greffe, Pascale Heurtel, Alexandre Sheldon-Duplaix, Florence Trystram

PERU: Raúl Hernández Asensio, Jorgé Ortiz-Sotelo, Eliecer Vilchez Ortega

SPAIN: Jorge Juan Guillén Salvetti, Antonio Lafuente García

UNITED STATES: Tamar Herzog, Jay Menaker, Elaine Protzman, David A. Taylor, Mary Terrall

Although I'm beholden to many for their help, I alone am responsible for any errors in fact, translation, or interpretation.

NOTES

Abbreviations Used in the Notes

AARS: Archives de l'Académie Royale des Sciences (Archives of the Royal Academy of Sciences), Paris, France

ADC: Archives Départementales du Cher (Departmental Archives of Cher), Bourges, France

ADLA: Archives Départementales de Loire-Atlantique (Departmental Archives of Loire-Atlantique), Nantes, France

AGI: Archivo General de Indias (General Archive of the Indies), Seville, Spain

ANEC: Archivo Nacional del Ecuador (National Archive of Ecuador), Quito, Ecuador

ANF: Archives Nationales de France (National Archives of France), Paris, France

AOP: Archives de l'Observatoire de Paris (Archives of the Observatory of Paris), Paris, France

BCIU: Bibliothèque Communautaire et Interuniversitaire (Community and Inter-University Library), Clermont-Ferrand, France

BIF: Bibliothèque de l'Institute de France (Library of the Institute of France), Paris, France

BNCBA: Biblioteca Nacional de Chile, Sala Medina, fondos Barros Ana (National Library of Chile, Medina Hall, Barros Ana Collection), Santiago, Chile

BNE: Biblioteca Nacional de España (National Library of Spain), Madrid, Spain

BNF: Bibliothèque Nationale de France (National Library of France), Paris, France

BNP: Biblioteca Nacional del Perú (National Library of Peru), Lima, Peru

DHAQ: *Documentos para la Historia de la Audiencia de Quito* (Documents for the History of the Audiencia of Quito), ed. José Rumazo, 8 vols. (Madrid: Afrodisio Aguado, 1948–1950)

DSB: *Dictionary of Scientific Biography*, ed. Charles Coulston Gillispie, 16 vols. (New York: Charles Scribner's Sons, 1970–1980)

HMAB: Histoire et Mémoires de l'Académie Royale des Sciences et des Belles Lettres de Berlin (History and Memoirs of the Royal Academy of Sciences and Literature of Berlin), Berlin, Prussia (Germany)
HMARS: Histoire et Mémoires de l'Académie Royale des Sciences de Paris (History and Memoirs of the Royal Academy of Sciences of Paris), Paris, France
MNHN: Muséum National d'Histoire Naturelle (National Museum of Natural History), Paris, France
PTRS: Philosophical Transactions of the Royal Society, London, Great Britain
PVARS: Procès-verbaux de l'Académie Royale des Sciences (Minutes of the Royal Academy of Sciences), Paris, France
RPRP: Recueils des Pieces qui ont Remporté les Prix de l'Académie Royale des Sciences (Collection of Works That Have Taken the Prizes of the Royal Academy of Sciences), Paris, France

Introduction The Baseline at Yaruquí

1. The descriptions of baseline operations at Yaruquí are from the published accounts of the members of the Geodesic Mission: Pierre Bouguer, "Relation abrégée du voyage fait au Pérou par Messieurs de l'Académie Royale des Sciences, pour mesurer les Degrés du Méridien aux environs de l'Equator, & en conclure la Figure de la Terre" (Abridged Relation of the Voyage Made to Peru by Gentlemen of the Royal Academy of Sciences, to Measure the Degrees of the Meridian Around the Equator, and to Deduce the Figure of the Earth), *HMARS 1744* (1748): 249–297 and *HMARS 1746* (1751): 569–606 (description on 279–282; Bouguer, *La Figure de la terre, déterminée par les observations de Messieurs Bouguer et de La Condamine, de l'Académie Royale des Sciences, envoyés par ordre du Roy au Pérou pour observer aux environs de l'équateur, avec une Relation abrégée de ce voyage qui contient la description du pays dans lequel les opérations ont été faites* (The Figure of the Earth, Determined by the Observations of the Messieurs Bouguer and La Condamine, of the Royal Academy of Sciences, Sent to Peru by Order of the King to Make Observations near the Equator, with an Abridged Relation of the Voyage Containing a Description of the Countries in Which the Observations Were Made) (Paris: Charles-Antoine Jombert, 1749), 37–44; Jorge Juan y Santacilia and Antonio de Ulloa y de la Torre-Guiral, *Relación histórica del Viaje á la América Meridional* (Historical Relation of the Voyage to South America), 2 vols. (Madrid: Antonio Marín, 1748; facsimile, ed. José Merino Navarro and Miguel Rodriguez San Vicente, Fundación universitaria española, 1978), 1:302–305; Jorge Juan and Ulloa, *Observaciones astronómicas y físicas hechas en los Reynos del Perú* (Astronomical and Physical Observations Made in the Kingdoms of Peru) (Madrid: Juan de Zúñiga, 1748), 144–155; Charles-Marie de La Condamine, *Journal du voyage fait à l'Equateur servant d'introduction historique à la Mesure des trois premiers degrés du Méridien* (Journal of the Voyage Made to the Equator, Serving as a Historical Introduction to the Measure of the First Three Degrees of the Meridian) (Paris: Imprimerie Royale, 1751), 19–20; La Condamine, *Mesure des trois premiers degrés du méridien dans l'hémisphère austral, tirée des observations de MM. de l'Académie royale des sciences envoyés par le roi sous l'Équateur* (Measure of the First Three Degrees of the Meridian in the Southern Hemisphere,

Drawn from the Observations of the Messieurs of the Royal Academy of Sciences, Sent by the King Below the Equator) (Paris: Imprimerie Royale, 1751), 80–85. See also Bouguer, *Justification des Mémoires de l'Académie royale des sciences de 1744, et du livre de la Figure de la terre déterminée par les observations faites au Pérou* (Justification of Memoirs of the Royal Academy of Sciences of 1744, and of the Book The Figure of the Earth Determined by Observations Made in Peru) (Paris: Charles-Antoine Jombert, 1752), 8; La Condamine, *Supplément au Journal historique du voyage a l'equateur, et au livre de la Mesure des trois premiers degrés du meridien* (Supplement to the Historical Journal of the Voyage to the Equator, and of the Book of the Measure of the First Three Degrees of the Meridian), 2 vols. (Paris: Durand and Pissot, 1752–1754), 1:36; Bouguer, *Lettre à M. *** dans laquelle on discute divers points d'astronomie pratique, et où l'on fait quelques remarques sur le supplément au Journal historique du voyage à l'équateur de M. de La C.* (Letter to Mr. *** in Which Is Discussed Diverse Points of Practical Astronomy, and Where Are Made Several Remarks on the Supplement of the Historical Journal of the Voyage to the Equator by Mr. de la Condamine) (Paris: Guérin et Delatour, 1754), 3.

Chapter I The Problem of the Earth's Shape

1. See Duane W. Roller, ed., *Eratosthenes' Geography* (Princeton, NJ: Princeton University Press, 2010), and J. L. Berggren and Alexander Jones, eds., *Ptolemy's Geography* (Princeton, NJ: Princeton University Press, 2001). Note that both Eratosthenes and Ptolemy used a figure of 252,000 *stade* for the circumference, but the exact length of a *stade* in those eras is still the subject of controversy.

2. Felipe Fernández-Armesto, *Pathfinders: A Global History of Exploration* (New York: Norton, 2006), 246.

3. Among the many histories of European maritime empires, several stand out: Jan Glete, *Navies and Nations: Warships, Navies and State Building in Europe and America 1500–1860*, 2 vols. (Stockholm: Academitryck AB Edsbruk, 1993); Glete, *Warfare at Sea, 1500–1650* (London: Routledge, 2000); John H. Elliot, *Empires of the Atlantic World: Britain and Spain in America, 1492–1830* (New Haven, CT: Yale University Press, 2006); and John R. McNeill, *Atlantic Empires of France and Spain: Louisbourg and Havana, 1700–1763* (Chapel Hill: University of North Carolina Press, 1985).

4. Bernard de Fontenelle, "Eloge de M. Abbé Gallois" (Eulogy for Abbot Gallois), *HMARS 1707* (1708): 178.

5. On the formation of the French Academy of Sciences, see especially Roger Hahn, *The Anatomy of a Scientific Institution: The Paris Academy of Sciences, 1666–1803* (Berkeley: University of California Press, 1971), 1–15, and David J. Sturdy, *Science and Social Status: The Members of the Académie des Sciences, 1666–1750* (Woodbridge, UK: Boydell Press, 1995), 63–79, 182.

6. Jean Richer, "Observations astronomiques et physiques faites en l'isle de Caïenne" (Astronomical and Physical Observations Made on the Island of Cayenne), *HMARS 1666–1699*, vol. 7: 1679 (1729): part 1, 231–326; John Olmsted, "The Scientific Expedition of Jean Richer to Cayenne (1672–1673)," *Isis* 24, no. 2:94 (Autumn 1942): 117–128.

7. More precisely, the seconds pendulum for Paris was 3 *pieds*, 8 3/5 *lignes*, and had to be shortened by 1 1/4 *lignes* in Cayenne to beat at exactly one-second intervals. On Huygens's theory of the pendulum in lower gravity, see John Greenberg, *The Problem of the Earth's Shape from Newton to Clairaut: The Rise of Mathematical Science in Eighteenth Century Paris and the Fall of "Normal" Science* (Cambridge, UK: Cambridge University Press, 1995), 16.

8. Giovanni-Domenico Cassini, "Les élémens de l'astronomie vérifiés par M. Cassini par le rapport de ses tables aux observations de M. Richer faites en l'isle de Caïenne" (Astronomical Elements Verified by Mr. Cassini in Relation to Tables of Observations of Mr. Richer Made on the Island of Cayenne), *HMARS 1666–1699*, vol. 8: 1684 (1730): 55–117.

9. Isaac Newton, *The Principia: Mathematical Principles of Natural Philosophy*, trans. I. Bernard Cohen and Anne Whitman (Berkeley: University of California Press, 1999).

10. I. Bernard Cohen, *Revolution in Science* (Cambridge, MA: Belknap Press, 1985), 127.

11. Cited in Eric J. Aiton, *The Vortex Theory of Planetary Motions* (London: MacDonald, 1972), 228.

12. See Greenberg, *The Problem of the Earth's Shape*, 1–14, for a succinct but mathematically precise description of Newton's explanation of Earth as a flattened sphere.

13. John Bennett Shank, *The Newton Wars and the Beginning of the French Enlightenment* (Chicago: University of Chicago Press, 2008), 114–115.

14. René Descartes, *Principles of Philosophy*, trans. Valentine Miller and Reese Miller (Boston: Kluwer, 1991). For a complete explanation of Descartes' vortex theories, see Aiton, *Vortex Theory*, and Danielle Fauque, "Tourbillons ou attractions; les physiciens du XVIIIe siècle entre un monde plein et un monde vide" (Vortices or Attractions: The Physicists of the 18th Century Between a Full World and an Empty World) in *De la nature: De la Physique Classique au Souci Écologique* (On Nature: From Classical Physics to Ecological Awareness), ed. Pierre Colin (Paris: Beauchesne, 1992), 205–235.

15. Bernard de Fontenelle, *Conversations on the Plurality of Worlds*, ed. Nina Gelbart, trans. Henry A. Hargreaves (Berkeley: University of California Press, 1990), 63.

16. James E. King, *Science and Rationalism in the Government of Louis XIV, 1661–1683* (Baltimore: Johns Hopkins University Press, 1949), 282–283; Christiane Aulaner, *Histoire du Palais et du Musée du Louvre* (History of the Palace and Museum of the Louvre) 9 vols. (Paris: Editions des Musées Nationaux, 1947–1971), 3: 51–53, 7:78; Sturdy, *Science and Social Status*, 283–285; Mary Terrall, *The Man Who Flattened the Earth: Maupertuis and the Sciences in the Enlightenment* (Chicago: University of Chicago Press, 2002), 29–31.

17. Giovanni-Domenico Cassini, "De la Méridienne de l'Observatoire Royal prolongée jusques aux Pyrenées" (Of the Meridian of the Royal Observatory Extended to the Pyrenees), *HMARS 1701* (1743): 171–184.

18. Bernard de Fontenelle, "Sur la Méridienne" (On the Meridian), *HMARS 1701* (1743): 96–97.

19. Jacques Cassini's final arc measurements were reported in his *De la grandeur et de la figure de la Terre* (Of the Size and Figure of the Earth), supplement to *HMARS 1718* (1720) (Paris: Imprimerie Royal, 1720). His historical examination of lengths of a degree of latitude was given in "De la Figure de la Terre" (Of the Figure of the Earth), *HMARS 1713* (1739): 187–198. Summarizing their decrease going from south to north:

Eratosthenes	230 BCE	Egypt	80 miles
Riccioli	1645	Italy	76 miles
Fernel, Picard	1525, 1670	France	69 miles
Norwood	1635	Britain	69 miles
Snellius	1615	Holland	67 miles

In-depth histories of determining the Earth's shape are given in Tom B. Jones, *The Figure of the Earth* (Lawrence, KS: Coronado Press, 1967); James R. Smith, *From Plane to Spheroid: Determining the Figure of the Earth from 3000 B.C. to the 18th Century Lapland and Peruvian Survey Expeditions* (Rancho Cordova, CA: Landmark Enterprises, 1986); and Michael R. Hoare, *The Quest for the True Figure of the Earth: Ideas and Explorations in Four Centuries of Geodesy* (London: Ashgate, 2005).

20. Aiton, *Vortex Theory*, 71–72, 84; Hoare, *The Quest for the True Figure of the Earth*, 41.

21. Greenberg, *The Problem of the Earth's Shape*, 51.

22. On the Anglo-French Alliance, see Jeremy Black, *The Collapse of the Anglo-French Alliance, 1727–1731* (New York: St Martin's Press, 1997). On the exchange of letters between the Royal Society and the French Academy, see Gavin de Beer, *The Sciences Were Never at War* (London: Thomas Nelson and Sons, 1960); and David J. Sturdy, "Scientific Exchange

in 'La République des Lettres': The Correspondence of Sir Hans Sloane and the Abbé Jean-Paul Bignon, 1709–1741," in *Franco-British Interactions in Science Since the Seventeenth Century*, ed. Robert Fox and Bernard Joly (London: King's College, 2010), 81–97.

23. Jean-Jacques d'Ourtous de Mairan, "Recherches géométriques sur la diminution des degrés terrestres, en allant de l'équateur vers les pôles: où l'on examine les conséquences qui en résultent, tant à l'égard de la figure de la Terre, que de la pesanteur des corps, de l'accourcissement du pendule" (Geometrical Researches on the Decrease in Terrestrial Degrees, Going from the Equator to the Poles, Examining the Consequences that Result Regarding the Figure of the Earth as Well as the Weight of the Bodies and the Shortening of the Pendulum), *HMARS 1720* (1722): 231–277; John Theophilus (Jean Théophile) Desaguliers, "A Dissertation Concerning the Figure of the Earth," *PTRS 1725*, no. 386 (Jan–Feb 1725): 201–222; no. 387 (Mar–Apr 1725): 239–255; no. 388 (May–June 1725): 277–304. Detailed analyses of the papers are given in Greenberg, *The Problem of the Earth's Shape*, 15–77.

24. Bernard de Fontenelle, "Sur l'inégalité des degrés de latitude terrestres, et sur celle du Pendule à secondes, ou sur la figure de la Terre" (On the Inequality of Terrestrial Degrees of Latitude, and on that of the Seconds Pendulum, or on the Figure of the Earth), *HMARS 1720* (1722): 77.

25. Voltaire's exile to London is described in Theodore Besterman, *Voltaire* (New York: Harcourt, Brace, 1969), 108–120.

26. Voltaire (François-Marie Arouet), *Letters Concerning the English Nation* (London: C. Davis and A. Lyon, 1733), 139–140. The French edition was published in Amsterdam in 1734 as *Lettres philosophiques* (Philosophical Letters). See also William Johnson and Srinivasan Chandrasekar, "Voltaire's Contribution to the Spread of Newtonianism: Part I, Letters from England: Les Lettres Philosophiques," *International Journal of Mechanical Sciences* 32, no. 5 (1990): 423–453.

27. Terrall, *The Man Who Flattened the Earth*, 41–47.

28. Pierre-Louis Moreau de Maupertuis, "De figuris quas fluida rotata induere possunt problemata duo; cum conjectura de stellis quae aliquando prodeunt vel deficieunt; et de annulo Saturni" (On Rotating Fluid Figures Which Present Two Problems; When Stars Grow and Diminish; and the Rings of Saturn), *PTRS 1732* (1733): 240–256; and *Discours sur les différentes figures des astres; d'où l'on tire des conjectures sur les étoiles qui paraissent changer de grandeur; et sur l'anneau de Saturne avec une exposition abbrégée avec une exposition des Systèmes de M. Descartes et M. Newton* (Discourse on the Different Figures of Celestial Bodies; Where One Draws Conjectures on Stars That Appear to Change Size; and on the Rings of Saturn with an Abbreviated Exposition of the systems of Mr. Descartes and Mr. Newton) (Paris: Imprimerie Royal, 1732). See also Robert Iliffe, "'Aplatisseur du monde et de Cassini': Maupertuis, Precision Measurement and the Shape of the Earth in the 1730s," *History of Science* 31 (1993): 335–375 (Maupertuis quote on 339); and Terrall, *The Man Who Flattened the Earth*, 53–71.

29. Terrall, *The Man Who Flattened the Earth*, 83.

30. See William Johnson and Srinivasan Chandrasekar, "Voltaire's Contribution to the Spread of Newtonianism," *International Journal of Mechanical Sciences* 32, no. 8 (1990): 423–453, 521–546; and David Bodanis, *Passionate Minds* (New York: Crown, 2006).

31. See Black, *The Collapse of the Anglo-French Alliance*. Newton elevated to level of Descartes: Bernard de Fontenelle, "Eloge de M. Neuton" (Eulogy of Mr. Newton), *HMARS 1727* (1729): 165.

32. Drop-off in correspondence between French and British scientists: Sturdy, "Scientific Exchange in 'La République des Lettres,'" 96. Quotes: Greenberg, *The Problem of the Earth's Shape*, 86–87.

33. Voltaire to La Condamine, Paris, June 22, 1734, in Voltaire, *The Complete Works of Voltaire*, ed. Theodore Besterman, et al., 135+ vols. (Toronto: University of Toronto Press, 1968–present), 87:37.

34. Johann Bernoulli, *Essai d'une nouvelle Physique céleste, Servant à expliquer les principaux Phénomènes du Ciel, & en particulier la cause physique de l'inclinaison des*

Orbites des Planètes par rapport au plan de l'équateur du Soleil (Essay on a New Celestial Physics to Explain the Principal Phenomena of the Heavens, and in Particular the Physical Cause of the Inclination of the Orbits of Planets with Respect to the Solar Equator) 1734 Prize, *RPRP*, vol. 3: 1734–1737 (1752); reprinted in *Johannis Bernoulli Opera Omnia* (Johann Bernoulli's Complete Works), 4 vols. (Geneva: Cramer, 1742), 3:261–364 (facsimile by Olms, Hildesheim 1968). On Bernoulli's proof, see Aiton, *Vortex Theory*, 228–235. Bernoulli had previously examined the problem of the drift of ships in 1714, when he became enmeshed in a very public dispute with a French naval officer over the proper way to analyze the course and speed of a sailing ship. A ship with an elongated shape blown obliquely by the wind would not move exactly in the direction of the wind, nor straight along the line of its keel, but rather at some resultant angle (called the drift) between the direction of the wind and the line of its keel. Bernoulli argued that an elongated Earth would do the same. See Larrie D. Ferreiro, *Ships and Science: The Birth of Naval Architecture in the Scientific Revolution, 1600–1800* (Cambridge, MA: MIT Press, 2007), 92–94.

35. For biographical information on Bouguer, see Jean-Paul Grandjean de Fouchy, "Eloge de M. Bouguer" (Eulogy of Mr. Bouguer), *HMARS 1760* (1766): 181–194; Roland Lamontagne, *La vie et l'oeuvre de Pierre Bouguer* (The Life and Work of Pierre Bouguer) (Montreal: Presses de l'Université de Montréal, 1964); Danielle Fauque, "Du bon usage de l'éloge: Cas de celui de Pierre Bouguer" (On the Correct Usage of the Eulogy: The Case of Pierre Bouguer), *Revue d'histoire des sciences* 54, no. 3 (2001): 351–382; René Chesnais, "Les trois Bouguer et Le Croisic" (The Three Bouguers and Le Croisic), in *Pierre Bouguer, un savant Breton au XVIIIe siècle* (Pierre Bouguer, a Breton Scholar in the 18th Century) (Vannes: Institut culturel de Bretagne, 2002), 9–31; and *DSB*, 2:343–344.

36. Pierre Bouguer, *Entretiens sur la cause de l'inclination des orbites des planètes* (Dialogues on the Cause of the Inclination of the Orbits of Planets) (Paris: Claude Jombert, 1734); and "Comparaison des deux Loix que la Terre & les autres Planètes doivent observer dans la figure que la pesanteur leur fait prendre" (Comparison of Two Laws that the Earth and Other Planets Must Observe in the Shape Caused by Their Weight), *HMARS* 1734 (1736): 21–40; Greenberg, *The Problem of the Earth's Shape*, 89–106, 108, 113.

37. Terrall, *The Man Who Flattened the Earth*, 83–84.

38. Biographical details of La Condamine are from Marie Jean Antoine Nicolas de Caritat, Marquis de Condorcet, "Eloge de M. de La Condamine" (Eulogy of M. de La Condamine), *HMARS 1774* (1778): 85–112, quote at 87; Achille Le Sueur, *La Condamine d'après ses papiers inédits* (La Condamine, from His Unpublished Papers) (Paris: Picard et fils, 1911); and *DSB*, 15:269–273.

39. Théophile-Imarigeon Duvernet, *La vie de Voltaire* (The Life of Voltaire) (Geneva, 1786; Paris: Hachette, 1978), 76; Jacques Donvez, *De quoi vivait Voltaire?* (What Did Voltaire Live On?) (Paris: Editions de Deux-Rives, 1949), 37–55.

40. La Condamine, "Observations mathématiques et physiques faites dans un Voyage de Levant en 1731 & 1732" (Mathematical and Physical Observations Made on a Voyage to the Levant in 1731 & 1732), *HMARS 1732* (1735): 295–322.

41. Biographical details for Louis Godin can be found in Jean-Paul Grandjean de Fouchy, "Eloge de M. Godin" (Eulogy of Mr. Godin), *HMARS 1760* (1766): 181–194, quotes on 181–182; AARS Biographical Dossier Louis Godin; and *DSB*, 5:434–436. The Royal College (Collège Royal, today the Collège de France) was an independent teaching and research institute in Paris, not part of the traditional French university system. See David J. Sturdy, *Science and Social Status: The Members of the Académie des Sciences, 1666–1750* (Woodbridge, UK: Boydell Press, 1995), 10.

42. Maupertuis, "Sur la Figure de la Terre, & sur les moyens que l'Astronomie & la Géographie fournissent pour la déterminer" (On the Figure of the Earth and on the Means that Astronomy and Geography Furnish to Determine It), *HMARS 1733* (1735): 153–164; Louis Godin, "Méthode pratique de tracer sur Terre un Parallèle par un degré de latitude donné" (Practical Method to Trace on the Earth a Parallel Along a Given Degree of Latitude), *HMARS 1733* (1735): 223–232; La Condamine, "Description d'un Instrument qui peut servir à déterminer, sur la surface de la Terre, tous les points d'un Cercle parallèle à l'Équa-

teur" (Description of an Instrument that Can Serve to Determine, on the Surface of the Earth, All the Points of a Circle Parallel to the Equator), *HMARS 1733* (1735): 294–301. See also Greenberg, *The Problem of the Earth's Shape*, 79–88, and Terrall, *The Man Who Flattened the Earth*, 90–94.

Chapter II Preparations for the Mission

1. Marie Jean Antoine Nicolas de Caritat, Marquis de Condorcet, "Eloge de M. de La Condamine" (Eulogy of M. de La Condamine), *HMARS 1774* (1778): 92. See also Elisabeth Badinter, *Les passions intellectuelles*, vol. 1, *Désirs de gloire, 1735–1751* (Intellectual Passions: Desires of Glory, 1735–1751) (Paris: Fayard, 1999), 72–73. My thanks to Mary Terrall for her observations on the politics of this dispute.

2. Jean-Paul Grandjean de Fouchy, "Eloge de M. Godin" (Eulogy of Mr. Godin), *HMARS 1760* (1766): 187. On Jean-Nicolas de Rarécourt de La Vallée, chevalier de Pimodan (his full name), see Alphonse Roserot, *Titres de la Maison de Rarécourt de La Vallée de Pimodan* (Titles of the House of Rarécourt de La Vallée de Pimodan) (Paris: Librairie Plon, 1903), 259. Florence Greffe of the French Academy of Sciences kindly provided me the location of the residences of Godin and Grandjean de Fouchy.

3. Grandjean de Fouchy, "Eloge de M. Godin," 187; *PVARS*, Dec. 23, 1733, 237. The minutes were not in Fontenelle's distinctive hand.

4. Quoted in Mary Terrall, *The Man Who Flattened the Earth: Maupertuis and the Sciences in the Enlightenment* (Chicago: University of Chicago Press, 2002), 93.

5. Royal Society of London Archives EL/G2/22, Letters from Louis Godin, Jan.–Feb. 1734.

6. Biographical information for Maurepas from Condorcet, "Eloge de M. le Comte de Maurepas" (Eulogy of Monsieur the Count of Maurepas), *HMARS 1781* (1784): 79–102; Maurice Filion, *Maurepas, Ministre de Louis XV* (Maurepas, Minister of Louis XV) (Montreal: Editions Leméac, 1967); and André Piccola, *Le Comte de Maurepas: Versailles et l'Europe á la fin de l'ancien Régime* (The Count of Maurepas: Versailles and Europe at the End of the Old Regime) (Paris: Perrin, 1999).

7. Quoted in Piccola, *Le Comte de Maurepas*, 153.

8. Carlos Daniel Malamud Rikles, *Cádiz y Saint Malo en el comercio colonial peruano (1698–1725)* (Cadiz and Saint-Malo in Colonial Peruvian Commerce, 1698–1725) (Cadíz: Diputación de Cádiz, 1986), 62–67.

9. Erik Dahlgren, *Les relations commerciales et maritimes entre la France et les côtes de l'océan Pacifique (commencement du XVIIIe siècle)* (Commercial and Maritime Relations Between France and the Coasts of the Pacific Ocean, Beginning of the 18th Century) (Paris: H. Champion, 1909).

10. The correspondence between Maurepas and Patiño was mediated through the chief of the French embassy in Madrid, Gérard Lévesque de Champeaux, and the French ambassador to the Spanish court, Conrad Alexandre, Comte de Rottembourg. Rottembourg was a highly respected figure in Madrid, having personally negotiated the Treaty of Escorial between France and Spain, and was instrumental in getting the two nations to agree on the terms of the Geodesic Mission. Quote from Antonio Lafuente García, "Una ciencia para el estado: La expedición geodesica hispano-francesa al virreinato del Perú (1734–1743)" (Science for the State: The French-Spanish Geodesic Mission to the Viceroyalty of Peru [1734–1743]), *Revista de Indias* 43 (1983): 555. See also Robert Finn Erickson, "The French Academy of Sciences Expedition to Spanish America, 1735–1744," PhD diss., University of Illinois, Champaign-Urbana, 1955, 70–72; Jorge Juan y Santacilia, Antonio de Ulloa y de la Torre-Guiral, and Luís Javier Ramos Gómez, *Epoca, génesis y texto de las 'Noticias secretas de América,' de Jorge Juan y Antonio de Ulloa, 1735–1745* (Epoch, Genesis and Text of the "Secret Notices of America" of Jorge Juan and Ulloa, 1735–1745), vol. 1, *El viaje a América* (Voyage to America) (Madrid: CSIC, 1985), 5–11; Antonio Lafuente García and Antonio Mazuecos, *Los Caballeros de Punto Fijo; Ciencia, politica y aventura en la expedición geodésica hispanofrancesa al virreinato del Perú en el siglo*

XVIII (The Gentlemen of the Fixed Point: Science, Politics and Adventure in the Hispanic/French Geodesic Expedition to the Viceroyalty of Peru in the 18th Century) (Barcelona: Serbal/CSC, 1987), 84–88.

11. Lafuente, "Una ciencia para el estado," 558.

12. Ibid., 560.

13. Voltaire, *Alzire, ou Les américains* (Alzire, or The Americans) (Paris: Claude Bauche, 1736); Theodore Besterman, *Voltaire* (New York: Harcourt, Brace, 1969), 187–189.

14. James E. McClellan III and Francois Regourd, "The Colonial Machine: French Science and Colonization in the Ancién Regime," *Osiris*, 2nd ser., 15 (2000): 40–44.

15. Biographical information on Joseph de Jussieu comes from three primary sources: his eulogy by Condorcet, "Eloge de M. de Jussieu" (Eulogy of Mr. de Jussieu), *HMARS 1779* (1782): 44–53; posthumous notes by his nephew Antoine-Laurent de Jussieu in the French Academy of Sciences (AARS Biographical Dossier Joseph de Jussieu); and from the encyclopedia article by Antoine-Laurent de Jussieu, "Jussieu," in *Encyclopédie Méthodique: Médicine* (Methodical Encyclopedia: Medicine), ed. Henri Agasse (Paris: Panckoucke, 1798), 7:765–772. See also two works by Gaston Lehir: "Joseph de Jussieu et son exploration en Amérique Méridionale (1735–1769)" (Joseph de Jussieu and His Exploration in South America, 1735–1769), master's thesis, University of Montreal, 1976, and "L'oeuvre de Joseph de Jussieu (1704–1779) en Amérique méridionale" (The Work of Joseph de Jussieu [1704–1779] in South America), in *Les naturalistes français en Amérique du Sud XVIe–XIXe siècles* (French Naturalists in South America 16th–19th Centuries), ed. Yves Lassius (Paris: Editions du CTHS, 1995),121–135. Patrick Drevet's novel *Le Corps du Monde* (The Body of the World) (Paris: Seuil, 1997) gives an intimate portrayal of Jussieu based on his correspondence.

16. Erickson, "The French Academy of Sciences Expedition," 72. The original decrees are in AGI Indiferente General, legajos 333 and 956. They are translated and referred to as "passports" in Charles-Marie de La Condamine, *Journal du voyage fait à l'Équateur servant d'introduction historique à la Mesure des trois premiers degrés du Méridien* (Journal of the Voyage Made to the Equator, Serving as a Historical Introduction to the Measure of the First Three Degrees of the Meridian) (Paris: Imprimerie Royale, 1751), 272–274.

17. On French trade with Spanish colonies, see Carlos Daniel Malamud Rikles, *Cádiz y Saint Malo en el comercio colonial peruano (1698–1725)* (Cadiz and Saint-Malo in Colonial Peruvian Commerce, 1698–1725) (Cadíz: Diputación de Cádiz, 1986), 104–105; and Susy Sánchez Rodríguez, "Temidos o admirados: negocios franceses en la ciudad de Lima a fines del siglo XVIII" (Fear Them or Admire Them: French Businessmen in the City of Lima at the End of the 18th Century), in *Passeurs, mediadores culturales y agentes de la primera globalización en el mundo Ibérico, siglos XVI-XIX* (Visitors, Cultural Mediators and Agents of the First Globalization in the Iberian World, 16th–19th Centuries), ed. Scarlett O'Phelan Godoy and Carmen Salazar-Soler (Lima: IFEA/PUCP, 2005), 441–469. On the financial arrangements negotiated through the French consul at Cadiz (Jean Partyet), see Jules Antoine René Maillard de La Gournerie, "Recherche de documents relatifs à l'Expédition scientifique faite au Pérou, de 1735 à 1743" (Research of Documents Relative to the Scientific Expedition Made in Peru, from 1735 to 1743), *Comptes rendus hebdomadiares des séances de l'Académie des Sciences* (Weekly Proceedings of the Sessions of the Academy of Sciences) 85 (1877): 424; and Erickson, "The French Academy of Sciences Expedition," 73–78.

18. Jorge Juan and Ulloa to Patiño, Tababela (Puembo), Nov. 18, 1736, BNCBA 4/19, 152–166; La Condamine, *Supplément au Journal historique du voyage a l'équateur, et au livre de la Mesure des trois premiers degrés du meridien* (Supplement to the Historical Journal of the Voyage to the Equator, and of the Book of the Measure of the First Three Degrees of the Meridian), 2 vols. (Paris: Durand and Pissot, 1752–1754), 2:193; Friedrich Melchior Baron von Grimm, Denis Diderot, et al., *Correspondance littéraire, philosophique et critique de Grimm et de Diderot, depuis 1753 jusqu'en 1790* (Literary, Philosophical and Critical Correspondance of Grimm and Diderot, 1753–1790), vol. 8, 1772–1776 (Paris: Furne, 1830), 285.

19. On La Grive's withdrawal: Erickson, "The French Academy of Sciences Expedition," 68. On Pimodan's withdrawal: Claude-Emmanuel-Henri, Comte de Pimodan, *Louise-Elisabeth*

d'Orléans, reine d'Espagne (1709–1742) (Louise-Elisabeth d'Orléans, queen of Spain 1709–1742) (Paris: Plon-Nourrit, 1923), 320. Grandjean de Fouchy stated that his withdrawal was due to a "long and dangerous malady"; Grandjean de Fouchy, "Eloge de M. Bouguer" (Eulogy for M. Bouguer), *HMARS 1758* (1763): 131. His son identified this malady as "obstructions," which at the time often referred to the urinary system; Charles Joseph Léopold Grandjean de Fouchy, "Notes for Condorcet, 1788," Groupe D'Alembert website, dalembert.obspm.fr/Fouchy-notes-JP.php, accessed June 2010). Elisabeth Badinter, however, argues that Grandjean de Fouchy was simply reluctant to leave his fiancée (*Les passions intellectuelles*, 74).

20. Morainville's date of birth is from his testimony at the Seniergues trial, BNF Cotes Espagnols 51 fol. 41. Morainville's marital status comes from a notice of letters sold at auction in July 1991, alluding to an undated death certificate for "Mme. de Morainville" (AARS Biographical Dossier La Condamine, 50J). On Verguin, see Bernard Cros, "Verguin," *Neptunia* 250 (2008): 29–38. Maurepas' letter to Caylus: Henry Medley et al., *Report on the Manuscripts of Lady Du Cane* (London: HM Stationary Office, 1905), 254.

21. Robert Whitaker, *The Mapmaker's Wife: A True Tale of Love, Murder and Survival in the Amazon* (New York: Perseus Books, 2004), 52–53.

22. Couplet's full name was never mentioned in standard works about the Geodesic Mission. It is reconstructed from La Condamine, *Journal du voyage*, 272 (the official passports naming him as "Couplet-Viguier"), and Jorge Juan y Santacilia and Antonio de Ulloa y de la Torre-Guiral, *Relación histórica del Viaje á la América Meridional* (Historical Relation of the Voyage to South America), 2 vols. (Madrid: Antonio Marín, 1748; facsimile, ed. José Merino Navarro and Miguel Rodriguez San Vicente, Fundación universitaria española, 1978), 2:652 (in Couplet's index entry, his first name "Santiago" is the Spanish version of the French "Jacques"). The Academy treasurer's given name was Nicolas Couplet, although he was inexplicably called "Pierre" in French Academy records. On Couplet's father, also named Jacques Couplet (a military engineer), see David J. Sturdy, *Science and Social Status: The Members of the Académie des Sciences, 1666–1750* (Woodbridge, UK: Boydell Press, 1995), 134–135, 242. Rose Godin as Nicolas Couplet's beneficiary: AARS Biographical Dossier Pierre Couplet de Tartreaux alias Tartereaux.

23. Hugo's lack of experience in building astronomical instruments is from La Condamine, *Supplément*, 2:254. Seniergues' birthplace in Bonneval, Quercy, is from the last testament of Jean Seniergues in Octavio Cordero Palacios, "La muerte de don Juan Seniergues" (The Murder of Mr. Jean Seniergues), *Revista del Centro de Estudios Históricos y Geograficos de Cuenca* (1924–1928): vol. 11, 142–184; vol. 12, 206–298; vol. 13, 302–317, vol. 12, 254. "Bonneval" may have been one of several hamlets with that name in the ancient province of Quercy (nowadays encompassing the *departements* of Lot and Tarn-et-Garonne), but the few remaining archival records cannot specify which.

Hugo, in a letter to Joseph de Jussieu written many years later, ruefully mentioned his early thoughts of reward that would never be realized (Quito, Nov. 21, 1749, MNHN Ms 179, no. 47). Seniergues' remarks are from his letter to Antoine and Bernard Jussieu, Feb. 18, 1736, MNHN Ms 179, fol. 21. The correspondence between Joseph and his brothers indicates they knew Seniergues and Hugo prior to the mission.

24. For a complete account of these developments in ship design, see Larrie D. Ferreiro, *Ships and Science: The Birth of Naval Architecture in the Scientific Revolution, 1600–1800* (Cambridge, MA: MIT Press, 2007).

25. The aversion of some (but certainly not most) Academicians to marriage and even sex is described in Badinter, *Les passions intellectuelles*, 1:30.

26. Bouguer's initial reluctance to go on the mission and his hatred of ocean voyages: Pierre Bouguer, *La Figure de la terre, déterminée par les observations de Messieurs Bouguer et de La Condamine, de l'Académie Royale des Sciences, envoyés par ordre du Roy au Pérou pour observer aux environs de l'équateur, avec une Relation abrégée de ce voyage qui contient la description du pays dans lequel les opérations ont été faites* (The Figure of the Earth, Determined by the Observations of the Messieurs Bouguer and La Condamine, of the Royal Academy of Sciences, Sent to Peru by Order of the King to Make Observations near the Equator, with an Abridged Relation of the Voyage Containing a

Description of the Countries in Which the Observations Were Made) (Paris: Charles-Antoine Jombert, 1749), iv. Bouguer and astronomical instruments: Bouguer to Nicolas Couplet, Le Havre, Sept. 12, 1734 and Feb. 25, 1735, BCIU Ms 337 folios 94–97; letter from Bouguer to Maurepas, Le Havre, Nov. 21, 1734, ANF Marine C7 40 Dossier Bouguer, folios 18–19. Nomination of Bouguer to the Geodesic Mission: Bignon to Maurepas, Paris, Dec. 18–29, 1734, BNF Manuscrits occidentaux, 22235, folios 289–292 (quote from folio 292, Dec. 29, 1734).

27. José Augustín Pardo de Figueroa (later the Marqués de Valleumbroso), having lived in both Spain and Peru, was intimately familiar with the politics of both the mother country and its colonies; in particular he was a close friend of the newly appointed president of Quito, José Araujo y Rio. See La Condamine, *Journal du voyage,* 268–269.

28. BNE Ms 8428 folio 38.

29. Ulloa was not the first choice; he was selected after another cadet, Juan García del Postigo, did not return from sea duty in time. The royal orders assigning Jorge Juan and Ulloa to the Geodesic Mission, dated Jan. 4 and Apr. 1, 1735, are in AGI Lima 590 folios 68 and 48–50, respectively. Biographical details of Jorge Juan and Ulloa are found in Arthur P. Whitaker, "Antonio de Ulloa," *Hispanic American Historical Review* 15, no. 2 (1935): 155–194; Julio Fernando Guillén Tato, *Los Tenientes de navío Jorge Juan y Santacilia y Antonio de Ulloa y de la Torre-Guiral, y la medición del Meridiano* (Lieutenants Jorge Juan y Santacilia and Antonio de Ulloa y de la Torre-Guiral, and the Measure of the Meridian) (Madrid: Galo Sáez, 1936; Novelda, Alicante: Caja de Ahorros, 1973); Miguel Losada and Consuelo Losada, eds., *Actas de II Centenario de Don Antonio de Ulloa* (Acts of the Bicentennial of Antonio de Ulloa) (Seville: CSIC/AGI, 1995); Francisco Solano Pérez-Lila, *La pasión de Reformar: Antonio de Ulloa, marino y científico, 1716–1795* (A Passion for Reforming: Antonio de Ulloa, Sailor and Scientist) (Cadiz: Universidad de Cádiz, 1999) ("Adventurer" quote, 43); and Emilio Soler Pascual, *Viajes de Jorge Juan y Santacilia: Ciencia y política en la España del siglo XVIII* (Voyages of Jorge Juan y Santacilia: Science and Politics in 18th-Century Spain) (Barcelona: Ediciones B, 2002). As for Juan del Postigo, he went on to have a distinguished naval career, rising to the rank of admiral in the Spanish navy.

30. An account of a proposed expedition to the equator by Louis Godin, Dec. 19, 1734, Royal Society of London Archives MM/20/29; Grandjean de Fouchy, "Eloge de M. Godin," 187–188.

31. Halley's fluency in French: Alan Cook, *Edmond Halley: Charting the Heavens and the Seas* (Oxford: Clarendon Press, 1998), 17, 152. AARS Biographical Dossier Louis Godin contains two notes from a David A. Taylor (Nov. 1965 and Mar. 1966) stating that none of Halley's personal records describe a meeting with Godin. Godin's election to the Royal Society: He was nominated Jan. 9, 1735, and elected Mar. 27, 1735, Royal Society of London Archives EC/1735/02.

32. *PVARS*, Jan. 1735, pp. 1,9,17,56; James R. Smith, *From Plane to Spheroid: Determining the Figure of the Earth from 3000 B.C. to the 18th Century Lapland and Peruvian Survey Expeditions* (Rancho Cordova, CA: Landmark Enterprises, 1986), 78–79.

33. Bouguer, *La Figure de la terre*, 60–61; Charles-Marie de La Condamine, *Mesure des trois premiers degrés du méridien dans l'hémisphère austral, tirée des observations de MM. de l'Académie royale des sciences envoyés par le roi sous l'Équateur* (Measure of the First Three Degrees of the Meridian in the Southern Hemisphere, Drawn from the Observations of the Messieurs of the Royal Academy of Sciences, Sent by the King Below the Equator) (Paris: Imprimerie Royale, 1751), 75–76; Lafuente, "Una ciencia para el estado," 583; Anthony J. Turner, "The Observatory and the Quadrant in Eighteenth-Century Europe," *Journal for the History of Astronomy* 33, part 4, no. 113 (2002): 381.

34. Louis Econches Feuillée, *Journal des observations physiques, mathématiques et botaniques, faites par l'ordre du Roy sur les côtes orientales de l'Amérique méridionale, & dans les Indes occidentales, depuis l'année 1707 jusques en 1712* (Journal of Physical, Mathematical and Botanical Observations, Made at the Order of the King Along the Eastern Coasts of South America and in the West Indies, Since 1707 Until 1712), 2 vols. (Paris: Giffart, 1714). An analysis of Feuillée's voyage is in Jordan Kellman, "Discovery and En-

lightenment at Sea: Maritime Exploration and Observation in the 18th-Century French Scientific Community," PhD diss., Princeton University, 1998, 209–253.

35. François Amedée Frézier, *Relation du voyage de la mer du sud aux côtes du Chily et du Pérou, fait pendant les années 1712, 1713 & 1714* (Relation of the voyage from the southern sea to the coasts of Chile and Peru, made during the years 1712, 1713, & 1714) (Paris: Nyon, 1716), 230; Kellman, "Discovery and Enlightenment at Sea," 264–284.

36. Bouguer's possible connection with Frézier had nothing to do with botany. In early 1734 both Bouguer (as a mathematician) and Frézier (in his role as military engineer) became interested in the stability of arched vaults of stone, and they simultaneously developed a novel "slicing technique" for their analyses, which was likely arrived at through close communication. Since this was the only foray Bouguer ever made into civil engineering, it is likely that he had discussed the problem directly with Frézier. See Santiago Huerta, "Mechanics of Masonry Vaults: The Equilibrium Approach," in *Historical Constructions 2001: Possibilities of Numerical and Experimental Techniques, Proceedings of the 3rd International Seminar*, ed. Paulo B. Lourenço and Pere Roca (Guimarães, Portugal: University of Minho, 2001), 58.

37. Maurepas to Fayet, Versailles, Jan. 25, 1735, ANF Colonies B63 fol. 366–377. See also Erickson, "The French Academy of Sciences Expedition," 77; and Lafuente, "Una ciencia para el estado," 568–569.

38. See Guillaume de Meschin, *Journal de la campagne du Portefaix* (Logbook of Portefaix), March 23, 1735–Feb. 25, 1736, ANF Marine 4JJ 28 fol. 37. The *Portefaix* was launched in Toulon in 1717. See Alain Demerliac, *La Marine de Louis XV: Nomenclature des navires français de 1715 à 1774* (Louis XV's Navy: Nomenclature of French Ships from 1715 to 1774) (Nice: Editions Omega, 1995), 90.

39. Saint-Laurent the cook: passenger list in AARS Biographical Dossier La Condamine, 50J fol. 14/15. For a description of the Trois Marchands, see Philippe Duprat and Robert Fontaine, "Deux auberges à Rochefort au XVIIe siècle" (Two Inns in Rochefort in the 17th Century), *Roccafortis, Bulletin de la Société de Géographie de Rochefort*, 3rd ser., 2, no. 16 (Sept. 1995): 351–356.

40. Maurepas to Champeaux, Versailles, Apr. 2, 1735, ANF Marine B7 150 fol. 565; Joseph de Jussieu to Antoine de Jussieu, Rochefort, Apr. 22, 1735, MNHN Ms 179-3.

41. See Kellman, "Discovery and Enlightenment at Sea," for a description of French scientific voyages in the seventeenth and eighteenth centuries.

42. Maurepas' reply to Beauharnais from Erickson, "The French Academy of Sciences Expedition," 82. His reply to Bouguer is quoted in Jean-Jacques Levallois, *Mesurer la terre: 300 ans de géodésie française, de la toise du Châtelet au satellite* (300 Years of French Geodesy, from the Toise of Châtelet to the Satellite) (Paris: Presses de Pontes et Chaussées, 1988), 36.

43. Meschin, *Journal de la campagne du Portefaix,* entry for May 10, 1735; Bouguer, "Expédition au Pérou" (Expedition to Peru), nonpaginated and undated (but circa 1747), AOP A.C. 2.7; Lafuente, "Una ciencia para el estado," 575.

44. Antonio de Ulloa y de la Torre-Guiral, *Diario de theniente Ulloa* (Personal Logbook of Lt. Ulloa), entry for May 26, 1735, May 26–July 9, 1735, BNCBA 4/37 fol. 49–99; Jorge Juan and Ulloa, *Relación histórica*, 1:10.

Chapter III Finding Quito

1. Pierre Bouguer, *La Figure de la terre, déterminée par les observations de Messieurs Bouguer et de La Condamine, de l'Académie Royale des Sciences, envoyés par ordre du Roy au Pérou pour observer aux environs de l'équateur, avec une Relation abrégée de ce voyage qui contient la description du pays dans lequel les opérations ont été faites* (The Figure of the Earth, Determined by the Observations of the Messieurs Bouguer and La Condamine, of the Royal Academy of Sciences, Sent to Peru by Order of the King to Make Observations near the Equator, with an Abridged Relation of the Voyage Containing a Description of the Countries in Which the Observations Were Made) (Paris: Charles-Antoine

Jombert, 1749), vi; La Condamine, *Journal du voyage fait à l'Équateur servant d'introduction historique à la Mesure des trois premiers degrés du Méridien* (Journal of the Voyage Made to the Equator, Serving as a Historical Introduction to the Measure of the First Three Degrees of the Meridian) (Paris: Imprimerie Royale, 1751), 8–9. On Hadley's octant, see Danielle Fauque, "The Introduction of the Octant in Eighteenth-Century France," in *Koersvast: Vijf eeuwen navigatie op zee* (On a Steady Course: Five Centuries of Navigation at Sea), ed. Leo Akveld (Zaltbommel, Netherlands: Uitgeverij Aprilis, 2005), 95–98.

2. Joseph de Jussieu to Antoine de Jussieu, Fort Royal, Martinique, Jul. 8, 1735, MNHN Ms 179/6.

3. Seniergues to Antoine de Jussieu, Fort Royal, Martinique, Jul. 4, 1735, MNHN Ms 179/5.

4. Gustave Desnoiresterres, *Voltaire et la société française au XVIIIe siècle* (Voltaire and French Society in the 18th Century), 8 vols. (Paris: Didier, 1867–1876), 2:141–144.

5. Bouguer, "Expédition au Pérou" (Expedition to Peru), nonpaginated and undated (but circa 1747), AOP A.C. 2.7.

6. Meschin, *Journal de la campagne du Portefaix,* entries for June 11–22, 1735. Bouguer's lodging: Bouguer, "Expédition au Pérou." Jussieu's lodging: Gaston Lehir, "Joseph de Jussieu et son exploration en Amérique Méridionale (1735–1769)" (Joseph de Jussieu and His Exploration in South America, 1735–1769), master's thesis, University of Montreal, 1976, 63.

7. Joseph de Jussieu to Antoine de Jussieu, Fort Royal, Martinique, Jul. 8, 1735; Bouguer, "Expédition au Pérou."

8. Meschin, *Journal de la campagne du Portefaix*, entries for July 9–21, 1735; La Condamine, *Journal du voyage*, 3–4. The "cures" (bleeding and purging) for yellow fever of course did nothing; a fair percentage of yellow fever victims recover spontaneously, as happened with La Condamine.

9. Quoted in Antonio Lafuente García, "Una ciencia para el estado: La expedición geodesica hispano-francesa al virreinato del Perú (1734–1743)" (Science for the State: The French-Spanish Geodesic Mission to the Viceroyalty of Peru, 1734–1743), *Revista de Indias* 43 (1983): 569.

10. Bouguer, *La Figure de la terre*, 56–57; Charles-Marie de La Condamine, *Mesure des trois premiers degrés du méridien dans l'hémisphère austral, tirée des observations de MM. de l'Académie royale des sciences envoyés par le roi sous l'Équateur* (Measure of the First Three Degrees of the Meridian in the Southern Hemisphere, Drawn from the Observations of the Messieurs of the Royal Academy of Sciences, Sent by the King Below the Equator) (Paris: Imprimerie Royale, 1751), 14. These accounts clearly show that Grangier worked under Bouguer's tutelage. On the slave population, see James E. McClellan III, *Colonialism and Science: Saint Domingue in the Old Regime* (Baltimore: Johns Hopkins University Press, 1992), 120.

11. For more complete treatments of the French slave trade, see Sue Peabody, *"There Are No Slaves in France": The Political Culture of Race and Slavery in the Ancien Régime* (New York: Oxford University Press, 1996); and Robert Harms, *The Diligent: A Voyage Through the Worlds of the Slave Trade* (New York: Basic Books, 2002).

12. Quoted in McClellan, *Colonialism and Science*, 60.

13. Joseph de Jussieu to Antoine de Jussieu, Petit Goâve, Oct. 27, 1735, MNHN Ms 179/15.

14. Godin to Antoine de Jussieu, Guayaquil, Apr. 20, 1736, MNHN Ms 179/22; Seniergues to Antoine de Jussieu, Fort Royal, Martinique, Jul. 4, 1735, MNHN Ms 179/5.

15. Quoted in McClellan, *Colonialism and Science*, 121.

16. Robert Finn Erickson, "The French Academy of Sciences Expedition to Spanish America, 1735–1744," PhD diss., University of Illinois, Champaign-Urbana, 1955, 85–87.

17. Godin to Antoine de Jussieu, Petit Goâve, Aug. 11, 1735, MNHN Ms 179/9; Seniergues to Antoine and Bernard de Jussieu, Panama City, Feb. 18, 1736, MNHN Ms 179/21.

18. On *Vautour*: Alain Demerliac, *La Marine de Louis XV: Nomenclature des navires français de 1715 à 1774* (Louis XV's Navy: Nomenclature of French Ships from 1715 to

1774) (Nice: Editions Omega, 1995), 93. On the passage: La Condamine, *Journal du voyage*, 7; La Condamine to Voltaire, Portobelo, Dec. 15, 1735, in Voltaire, *The Complete Works of Voltaire*, ed. Theodore Besterman, et al., 135+ vols. (Toronto: University of Toronto Press, 1968–present), 87:275–277 (quote at 276).

19. Antonio de Ulloa y de la Torre-Guiral, *Diario de theniente Ulloa* (Personal Logbook of Lt. Ulloa), entries for June 7 and 21 and July 2–7, 1735, BNCBA 4/37 fol. 49–99.

20. Ulloa, *Diario*, entry for July 9, 1735.

21. Antonio de Ulloa y de la Torre-Guiral, *Prosigue del diario después de Cartagena* (Continuation of Personal Logbook after Cartagena), July 9–Oct. 15, 1735, BNCBA 4/36 fol. 100–139.

22. See Lance Grahn, "Political Corruption and Reform in Cartagena Province, 1700–1740," North Central Council of Latin Americanists (NCCLA) Discussion Paper #88, Feb. 1995.

23. Rubén Vargas Ugarte, *Historia del Perú virreinato (siglo XVIII) 1700–1790* (History of Viceroyal Peru, 18th Century 1700–1790) (Lima: Hernandus Vega Centeno. Vicaris Generales, 1956), 182–183.

24. Jorge Juan and Ulloa to Patiño, Cartagena de Indias, Oct. 25, 1735, and Nov. 4, 1735, BNCBA 4/31 fol. 140–143 and 4/17 fol. 332–335.

25. Jorge Juan y Santacilia, Antonio de Ulloa y de la Torre-Guiral, and Luís Javier Ramos Gómez, *Epoca, génesis y texto de las 'Noticias secretas de América,' de Jorge Juan y Antonio de Ulloa, 1735–1745* (Epoch, Genesis and Text of the "Secret Notices of America" of Jorge Juan and Ulloa, 1735–1745), vol. 1, *El viaje a América* (Voyage to America) (Madrid: CSIC, 1985), 23–24, 30.

26. Kenneth J. Andrien, "The *Noticias secretas de América* and the Construction of a Governing Ideology for the Spanish American Empire," *Colonial Latin American Review* 7, no. 2 (1998): 184–185.

27. Jorge Juan y Santacilia and Antonio de Ulloa y de la Torre-Guiral, *Relación histórica del Viaje á la América Meridional* (Historical Relation of the Voyage to South America), 2 vols. (Madrid: Antonio Marín, 1748; facsimile, ed. José Merino Navarro and Miguel Rodriguez San Vicente, Fundación universitaria española, 1978), 1:40.

28. Ibid.

29. Jean Descola, *Daily Life in Colonial Peru, 1710–1820* (New York: Macmillan, 1968), 25–34; Marcelin Defourneaux, *Daily Life in Spain in the Golden Age* (Stanford, CA: Stanford University Press, 1970), 83–84; Enrique Ayala Mora, ed., *Nueva Historia del Ecuador* (New History of Ecuador), 15 vols. (Quito: Corporación Editora Nacional—Grijalbo, 1983–1995), 5:45.

30. Bartolomé de las Casas, *A Short Account of the Destruction of the Indies*, trans. Nigel Griffins (London: Penguin, 1999).

31. Joseph de Jussieu to Antoine de Jussieu, Portobelo, Dec. 16, 1735, MNHN Ms 179/16; La Condamine to Voltaire, Portobelo, Dec. 15, 1735, in Voltaire, *Complete Works*, 87:275–277; Jorge Juan and Ulloa to Patiño, Portobelo, Dec. 17, 1735, BNCBA 4/32 fol. 148–151.

32. La Condamine, *Journal du voyage*, 6–7; Jorge Juan, Ulloa, and Ramos Gómez, *Las 'Noticias secretas de América'*, 1:34.

33. The complete inventory recorded by the Panama authorities was carried by the expedition all the way to Quito, where it was carefully rechecked; a tally of the personnel shows they had bought three more slaves in either Cartagena or Portobelo in addition to the three they bought in Saint Domingue. See "Inventario y reconocimiento del equipaje de los Académicos Franceses" (Inventory and Acknowledgement of Baggage of the French Academicians), Quito, Jun. 1, 1736, AGI Quito 134 fol. 397–400, and *DHAQ* 6:17–23.

34. Joseph de Jussieu to Antoine de Jussieu, Portobelo, Dec. 16, 1735, MNHN Ms 179/16.

35. Joseph de Jussieu to Antoine de Jussieu, Panama City, Feb. 5, 1736, MNHN Ms 179/18; La Condamine, *Journal du voyage*, 8–9.

36. See "Informe de Dionisio de Alsedo y Herrera," Madrid, Nov. 18, 1740, AGI Quito 134 fol. 379–380, and *DHAQ* 6:49–50; and Jorge Juan, Ulloa, and Ramos Gómez, *Las 'Noticias secretas de América'*, 1:38. Quote in Lafuente, "Una ciencia para el estado," 569–570.

37. Jorge Juan, Ulloa, and Ramos Gómez, *Las 'Noticias secretas de América'*, 1:34.

38. Seniergues to Antoine and Bernard de Jussieu, Panama City, Feb. 18, 1736, MNHN Ms 179/21.

39. La Condamine, *Journal du voyage*, 274.

40. Bouguer, "Memoire sur les avantages qu'il y a faire passer sur la coste la Meridienne que nous devons tracer" (Memoir on the Advantages of Tracing the Meridian Along the Coast), Monte Cristo, Mar. 12, 1736, AOP A.C. 2.7; Pierre Bouguer, "Expédition au Pérou"; Jorge Juan and Ulloa, *Relación histórica*, 1:190; La Condamine, *Journal du voyage*, 12.

41. Godin to Jorge Juan and Ulloa, Manta, Mar. 12, 1736, BNCBA 4/16 fol. 178–183; Jorge Juan and Ulloa to Patiño, Guayaquil, Apr. 12, 1736, BNCBA 4/20 fol. 168–176. Bouguer and La Condamine presumably retained two slaves and a servant, as this is the difference in head count between the December enumeration at Portobelo and the register of arrival in Quito of Godin's group on May 29.

42. Jorge Juan and Ulloa, *Relación histórica*, 1:219.

43. The description of the route from Guayaquil to Quito is from Antonio Bermeo, *La Primera Misión Geodésica Venida al Ecuador* (The First Geodesic Mission to Come to the Equator) (Guaranda, Ecuador: Casa de la Cultura Ecuatoriana Benjamín Carrion, 1983).

44. "Razón de los pagos hechos a Jorge Juan y Antonio de Ulloa por los gastos ocasionados en el traslado de sus tiendas al Virreynato en cumplimiento de su cometido científico" (Reasons for Payments Made to Jorge Juan and Antonio de Ulloa for the Expenses Incurred in the Transfer of Their Households to the Viceroyalty in Completion of the Scientific Duties), Quito, Aug. 18, 1736, BNP C77; "Comunicación que ordena a la Real Tesorería de Guayaquil prestar la ayuda necesaria a los académicos Franceses que vienen a hacer estudios astronómicos en Indias" (Memorandum Ordering the Royal Treasury of Guayaquil to Loan the Aid Necessary to the French Academicians Who Came to Study Astronomy in the Indies), BNP C100. Seniergues' surgery: La Condamine, *Journal du voyage*, 76.

45. Jorge Juan and Ulloa, *Relación histórica*, 1:281.

46. Bermeo, *La Primera Misión Geodésica Venida al Ecuador*, 17–34.

47. "Lettre de M. Verguin, datée de Quito, le 20 Juillet 1736," *Mercure de France* (March 1737): 551–552; Jorge Juan and Ulloa, *Relación histórica*, 1:293–294.

48. "Lettre de M. Verguin," 552.

49. Ibid. The document registering their arrival in Quito (AGI Quito 134 fol. 1) contains two significant errors that have so far defied explanation. First, there is an entry for "M. Couplet Desordonays, capitán de fragata, dibujante," clearly a conjoint of Couplet and Godin des Odonais. For unknown reasons it incorrectly lists the person as a draftsman (that was Morainville's job) and as a frigate captain, a high-ranking naval officer equivalent to commander, which neither were. They were both far too young for that rank, nor does either name appear in any contemporary record of French naval officers or officers of the Compagnie de L'Orient (East Indies Company). The second error is that the name of Théodore Hugo does not appear in this list, nor did he appear in the tallies taken in Cartagena and Portobelo. He was on the passenger list for *Portefaix* back in France, mentioned by name in a letter from Seniergues in Martinique, and it is unlikely that he was left behind in Saint Domingue. His position as instrument maker was absolutely vital to the expedition; he was needed to repair the equipment that would be damaged by the long journey. It is unclear whether he was disguised as a servant or, less likely, he took a different route than the others (perhaps to escape Godin) to arrive in Quito.

50. Bouguer, "Relation abrégée," 253. The actual westernmost point of South America is Punta Pariñas, one hundred fifty miles south of Guayaquil.

51. Bouguer sent his observations back to Paris, where they were published as "Sur les Réfractions Astronomiques dans la Zone Torride" (On Astronomical Refractions in the Torrid Zone), *HMARS 1737* (1739): 407–422.

52. The sojourn of Bouguer and La Condamine on the coast is recounted in Bouguer, *La Figure de la terre*, viii–xxv; and La Condamine, *Journal du voyage*, 11–13. The Palmar boulder that La Condamine inscribed has long since eroded into the ocean. In 1985, a group of historians attempted without success to precisely fix the site of the promontory described

by La Condamine, which could have been any of a number of outcroppings along that stretch of the coastline. The nearby town of Pedernales erected in the central plaza a replica of La Condamine's inscribed boulder. See Pierre Olivares, "La roca equinoccial del Punta Palmar grabada por La Condamine" (The Equatorial Rock Engraved by La Condamine), *Boletín de la Academia Nacional de Historia de Quito* 72, nos. 153–154 (1989): 154–163.

53. Bouguer, "Expédition au Pérou"; Bouguer, *La Figure de la terre*, xxvi–xxxi.

54. La Condamine, *Journal du voyage*, 13.

55. La Condamine, "Mémoire sur une résine élastique, nouvellement découverte à Cayenne par M. Fresneau; et sur l'usage des divers sucs laiteux d'arbres de la Guiane ou France équinoctiale" (Memoir on an Elastic Resin Newly Discovered in Cayenne by Mr. Fresneau, and on the Usage of Different Milky Tree Saps in Guyane or Equatorial France), *HMARS 1751* (1755): 319–333.

56. La Condamine, *Journal du voyage*, 14.

57. La Condamine, *Journal du voyage*, 14–15.

Chapter IV Degree of Difficulty

1. A short biography of Alsedo (which is how he spelled his own name in his letters and books, though many later works spell it "Alcedo") is provided in Carlos Manuel Larrea, *El Presidente de la Real Audiencia de Quito, Dn. Dionisio de Alsedo y Herrera* (The President of the Royal Audiencia of Quito, Mr. Dionisio de Alsedo y Herrera) (Quito: Casa de la Cultura Ecuatoriana, 1961). See also Victor Peralta Ruiz, "Un indiano en la corte de Madrid: Dionisio de Alsedo y Herrera y el *Memorial informativo* del Consulado de Lima, 1725" (An American in the Court of Madrid: Dionisio de Alsedo y Herrera and the *Informative Memoir* of the Consulate of Lima, 1725), *Revista Histórica del Pontificia Universidad Católica del Perú* 28, no. 2 (2003): 319–355.

2. Most of the legends of El Dorado focused on a gilded chieftain at Lake Guatavita, near modern-day Bogotá, or on the region of Canelos, southeast of Quito, supposedly rich in cinnamon and gold. But according to Ecuadorian historian Tamara Estupiñan Viteri, several Inca cities, including Tomebamba (present-day Cuenca), described by the Spanish conquistador Cieza de León as awash with gold, were the real origins of the legend of El Dorado. See Tamara Estupiñan Viteri, "La manipulación del Tirano llamado Rumiñahui" (The Manipulation of the Tyrant Rumiñahui), in *Mitos políticos en las sociedades andinas: Orígenes, invenciones y ficciones* (Political Myths in Andean Societies: Origins, Inventions and Fictions), ed. Germán Carrera Damas (Caracas: Editorial Equinoccio, 2006), 174; and Pedro de Cieza de León, *The Discovery and Conquest of Peru*, trans. Alexandra Parma Cook and Noble David Cook (Durham, NC: Duke University Press, 1998), 270–273, 392.

3. The story of Quito is told in Eliecer Enríquez Bermeo, *Quito a través de los siglos* (Quito Across the Centuries) (Quito: Imprenta Municipal, 1938); and Ximena Romero, *Quito en los ojos de los viajeros: El siglo de la ilustración* (Quito in the Eyes of Travelers: The Enlightenment) (Quito: Editorial Abya Yala, 2000).

4. Martin Minchom, *The People of Quito, 1690–1810: Change and Unrest in the Underclass* (Boulder, CO: Westview Press, 1994), 17–18, 45–47, 135.

5. *Real Cedula* (Royal Decree), San Ildefonso, August 14, 1734, AGI Quito 134 fol. 1; Alsedo to Villagarcía, Quito, Oct. 29, 1735; Villagarcía to Alsedo, Lima, Jan. 20 1736, AGI Quito 134 fol. 394. See also *DHAQ* 6:8–13.

6. *Informe de Dionisio de Alsedo y Herrera* (Alsedo's Report to the King), Madrid, Nov. 18, 1740, AGI Quito 134 fol. 379–380; see also *DHAQ* 6:49–50.

7. Kenneth J. Andrien, *The Kingdom of Quito, 1690–1830: The State and Regional Development* (Cambridge, UK: Cambridge University Press, 1995), 29–30, 56–59; Tamar Herzog, *Upholding Justice: Society, State, and the Penal System in Quito (1650–1750)* (Ann Arbor: University of Michigan Press, 2004), 131, 232–235.

8. See Luís Javier Ramos Gómez, "Enfrentamientos entre grupos de poder por el domino del Cabildo de Quito entre 1735 y 1739" (Confrontations Between Power Groups for the Control of the City Council of Quito Between 1735 and 1739), *Revista Complutense de Historia de América* 31 (2005): 53–77.

9. Bouguer, "Expédition au Pérou" (Expedition to Peru), nonpaginated and undated (but circa 1747), AOP A.C. 2.7.

10. Alsedo to Villagarcía, Quito, June 18, 1735, AGI Quito 134 fol. 401; see also *DHAQ* 6:28–32.

11. Herzog, *Upholding Justice*, 182–187.

12. Charles-Marie de La Condamine, *Journal du voyage fait à l'Équateur servant d'introduction historique à la Mesure des trois premiers degrés du Méridien* (Journal of the Voyage Made to the Equator, Serving as a Historical Introduction to the Measure of the First Three Degrees of the Meridian) (Paris: Imprimerie Royale, 1751), 15–16.

13. Jorge Juan and Ulloa to Patiño, Quito, July 15, 1736, BNCBA 4/18 fol. 250–265.

14. La Condamine, *Journal du voyage*, 16–17.

15. Letters between La Condamine and Alsedo, Quito, June 14–15, 1735, AGI Quito 134 fol. 401–402; see also *DHAQ* 6:28–32.

16. Quotes, in order, are from La Condamine, *Journal du voyage*, 114–115; Bouguer to Bignon, Quito, Feb. 10, 1737, BNF Collection Bréquigny 62 fol. 94; Jorge Juan and Ulloa, *Relación histórica*, 392; Pierre Bouguer, *La Figure de la terre, déterminée par les observations de Messieurs Bouguer et de La Condamine, de l'Académie Royale des Sciences, envoyés par ordre du Roy au Pérou pour observer aux environs de l'équateur, avec une Relation abrégée de ce voyage qui contient la description du pays dans lequel les opérations ont été faites* (The Figure of the Earth, Determined by the Observations of the Messieurs Bouguer and La Condamine, of the Royal Academy of Sciences, Sent to Peru by Order of the King to Make Observations near the Equator, with an Abridged Relation of the Voyage Containing a Description of the Countries in Which the Observations Were Made) (Paris: Charles-Antoine Jombert, 1749), lxiii.

17. Jorge Juan y Santacilia and Antonio de Ulloa y de la Torre-Guiral, *Relación histórica del Viaje á la América Meridional* (Historical Relation of the Voyage to South America), 2 vols. (Madrid: Antonio Marín, 1748; facsimile, ed. José Merino Navarro and Miguel Rodriguez San Vicente, Fundación universitaria española, 1978), 1:366–367.

18. Bouguer to Bignon, Quito, Feb. 10, 1737.

19. Jorge Juan and Ulloa, *Relación histórica*, 1:366–367. For Peruvian fashions see Jean Descola, *Daily Life in Colonial Peru, 1710–1820* (New York: Macmillan, 1968), 133–146. French fashions are described in Madeleine Delpierre, *Dress in France in the Eighteenth Century* (New Haven: Yale University Press, 1998).

20. On the education of Peruvian women see Enrique Ayala Mora, ed., *Nueva Historia del Ecuador* (New History of Ecuador), 15 vols. (Quito: Corporación Editora Nacional—Grijalbo, 1983–1995), 5:196–197; for that of French women see Guy Chaussinand-Nogaret, *La Noblesse au XVIIIe siècle: De la féodalité aux lumières* (The Nobility in the Eighteenth Century: From Feudalism to Enlightenment) (Paris: Hachette, 1976), 68–69.

21. Women's sexuality in France: Guy Chaussinand-Nogaret, *La vie quotidienne des français sous Louis XV* (Daily Life of the French Under Louis XV) (Paris: Hachette, 1979), 124–127. The use of the *manto* is discussed in Descola, *Daily Life in Colonial Peru*, 138–139.

22. Bouguer to Bignon, Quito, Feb. 10, 1737. On the Inquisition, see Jerry M. Williams, "A New Text in the Case of Ana de Castro: Lima's Inquisition on Trial," Museum of the Inquisition and Congress, Lima, Peru, http://www.congreso.gob.pe/museo/inquisicion/ana-de-castro.pdf, accessed July 2010.

23. Villagarcía to Alsedo, Lima, July 21, 1736, BNCBA 4/25 fol. 328–331. The *fiscal* or crown attorney was responsible for all financial matters associated with legal affairs.

24. Treaty: La Gournerie, "Remarques historiques et critiques sur les observations faites au Pérou . . . par Bouguer" (Historical Remarks and Criticisms on the Observations Made in Peru . . . by Bouguer), in *Annales de l'observatoire de Paris* 14, no. 1 (Paris: Gauthier-Villar, 1877), D30. Quote from Jorge Juan and Ulloa to Patiño, Tababela (Puembo), Nov. 18, 1736, BNCBA 4/19 fol.152–166. La Condamine mentioned that the loan was from a "Frenchman established in Quito"; that Frenchman was almost certainly the doctor and surgeon Raimundo Dablanc, with whom the expedition became good friends; *Supplément au Journal historique du voyage a l'équateur, et au livre de la Mesure des trois premiers*

degrés du meridien (Supplement to the Historical Journal of the Voyage to the Equator, and of the Book of the Measure of the First Three Degrees of the Meridian), 2 vols. (Paris: Durand and Pissot, 1752–1754), 2:193–194.

25. Jorge Juan and Ulloa to Patiño, Nov. 18, 1736; La Condamine, *Journal du voyage*, 18–19; Diego Bonifaz Andrade, *Guachalá: Historia de una hacienda en Cayambe* (Guachalá, History of a Hacienda in Cayambe) (Quito: Ediciones Abya-Yala, 1995), 15–18.

26. Jorge Juan and Ulloa to Patiño, Nov. 18, 1736; La Condamine, *Journal du voyage*, 19; Jorge Juan and Ulloa, *Relación histórica*, 304.

27. La Condamine, *Journal du voyage*, 12. Regarding quinine: Many Spanish Peruvians at the time believed in curing illnesses of specific qualities with medicines of the opposite quality. They considered malaria to be a "warm" disease, so a "warm" medicine like quinine should not be used. See Eduardo Estrella, "Ciencia ilustrada y saber popular en el conocimiento de la quina en el siglo XVIII" (Illustrated Science and Common Knowledge in the Understanding of Quinoa in the 18th Century), in *Saberes andinos: Ciencia y tecnología en Bolivia, Ecuador y Perú* (Andean Knowledge: Science and Technology in Bolivia, Ecuador and Peru), ed. Marcos Cueto (Lima: Instituto de estudios peruanos, IEP, 1995), 49–50.

28. *Informe de Dionisio de Alsedo y Herrera*, Madrid, Nov. 18, 1740. Alsedo stated that Couplet was "buried in the church of that town [Cayambe]." However, in May 2006 I met with the priest of Cayambe, Diego Brito, who informed me that the parochial church was only built around 1850 and that prior to that time, burials took place in local chapels, so it is not clear where Couplet may have been interred. His tomb has not yet been found.

29. Jorge Juan and Ulloa to Patiño, Nov. 18, 1736; Bouguer to Bignon, Quito, Feb. 10, 1737; Jacques Cassini, "Observation de l'Eclipse totale de la lune, faite à Thury le 20 Septembre 1736" (Observation of the Total Eclipse of the Moon, Performed at Thury on 20 September 1736), *HMARS 1736* (1739): 313–315. Godin sent word of Couplet's death in a letter to Maurepas in February 1737; Pierre Bouguer, *Justification des Mémoires de l'Académie royale des sciences de 1744, et du livre de la Figure de la terre déterminée par les observations faites au Pérou* (Justification of Memoirs of the Royal Academy of Sciences of 1744, and of the Book The Figure of the Earth Determined by Observations Made in Peru) (Paris: Charles-Antoine Jombert, 1752), 12–13. Maurepas presumably then informed Couplet's family.

30. Description of the Yaruquí plateau and the subsequent creation and measurement of the baseline from Bouguer, "Relation abregée du voyage fait au Pérou par Messieurs de l'Académie Royale des Sciences, pour mesurer les Degrés du Méridien aux environs de l'Equator, & en conclure la Figure de la Terre" (Abridged Relation of the Voyage Made to Peru by Gentlemen of the Royal Academy of Sciences, to Measure the Degrees of the Meridian Around the Equator, and to Deduce the Figure of the Earth), *HMARS 1744* (1748): 279–282, and *La Figure de la terre*, 37–44, 56–57; Jorge Juan and Ulloa, *Relación histórica*, 1:302–305, and *Observaciones astronómicas y físicas hechas en los Reynos del Perú* (Astronomical and Physical Observations Made in the Kingdoms of Peru) (Madrid: Juan de Zúñiga, 1748), 144–155, La Condamine, *Journal du voyage*, 19–20, and *Mesure des trois premiers degrés du méridien dans l'hémisphère austral, tirée des observations de MM. de l'Académie royale des sciences envoyés par le roi sous l'Équateur* (Measure of the First Three Degrees of the Meridian in the Southern Hemisphere, Drawn from the Observations of the Messieurs of the Royal Academy of Sciences, Sent by the King Below the Equator) (Paris: Imprimerie Royale, 1751), 4–10, 80–85. Quotes from Bouguer to Bignon, Quito, Feb. 10, 1737; and Bouguer, *La Figure de la terre*, 57.

31. Quote: Jorge Juan and Ulloa, *Relación histórica*, 1:317. Information on *ceques* is from Brian S. Bauer, *The Sacred Landscape of the Inca: The Cusco Ceque System* (Austin: University of Texas Press, 1998).

32. La Condamine, *Journal du voyage*, 20. A letter from Verguin dated Oct. 2, 1736, also mentioned the construction of a shelter on the summit (AARS Biographical Dossier La Condamine, 50J, in a notice of letters sold at auction in July 1991).

33. Bouguer, *La Figure de la terre*, 40.

34. A complete explanation of the calculation and reduction of the baseline to level is given in Antonio Lafuente García and Antonio J. Delgado, *La geometrización de la tierra:*

Observaciones y resultados de la expedición geodésica hispanofrancesa al virreinato del Perú (1735–1744) (The Geometrization of the Earth: Observations and Results of the Hispanic/French Geodesic Expedition to the Viceroyalty of Peru, 1735–1744) (Madrid: CSIC, 1984), 66–86.

35. On the Maupertuis expedition, see Robert Iliffe, "'Aplatisseur du monde et de Cassini': Maupertuis, Precision Measurement and the Shape of the Earth in the 1730s," *History of Science* 31 (1993); and Mary Terrall, *The Man Who Flattened the Earth: Maupertuis and the Sciences in the Enlightenment* (Chicago: University of Chicago Press, 2002).

36. Maupertuis to La Condamine, Thury, Sept. 8, 1735 (AARS Biographical Dossier La Condamine, 50J fol. 91); La Condamine, *Journal du voyage*, 44. La Condamine's scrawled note on the letter clearly records that he received it on Nov. 21, 1736, even though his *Journal* gives October 1736 as the date of arrival.

37. Jorge Juan y Santacilia, Antonio de Ulloa y de la Torre-Guiral, and Luís Javier Ramos Gómez, *Epoca, génesis y texto de las 'Noticias secretas de América,' de Jorge Juan y Antonio de Ulloa, 1735–1745* (Epoch, Genesis and Text of the "Secret Notices of America" of Jorge Juan and Ulloa, 1735–1745), 2 vols. (Madrid: CSIC, 1985), 2:245.

38. The descriptions of the Yaruquí mill complex and the harsh conditions of the *mita* are from Nicholas P. Cushner, *Farm and Factory: The Jesuits and the Development of Agrarian Capitalism in Colonial Quito, 1600–1767* (Albany: State University of New York Press, 1982), 89–123; and Ramos Gómez, "Algunos datos sobre los abusos e injusticias padecidas en 1737 por los indios de los obrajes de la ciudad de Quito" (Some Data on the Abuses and Injustices Suffered in 1737 by the Indians in the Mills of the City of Quito), *Revista Española de Antropología Americana* 27 (1997): 153–166.

39. Jorge Juan, Ulloa, and Ramos Gómez, *Las 'Noticias secretas de América'*, 2:219–221.

40. La Condamine, *Journal du voyage*, 21–22.

41. Bouguer, *La Figure de la terre*, 230–231.

42. Bouguer, "Relation des observations faites à Quito de l'obliquité de l'Ecliptique," AOP A.A. 2.4; Bouguer, *A Relation of the Observations Made at Quito, of the Obliquity of the Ecliptic at the Latter Solstice of 1736, and the Former of 1737, with an Instrument of 12 Feet Radius*, trans. Edmond Halley (?) (London: n.p., 1738).

43. Quotes from Stephen Peter Rigaud and Stephen Jordan Rigaud, eds., *Correspondence of Scientific Men of the Seventeenth Century*, 2 vols. (Oxford: Oxford University Press, 1841; rpt., Hildesheim: G. Olms, 1965), 2:333–335.

Chapter V City of Kings

1. Ruth Hill, *Hierarchy, Commerce and Fraud in Bourbon Spanish America* (Nashville: Vanderbilt University Press, 2005), 34–135; Angel Sanz Tapia, "El acceso a los cargos de gobierno de la audiencia de Quito, 1701–1750" (Access to the Positions of Government in the *Audiencia* of Quito, 1701–1750), *Anuario de Estudios Americanos* 63, no. 2 (2006): 49–73.

2. Jorge Juan y Santacilia, Antonio de Ulloa y de la Torre-Guiral, and Luís Javier Ramos Gómez, *Epoca, génesis y texto de las 'Noticias secretas de América,' de Jorge Juan y Antonio de Ulloa, 1735–1745* (Epoch, Genesis and Text of the "Secret Notices of America" of Jorge Juan and Ulloa, 1735–1745), vol. 1, *El viaje a América* (Voyage to America) (Madrid: CSIC, 1985), 55–60; Hill, *Hierarchy,* 134–135.

3. Unfortunately for Alsedo, by denouncing Araujo he had unknowingly ensnared himself, for during the three days he was still president, it had been within his authority to arrest Araujo and confiscate the smuggled goods. By not doing so he had himself violated the law. These denunciations would take several years to weave through the Spanish court system before being finally resolved against *both* Alsedo and Araujo. See Luís Javier Ramos Gómez, "La pugna por el poder local en Quito entre 1737 y 1745 según e proceso contra el Presidente de la Audiencia, José de Araujo y Río" ("The Fight for Local Power in Quito Between 1737 and 1745, According to the Trial Against the President of the Audiencia, José de Araujo y Río), *Revista Complutense de Historia de América* 18 (1992): 179–196.

4. Luís Javier Ramos Gómez, "Enfrentamientos entre grupos de poder por el domino del Cabildo de Quito entre 1735 y 1739" (Confrontations Between Power Groups for the Control of the City Council of Quito Between 1735 and 1739), *Revista Complutense de Historia de América* 31 (2005): 53–77, quote at 62.

5. Quote from Jorge Juan, Ulloa, and Ramos Gómez, *Las 'Noticias secretas de América'*, 1:71. On the hierarchy of salutations, it should be noted that in the exchange of letters between La Condamine and Alsedo in June 1736 described in Chapter 4, La Condamine addressed Alsedo as *vuesa señoría*, and Alsedo called La Condamine *vuesa merced*, showing a clear understanding of who outranked whom.

6. Charles-Marie de La Condamine, *Journal du voyage fait à l'Équateur servant d'introduction historique à la Mesure des trois premiers degrés du Méridien* (Journal of the Voyage Made to the Equator, Serving as a Historical Introduction to the Measure of the First Three Degrees of the Meridian) (Paris: Imprimerie Royale, 1751), 27; Neptali Zúñiga, *La expedición científica de Francia del siglo XVIII en la Presidencia de Quito* (The Scientific Expedition of France in the 18th Century in the Audiencia of Quito) (Quito: Instituto Panamericano de Geografía e Historia, Sección Nacional del Ecuador, 1977/1986), 30–31.

7. La Condamine, *Journal du voyage*, 22.

8. Jean Descola, *Daily Life in Colonial Peru, 1710–1820* (New York: Macmillan, 1968), 79–80, 114, 224.

9. La Condamine, *Journal du voyage*, 22–25.

10. Ibid., 25–27.

11. Ulloa described the incident to Patiño, though leaving out some crucial details, in a letter dated Feb. 12, 1737 (AGI Quito 133 fol. 206–223). The court case that followed is documented in AGI Quito 193, "Expediente sobre las quejas que dio el Presidente de Quito don José Araujo Río de haberle perdido el respeto don Antonio de Ulloa y don Jorge Juan" (Docket About the Complaints of President Araujo of Quito, Having Lost Respect for Ulloa and Jorge Juan), which is reproduced in Santiago Montoto, "El proceso contra Jorge Juan y Antonio de Ulloa en Quito, 1737" (The Trial Against Jorge Juan and Antonio de Ulloa in Quito, 1737), *Anuario de estudios americanos* 5 (1948): 747–780. The incident was of course glossed over in the published accounts of the expedition. It is critically analyzed in Tamar Herzog, *Upholding Justice: Society, State, and the Penal System in Quito (1650–1750)* (Ann Arbor: University of Michigan Press, 2004), 221–225; and Jorge Juan, Ulloa, and Ramos Gómez, *Las 'Noticias secretas de América'*, 1:70–81 (Ramos Gómez convincingly argues that Valparda was behind the whole incident).

12. Certificate of Godin et al., Quito, Feb. 11, 1737, BNCBA 4/7 fol. 398–401; Certificate of Bouguer, Quito, Feb. 17, 1737, AGI Quito 133 fols. 224–225; José de Zenitagoya to Patiño, Quito, Mar. 2, 1737, AGI Quito 133 fol. 167.

13. La Condamine, *Journal du voyage*, 26–27.

14. La Condamine's home (near the corner of Imbabura and Mideros) is still known locally as the "Observatory of the French Academics"; Fernando Jurado Noboa, *Calles, Casas y Gente del Centro Histórico de Quito* (Streets, Houses and People of Quito's Historic Center), 7 vols. (Quito: Fonsal, 2004–2009), 5:177–179. Lorenzo Nates was identified as the expedition's banker in a letter from Bouguer; Bouguer to unknown recipient, Quito, April(?) 1738, BNF *Nouvelles acquisitions francaises* 6197, folios 33–36. La Condamine stated (*Journal du voyage*, 127) that Bouguer lived with the "most famous lawyer in Quito" and commissioner of the Inquisition, later explaining that this was "Joseph Sánchez" (*Supplément* 2:xv). A brief biography of José Sánchez de Orellana is in Alfonso Anda Aguirre, *Los Marqueses de Solanda* (Quito: Editorial Casa de la Cultura Ecuatoriana, 1971), 57–58.

15. Zenitagoya to Patiño, Quito, Mar. 2, 1737. Ricardo Palma, the great Peruvian author and historian, explains that the folklore of *los caballeros del punto fijo* came from a now-lost manuscript originally in the National Library of Peru, titled *Viaje al globo de la luna* (Voyage to the Globe of the Moon), circa 1790, by an anonymous author, which touched on various technical and scientific themes around Peru, including a description of a flying machine. See *Tradiciones Peruanos* (Peruvian Traditions), 6 vols. (Madrid: Espasa Calpe, 1966–1967), 1:286, 2:87. "Knights of the exact" is from Robert Whitaker, *The Mapmaker's*

Wife: A True Tale of Love, Murder and Survival in the Amazon (New York: Perseus Books, 2004), 112.

Chapter VI The Triangles of Peru

1. The dispute between Bouguer and Godin over which measure to use is detailed in Robert Finn Erickson, "The French Academy of Sciences Expedition to Spanish America, 1735–1744," PhD diss., University of Illinois, Champaign-Urbana, 1955, 104–109; and in Antonio Lafuente García and Antonio J. Delgado, *La geometrización de la tierra: Observaciones y resultados de la expedición geodésica hispanofrancesa al virreinato del Perú (1735–1744)* (The Geometrization of the Earth: Observations and Results of the Hispanic/French Geodesic Expedition to the Viceroyalty of Peru, 1735–1744) (Madrid: CSIC 1984), 45–51.

2. Examples of these initial survey maps are in MNHN Ms 111.

3. Charles-Marie de La Condamine, *Journal du voyage fait à l'Équateur servant d'introduction historique à la Mesure des trois premiers degrés du Méridien* (Journal of the Voyage Made to the Equator, Serving as a Historical Introduction to the Measure of the First Three Degrees of the Meridian) (Paris: Imprimerie Royale, 1751), 30–31; Erickson, "The French Academy of Sciences Expedition," 118–119, 184–185.

4. Seniergues, while he was in Cartagena from 1737 to 1738, had explained to Alsedo that the group expected to be finished by May 1738: see *Informe de Dionisio de Alsedo y Herrera* (Alsedo's Report to the King), Madrid, Nov. 18, 1740, AGI Quito 134 fol. 379–380.

5. On the construction and use of the quadrant, see Danielle Fauque, "Un instrument essential de l'expédition pour la mesure de la terre: Le quart de cercle mobile" (An Essential Instrument of the Expedition for Measuring the Earth: The Portable Quadrant), in *La figure de la Terre du XVIIIe siècle à l'ère spatiale* (The Figure of the Earth from the 18th Century to the Space Age), ed. Henri Lacombe and Pierre Costable (Paris: Gauthier-Villers, 1988), 209–221.

6. James R. Smith, *From Plane to Spheroid: Determining the Figure of the Earth from 3000 B.C. to the 18th Century Lapland and Peruvian Survey Expeditions* (Rancho Cordova, CA: Landmark Enterprises, 1986), 151.

7. Erickson, "The French Academy of Sciences Expedition," 110.

8. Quote from Pierre Bouguer, *La Figure de la terre, déterminée par les observations de Messieurs Bouguer et de La Condamine, de l'Académie Royale des Sciences, envoyés par ordre du Roy au Pérou pour observer aux environs de l'équateur, avec une Relation abrégée de ce voyage qui contient la description du pays dans lequel les opérations ont été faites* (The Figure of the Earth, Determined by the Observations of the Messieurs Bouguer and La Condamine, of the Royal Academy of Sciences, Sent to Peru by Order of the King to Make Observations near the Equator, with an Abridged Relation of the Voyage Containing a Description of the Countries in Which the Observations Were Made) (Paris: Charles-Antoine Jombert, 1749), 262. See also La Condamine, *Journal du voyage*, 34–37; and Jorge Juan y Santacilia and Antonio de Ulloa y de la Torre-Guiral, *Relación histórica del Viaje á la América Meridional* (Historical Relation of the Voyage to South America), 2 vols. (Madrid: Antonio Marín, 1748; facsimile, ed. José Merino Navarro and Miguel Rodriguez San Vicente, Fundación universitaria española, 1978), 1:305–306.

9. Bouguer, *La Figure de la terre*, xxxvi–xxxvii.

10. La Condamine, *Journal du voyage*, 45–46.

11. Ibid., 52.

12. Jorge Juan and Ulloa, *Relación histórica*, 1:592–593; Bouguer, *La Figure de la terre*, xliii–xliv. For an explanation of this phenomenon, see David K. Lynch and Susan N. Futterman, "Ulloa's Observations of the Glory, Fogbow, and an Unidentified Phenomenon," *Applied Optics* 30 (1991): 3538–3541.

13. Jorge Juan and Ulloa, *Relación histórica*, 1:308–311.

14. Bouguer to Réaumur, Quito, April 12, 1738, AARS Dossier Pierre Bouguer. For a complete account of these developments in naval architecture, see Larrie D. Ferreiro,

Ships and Science: The Birth of Naval Architecture in the Scientific Revolution, 1600–1800 (Cambridge, MA: MIT Press, 2007).

15. Tamar Herzog, *Upholding Justice: Society, State, and the Penal System in Quito (1650–1750)* (Ann Arbor: University of Michigan Press, 2004), 113, 174–175.

16. Murder of Bouguer's slave: La Condamine, *Journal du voyage*, 55–56. Regarding *pishtacos*, see Mary J. Weismantel, *Cholas and Pishtacos: Stories of Race and Sex in the Andes* (Chicago: University of Chicago Press, 2001). The stories of *pishtacos* persisted even during the twentieth century. My wife's grandmother María Maraví Olivos, who grew up during the 1910s and 1920s in the Junín region of the central Peruvian highlands, related that her father, Ricardo Maraví Peña, was friendly with a *pishtaco* who supposedly sold body fat to the railroads to keep the engines lubricated. She said that the children would hide under the beds whenever the man came to their house but that her father wisely believed it was better to have the *pishtaco* as a friend rather than an enemy.

17. Jorge Juan and Ulloa, *Relación histórica*, 1:319–320.

18. Bouguer to unknown recipient, Quito, April(?) 1738. Maurepas' warning: Antonio Lafuente García, "Una ciencia para el estado: La expedición geodesica hispano-francesa al virreinato del Perú (1734–1743)" (Science for the State: The French-Spanish Geodesic Mission to the Viceroyalty of Peru, 1734–1743), *Revista de Indias* 43 (1983): 580. Maldonado's loan: Neptali Zúñiga, *La expedición científica de Francia del siglo XVIII en la Presidencia de Quito* (The Scientific Expedition of France in the 18th Century in the Audiencia of Quito) (Quito: Instituto Panamericano de Geografía e Historia, Sección Nacional del Ecuador, 1977/1986), 39.

19. Some excellent biographies of Pedro Vicente Maldonado and his extended family are Neptali Zúñiga, *Pedro Vicente Maldonado: Un científico de América* (Pedro Vicente Maldonado: An American Scientist) (Madrid: Publicaciones Españolas, 1951); Piedad Costales and Alfredo Costales, *Los Maldonado en la Real Audiencia de Quito* (The Maldonados in the Royal Audiencia of Quito) (Quito: Banco Central de Ecuador, 1987); and Carlos Ortiz Arellano, *Pedro Vicente Maldonado: Forjador de la Patria Ecuatoriana, 1704–1748* (Pedro Vicente Maldonado: The Man Who Forged the Ecuadorian Fatherland) (Quito: Casa de la Cultura Ecuatoriana Benjamín Carrion, 2004).

20. Jorge Juan y Santacilia and Antonio de Ulloa y de la Torre-Guiral, *Observaciones astronómicas y físicas hechas en los Reynos del Perú* (Astronomical and Physical Observations Made in the Kingdoms of Peru) (Madrid: Juan de Zúñiga, 1748), 132–143. For a modern examination of these experiments, see José Manuel Vaquero-Martínez et al., "Analysis of an Early Measurement of the Speed of Sound Propagation in the Atmosphere," *Applied Acoustics* 65, no. 1 (2004): 59–67.

21. Several authors provide highly detailed itineraries of the triangulation, compiled from the various primary sources: Lafuente and Delgado, *La geometrización de la tierra*, 119–123; Jorge Juan y Santacilia, Antonio de Ulloa y de la Torre-Guiral, and Luís Javier Ramos Gómez, *Epoca, génesis y texto de las 'Noticias secretas de América,' de Jorge Juan y Antonio de Ulloa, 1735–1745* (Epoch, Genesis and Text of the "Secret Notices of America" of Jorge Juan and Ulloa, 1735–1745), vol. 1, *El viaje a América* (Voyage to America) (Madrid: CSIC, 1985), 91–98; and Smith, *From Plane to Spheroid*, 111–117.

22. La Condamine, *Journal du voyage*, 59–60.

23. Bouguer to unknown recipient, Quito, April(?) 1738.

24. Ibid.

25. La Condamine, *Journal du voyage*, 65–71.

26. The evidence for the whereabouts of Jussieu, Seniergues, and Morainville is as follows: They were not in Cayambe in September 1736, when Couplet fell ill and died of malaria. In January 1737, Jussieu did not accompany La Condamine on his trip south so as to view the all-important cinchona tree, instead giving him detailed instructions on what to look for. In February 1737, Jussieu, Seniergues, and Morainville were in Quito when they signed the testimonial supporting Jorge Juan and Ulloa against Araujo. In July of that year, Jussieu, still in Quito, penned a four-page memoir on the cinchona tree, based on La Condamine's observations after he had just returned from Lima (Joseph de Jussieu, untitled notes on

the cinchona tree, Quito, July 20, 1737, MNHN Ms 1626–2). In 1737 Seniergues trekked overland, alone, to Cartagena de Indias, returning to Quito by the beginning of 1739. Apart from these few facts, we know nothing of their activities.

27. *Informe de Dionisio de Alsedo y Herrera*, Madrid, Nov. 18, 1740. Alsedo lived in Cartagena de Indias from 1737 until 1739, after which he returned to Spain.

28. Zúñiga, *La expedición científica de Francia*, 31; Enrique Ayala Mora ed., *Nueva Historia del Ecuador* (New History of Ecuador), 15 vols. (Quito: Corporación Editora Nacional—Grijalbo, 1983–1995), 5:207.

29. Jorge Juan and Ulloa, *Relación histórica*, 1:433. Comparative mortality rates from Paolo Malanima, *Pre-Modern European Economy* (Leiden: Brill, 2009), 39; and Victor Bulmer-Thomas et al. eds., *The Cambridge Economic History of Latin America: The Colonial Era* (Cambridge UK: Cambridge University Press, 2006), 170.

30. The influence of disease transmission from Europe to the Americas is highlighted in two recent works: Diamond Jared, *Guns, Germs and Steel* (New York: Norton, 2005) and Charles C. Mann, *1491* (New York: Knopf, 2005). For a comprehensive examination of the impact of disease, see Suzanne Alchon, *Native Society and Disease in Colonial Ecuador* (Cambridge: Cambridge University Press, 1991). Data on the 1718–1723 influenza epidemic is from Adrian J. Pearce, "The Peruvian Population Census of 1725–1740," *Latin American Research Review* 36, no. 3 (2001): 70.

31. Seniergues gave Jussieu 1,120 pesos (about $50,000) before he left for Cartagena de Indias. See "Testamento de don Juan Seniergues" (Testament of Mr. Jean Seniergues), in Octavio Cordero Palacios, "La muerte de don Juan Seniergues" (The Murder of Mr. Jean Seniergues), *Revista del Centro de Estudios Históricos y Geograficos de Cuenca* 12 (Dec. 1925): 254–259. His cataract surgery and visit to Cartagena are described in La Condamine, *Journal du voyage*, 59, 76, 131, and *Informe de Dionisio de Alsedo y Herrera*. His slave buying is inferred from the fact that he had none upon his entry to Quito in 1736 but he had two slaves in Cuenca in 1739.

32. La Condamine, *Journal du voyage*, 76; "Testamento de Don Juan Seniergues."

33. A complete history of quinine is given in Fiammetta Rocco, *The Miraculous Fever-tree: Malaria and the Quest for a Cure That Changed the World* (New York: Harper Collins, 2003).

34. La Condamine, "Sur l'arbre de quinquina" (On the cinchona tree), HMARS 1738 (1742): 226–244.

35. Joseph de Jussieu, "Descripto arboris Kina Kina" (Description of the Cinchona Tree), c. 1739, MNHN Ms 1626–6. It was published in translation almost two hundred years later as *Description de l'Arbre a Quinquina* (Paris: Société du Traitement des Quinquinas, 1936). Morainville's drawings and aquarelles are in two separate folios, MNHN Ms 111 and Ms 1626. See also Gaston Lehir, "Joseph de Jussieu et son exploration en Amérique Méridionale (1735–1769)" (Joseph de Jussieu and His Exploration in South America, 1735–1769), master's thesis, University of Montreal, 1976, 102–108; and Germán Rodas Chaves, "J. de Morainville y el primer dibujo universal de la quina o cascarilla" (J. de Morainville and the First Universal Drawing of Cinchona or Cascarilla), *Bulletin de l'Institut Français d'Études Andines* 32, no. 3 (2003): 431–440.

36. La Condamine, *Journal du voyage*, 66–67. On the Dávalos family, see Elmer Carvajal, *Riobamba: Personajes Ilustres de la Colonia* (Riobamba: Illustrious Personages of the Colony) (Riobamba, Ecuador: Editorial Pedagógica Freire, 1999), 32–37.

37. La Condamine, *Journal du voyage*, 62. Maupertuis' expedition is detailed in Robert Iliffe, "'Aplatisseur du monde et de Cassini': Maupertuis, Precision Measurement and the Shape of the Earth in the 1730s," *History of Science* 31 (1993); and Mary Terrall, *The Man Who Flattened the Earth: Maupertuis and the Sciences in the Enlightenment* (Chicago: University of Chicago Press, 2002).

38. Bouguer's memoir on his gravity experiment is found in his *La Figure de la terre*, 367–394. Further details are from La Condamine, *Journal du voyage*, 68–70. A recent re-examination of his results shows that deep crustal features were responsible for his anomalous results. See John R. Smallwood, "Bouguer Redeemed: The Successful 1737–1740

Gravity Experiments on Pichincha and Chimborazo," *Earth Sciences History* 29, no. 1 (2010): 1–25.

39. Comparative height data from Deborah Poole, *Vision, Race and Modernity: A Visual Economy of the Andean Image World* (Princeton, NJ: Princeton University Press, 1997), 59; and Richard Steckel, "Health and Nutrition in the Pre-Industrial Era: Insights from a Millennium of Average Heights in Northern Europe" in *Living Standards in the Past: New Perspectives on Well-Being in Asia and Europe*, ed. Robert C. Allen et al. (Oxford: Oxford University Press, 2005), 237. My thanks to Nelson Gómez Espinoza for explaining the contributions of the regional Indians to the Geodesic Mission.

40. La Condamine, *Journal du voyage*, 52; Bouguer, *La Figure de la terre*, xcix. Quote from Jorge Juan and Ulloa, *Relación histórica*, 1:544–545.

41. Juan de Velasco, *Historia del reino de Quito en la América meridional, 1789* (History of the Kingdom of Quito in South America, 1789), vol. 1 (Quito: Editorial Casa de la Cultura Ecuatoriana, 1977), 342.

42. La Condamine, *Journal du voyage*, 81; La Condamine, "Mémoire sur quelques anciens monumens du Pérou, du tems des Incas" (Memoir on Some Ancient Monuments of Peru in the Time of the Incas), *HMAB 1746* (1748): 436–456, quote at 441. See also Monica Barnes and David Fleming, "Charles-Marie de La Condamine's Report on Ingapirca and the Development of Scientific Field Work in the Andes, 1735–1744," *Andean Past* 2 (1989): 175–236.

43. La Condamine, *Journal du voyage*, 76–80, quote at 78.

44. Erickson, "The French Academy of Sciences Expedition," 125–128.

45. Charles-Marie de La Condamine, *Mesure des trois premiers degrés du méridien dans l'hémisphère austral, tirée des observations de MM. de l'Académie royale des sciences envoyés par le roi sous l'Équateur* (Measure of the First Three Degrees of the Meridian in the Southern Hemisphere, Drawn from the Observations of the Messieurs of the Royal Academy of Sciences, Sent by the King Below the Equator) (Paris: Imprimerie Royale, 1751), 71–75; Jorge Juan and Ulloa, *Observaciones astronómicas*, 166–167.

46. Examples of the field calculations made by La Condamine are found in AARS Biographical Dossier La Condamine 50J, fols. 17,18.

47. A complete record and painstaking recalculation of the triangulation, including explanations for sources of error, is given in Smith, *From Plane to Spheroid*, 118–148. Godin left no notes of his initial measurements, and Jorge Juan did not provide any information on them in his published accounts.

48. La Condamine, *Journal du voyage*, 85; Jorge Juan and Ulloa to Mateo Pablo Díaz de Lavandero, Marqués de Torrenueva, Cuenca, August 26, 1739, BNCBA 4/5 fol. 430–437. Torrenueva had taken over the Ministry of the Navy portfolio upon the death of Patiño in 1736.

49. La Condamine, "Mémoire sur quelques anciens monumens," 439; La Condamine, *Journal du voyage*, 32, 97, 104.

50. Extracts from *Gazette d'Amsterdam* 1738, dated January 14, April 1, and July 8, in AOP A.A. 7.7. See also Anne-Marie Chouillet, "Rôle de la presse périodique de langue française dans la diffusion des informations concernant les missions en Laponie ou sous l'équateur" (Role of the French-Language Newspapers in the Diffusion of Information Concerning the Missions in Lapland or Under the Equator), in *La figure de la Terre du XVIIIe siècle à l'ère spatiale* (The Figure of the Earth from the 18th Century to the Space Age), ed. Henri Lacombe and Pierre Costable (Paris: Gauthier-Villers, 1988), 174.

51. Quoted in Michael R. Hoare, *The Quest for the True Figure of the Earth: Ideas and Explorations in Four Centuries of Geodesy* (London: Ashgate, 2005), 165.

Chapter VII Death and the Surgeon

1. The information in this chapter comes principally from two sources: first, the depositions of witnesses recorded in two handwritten sets of court documents found in the archives of Quito and Cuenca; second, a famous account published by Charles-Marie de

La Condamine in Paris seven years after the events. The first set of court documents is located in Quito: *Autos criminales seguidos para descubrir autores, cómplices y encubridores del asesinato de Senierges* (Criminal Case Pursued to Discover the Authors, Accomplices and Accessories After the Fact of the Assassination of Seniergues), Aug. 30, 1739, to Sept. 2, 1768, *ANEC Criminales 14*, 30, no. 5. (My thanks to María Antonieta Vásquez Hahn and José Lorenzo Saa Bernstein for providing me a copy of the originals and a transcription.) The second set is in Cuenca: *Documentos sobre la muerte de Seniergues* (Documents on the Murder of Seniergues), Sept. 2, 1739, to Nov. 10, 1744, Universidad de Cuenca, Biblioteca Juan Bautista Vazquez, Código 340.986.6. These latter documents were largely transcribed by the great Ecuadorian historian Octavio Cordero Palacios in a series of articles: "La muerte de don Juan Seniergues" (The Murder of Mr. Jean Seniergues), *Revista del Centro de Estudios Históricos y Geograficos de Cuenca*, vol. 11 (Dec. 1924): 42–184; vol. 12 (Dec. 1925): 206–298; and vol. 13 (Apr. 1928): 302–317. Dr. Cordero died before he was able to provide his promised analysis of the affair.

La Condamine's account, *Lettre à Mme ***, sur l'émeute populaire excitée en la ville de Cuenca, au Pérou, le 29 d'août 1739, contre les académiciens des sciences envoyés pour la mesure de la terre* (Letter to Mrs. X, on the Popular Uprising Raised in the City of Cuenca, in Peru, 29 August 1739, Against the Scientific Academicians Sent to Measure the Earth) (Paris: Veuve Pissot, 1745), was first appended to the account of his journey down the Amazon and later included in the second edition of his *Journal du voyage fait à l'Équateur servant d'introduction historique à la Mesure des trois premiers degrés du Méridien* (Journal of the Voyage Made to the Equator, Serving as a Historical Introduction to the Measure of the First Three Degrees of the Meridian) (Paris: Imprimerie Royale, 1751). His transcribed documents from the original court case, *Copie authentique du procès introduit à l'audience de Quito, à l'occasion de l'assassinat commis sur la personne de Jean Seniergues, à Quito 7 Septembre 1742* (Authentic Copy of the Trial Introduced at the Audiencia of Quito on the Occasion of the Assassination Committed Against the Person of Jean Seniergues), are found in BNF *cotes Espagnols* 51.

In reconstructing these events, I have only used those accounts that were corroborated by multiple witnesses. All quotes are directly from the testimonies. My thanks to Raúl Hernández Asensio, who provided me his unpublished analysis on the politics surrounding the events.

2. René Taton, ed., *Enseignement et diffusion des sciences en France au XVIIIe siècle* (Instruction and Diffusion of Sciences in France in the 18th Century) (Paris: Hermann, 1964), 190.

3. Testimony of Manuel Astudillo, *Autos criminales seguidos*, fol. 6v; testimony of Andrés Cubillús, *Documentos sobre la muerte*, fol. 53–54, and Cordero Palacios, "Muerte de Seniergues," 246–247.

4. Testimony of Nicholás de Neyra, *Autos criminales seguidos*, fol. 9v.

5. Maximiliano Borrero Crespo, *Orígenes Cuencanos* (Cuencan Lineages), 2 vols. (Cuenca: Universidad de Cuenca, 1962), 2:18, 104, 238–239. La Condamine mistakenly identified León's wife, for whom he jilted Manuela, as Serrano's daughter (*Lettre à Mme ****, 6, 10). This error was repeated ever since by other historians. Serrano, however, was just thirty-five at the time and married with three young children, the oldest of whom was six.

6. Testimony of Tomas Nugent, *Documentos sobre la muerte*, fol. 7v; Cordero Palacios, "Muerte de Seniergues," 179.

7. Romero, *Quito en los ojos de los viajeros*, 192.

8. Testimony of Nicholás de Neyra, *Autos criminales seguidos*, fol. 9v.

9. Testimonies of Sebastián Serrano, *Autos criminales seguidos*, fol. 6, and Nicolás de Neyra, *Autos criminales seguidos*, fol. 8v.

10. Testimony of Nicolás de Neyra, *Autos criminales seguidos*, fol. 9.

11. Testimony of Sebastián Serrano, *Documentos sobre la muerte*, fol. 14; Cordero Palacios, "Muerte de Seniergues," 162–163.

12. La Condamine, *Lettre à Mme ****, 28.

13. Testimony of Sebastián de la Madris y Cosio, *Documentos sobre la muerte*, fol. 3v; Cordero Palacios, "Muerte de Seniergues," 172.

14. See Michael F. Brown and Eduardo Fernández, *War of Shadows: The Struggle for Utopia in the Peruvian Amazon* (Berkeley: University of California Press, 1993). The more famous Juan Santos Atahualpa led his rebellion just a few years after Torote, in 1742.

15. Joseph de Jussieu to Antoine de Jussieu, Cuenca, Aug. 31, 1739, *MNHN* Ms 179/25. I would like to thank Dr. Jay Menaker, of the R. Adams Cowley Shock Trauma Center in Baltimore, for explaining the progression of wounds like Seniergues'.

16. "Declaration et Testamento de don Juan Seniergues" (Declaration and Testament of Mr. Jean Seniergues), in Corderos Palacio, "Muerte de Seniergues," 244, 254–259.

17. See Tamar Herzog, *Upholding Justice: Society, State, and the Penal System in Quito (1650–1750)* (Ann Arbor: University of Michigan Press, 2004), for a fascinating and detailed look at the legal system of colonial Quito.

18. The actions of the magistrates are analyzed in Juan Chacón Zhapán, *Historia del corregimiento de Cuenca, 1557–1777* (History of the Municipality of Cuenca, 1557–1777) (Quito: Banco Central del Ecuador, 1990), 320–325.

19. *Autos criminales enviados desde Cuenca al tribunal de la Audiencia contra el capitán Diego de León, quien huyó de la cárcel en donde se hallaba desde hace unos dos años atrás implicado en el tumulto que se originó en dicha ciudad y causó la muerte del francés Juan Sénierges* (Criminal Case Against Captain Diego de León, Who Fled the Prison Where He Was Located Since One or Two Years Earlier, Implicated in the Tumult in Said City and Caused the Death of the Frenchman Jean Seniergues), April 6,1742, *ANEC Criminales* 14, 31, no. 12. My thanks to Raúl Hernández Asensio for bringing this source to my attention.

20. Voltaire, *The Complete Works of Voltaire*, ed. Theodore Besterman, et al., 135+ vols. (Toronto: University of Toronto Press, 1968–present), 130:82.

21. La Condamine, *Lettre à Mme ****, 46. Two works of fiction carry on the legend of an indiscreet, and fatal, love triangle: Manuel Coronel, *La muerte de Seniergues: Leyenda histórica* (The Death of Seniergues: Historical Legend) (Cuenca, Ecuador: Imprimiera de la Alianza Obrera, 1906); and Gonzalo Humberto Mata, *Cusinga: Capulí en Lis: Exégesis y poema, romance histórico* (Cusinga [Happy One]: Cherry Blossom: Interpretation and Poem, Historical Romance) (Cuenca, Ecuador: Editorial Biblioteca Cenit, 1944).

22. Alexander von Humboldt, *Reise auf dem Río Magdalena, durch die Anden und Mexico* (Journey on the Magdalena River, Through the Andes and Mexico), ed. Margot Faak, 2 vols. (Berlin: Akademie-Verlag, 1986), 1:234.

23. Borrero Crespo, *Orígenes Cuencanos*, 2:163. Humboldt claimed that the woman at the center of the Seniergues dispute (her married name would have been Manuela Ullauri de Quesada) had died just shortly before he arrived in Cuenca in 1802 (*Reise auf dem Río Magdalena*, 1:234).

24. In 1866 the Inmaculada Concepción cathedral was erected on the site of the Jesuit church. See Ross Jamieson, *Domestic Architecture and Power: The Historical Archaeology of Colonial Ecuador* (New York: Kluwer Academic/Plenum, 2000), 46. Thanks to Patrick Drevet for pointing out the connection with El Dorado.

Chapter VIII The War of Jenkins's Ear

1. Antonio Lafuente García and Antonio J. Delgado, *La geometrización de la tierra: Observaciones y resultados de la expedición geodésica hispanofrancesa al virreinato del Perú (1735–1744)* (The Geometrization of the Earth: Observations and Results of the Hispanic/French Geodesic Expedition to the Viceroyalty of Peru, 1735–1744) (Madrid: CSIC 1984), 213–216.

2. Juan de Velasco, *Historia del reino de Quito en la América meridional, 1789* (History of the Kingdom of Quito in South America, 1789), vol. 2 (Caracas: Biblioteca Ayacucho, 1981), 397. Velasco mistakenly believed they were mapping the city of Cuenca.

3. *Procès-verbal des observations astronomiques à Tarqui (Statement of astronomical observations at Tarqui)*, Cuenca, Jan. 13, 1740, *AARS* Biographical Dossier La Condamine, 50J fol. 19. See also Jorge Juan y Santacilia and Antonio de Ulloa y de la Torre-Guiral, *Observaciones astronómicas y físicas hechas en los Reynos del Perú* (Astronomical and

Physical Observations Made in the Kingdoms of Peru) (Madrid: Juan de Zúñiga, 1748), 270–283; Charles-Marie de La Condamine, *Mesure des trois premiers degrés du méridien dans l'hémisphère austral, tirée des observations de MM. de l'Académie royale des sciences envoyés par le roi sous l'Équateur* (Measure of the First Three Degrees of the Meridian in the Southern Hemisphere, Drawn from the Observations of the Messieurs of the Royal Academy of Sciences, Sent by the King Below the Equator) (Paris: Imprimerie Royale, 1751), 128–158.

4. Charles-Marie de La Condamine, *Journal du voyage fait à l'Equateur servant d'introduction historique à la Mesure des trois premiers degrés du Méridien* (Journal of the Voyage Made to the Equator, Serving as a Historical Introduction to the Measure of the First Three Degrees of the Meridian) (Paris: Imprimerie Royale, 1751), 88.

5. Ibid., 90–91. The marriage was probably of either Liberata or Mariana, the daughters of the hacienda owners Julián Mancheno and Rosa Maldonado; Carlos Ortiz Arellano, *Pedro Vicente Maldonado: Forjador de la Patria Ecuatoriana, 1704–1748* (Pedro Vicente Maldonado: The Man Who Forged the Ecuadorian Fatherland) (Quito: Casa de la Cultura Ecuatoriana Benjamín Carrion, 2004), 204.

6. La Condamine, *Mesure des trois premiers degrés du méridien*, 159–167, quote at 165, emphasis in the original. See also Pierre Bouguer, *La Figure de la terre, déterminée par les observations de Messieurs Bouguer et de La Condamine, de l'Académie Royale des Sciences, envoyés par ordre du Roy au Pérou pour observer aux environs de l'équateur, avec une Relation abrégée de ce voyage qui contient la description du pays dans lequel les opérations ont été faites* (The Figure of the Earth, Determined by the Observations of the Messieurs Bouguer and La Condamine, of the Royal Academy of Sciences, Sent to Peru by Order of the King to Make Observations near the Equator, with an Abridged Relation of the Voyage Containing a Description of the Countries in Which the Observations Were Made) (Paris: Charles-Antoine Jombert, 1749), 114–115.

7. La Condamine, *Journal du voyage*, 99.

8. Ortiz Arellano, *Pedro Vicente Maldonado*, 199. On the construction and impact of the Esmeraldas road, see Neptali Zúñiga, *Pedro Vicente Maldonado: Un científico de América* (Pedro Vicente Maldonado: An American Scientist) (Madrid: Publicaciones Españolas, 1951), 149–200.

9. Bouguer to La Condamine, Isla de Linga, July 18, 1740, AARS Biographical Dossier La Condamine, 50J fol. 39; Bouguer, "Relation du Voyage fait à la rivière des Emeraudes pour déterminer au dessus du niveau de la mer la hauteur absolue des Montagnes qui ont servi à la Méridienne" (Relation of Voyage to the Esmeraldas River to Determine the Absolute Height Above Sea Level of the Mountains for the Meridian), undated, AOP A.C 2.7. Quotes from Pierre Bouguer, "Relation abrégée du voyage fait au Pérou par Messieurs de l'Académie Royale des Sciences, pour mesurer les Degrés du Méridien aux environs de l'Equator, & en conclure la Figure de la Terre" (Abridged Relation of the Voyage Made to Peru by Gentlemen of the Royal Academy of Sciences, to Measure the Degrees of the Meridian Around the Equator, and to Deduce the Figure of the Earth), *HMARS 1744* (1748): 290. For an examination of Bouguer's observation of Illiniza, see Bouguer, *La Figure de la terre*, 159–163, and James R. Smith, *From Plane to Spheroid: Determining the Figure of the Earth from 3000 B.C. to the 18th Century Lapland and Peruvian Survey Expeditions* (Rancho Cordova, CA: Landmark Enterprises, 1986), 139–141.

10. La Condamine, *Journal du voyage*, 95; Jorge Juan y Santacilia, Antonio de Ulloa y de la Torre-Guiral, and Luís Javier Ramos Gómez, *Epoca, génesis y texto de las 'Noticias secretas de América,' de Jorge Juan y Antonio de Ulloa, 1735–1745* (Epoch, Genesis and Text of the "Secret Notices of America" of Jorge Juan and Ulloa, 1735–1745), vol. 1, *El viaje a América* (Voyage to America) (Madrid: CSIC, 1985), 1:130–133.

11. Kenneth J. Andrien, *The Kingdom of Quito, 1690–1830: The State and Regional Development* (Cambridge, UK: Cambridge University Press, 1995), 177–178.

12. Anson's voyage is memorably told in Glyn Williams, *The Prize of All the Oceans: Commodore Anson's Daring Voyage and Triumphant Capture of the Spanish Treasure Galleon* (New York: Penguin Books, 1999). A highly detailed account of the Spanish de-

fensive plans is given in Jorge Juan, Ulloa, and Ramos Gómez, *Las 'Noticias secretas de América,'* 103–181.

13. La Condamine, *Journal du voyage*, 95–103 (quote at 103).

14. Jorge Juan y Santacilia and Antonio de Ulloa y de la Torre-Guiral, *Relación histórica del Viaje á la América Meridional* (Historical Relation of the Voyage to South America), 2 vols. (Madrid: Antonio Marín, 1748; facsimile, ed. José Merino Navarro and Miguel Rodriguez San Vicente, Fundación universitaria española, 1978), 2:2–35 (quote at 4).

15. Robert Whitaker, *The Mapmaker's Wife: A True Tale of Love, Murder and Survival in the Amazon* (New York: Perseus Books, 2004), 157–158.

16. Godin des Odonais to Mathias Dávila, Cartagena, May 2, 1741, *Documentos sobre la muerte de Seniergues* (Documents on the Murder of Seniergues), Sept. 2, 1739, to Nov. 10, 1744, Universidad de Cuenca, Biblioteca Juan Bautista Vazquez, Código 340.986.6, fol. 19v.

17. La Condamine, *Journal du voyage*, 138.

18. Whitaker, *The Mapmaker's Wife*, 31–32, 160–161.

19. The story of Harrison's clock is brilliantly told in Dava Sobel's *Longitude* (New York: Walker and Company, 1995).

20. Details of Jorge Juan's and Ulloa's travels to Guayaquil, to Lima, and on the ocean are found in Jorge Juan and Ulloa, *Relación histórica*, 2:257–380 (quote at 267); and analyzed in Jorge Juan, Ulloa and Ramos Gómez, *Las 'Noticias secretas de América'*, 1:206–260.

21. Lawrence Clayton, *Caulkers and Carpenters in a New World: The Shipyards of Guayaquil* (Athens: Center for International Studies, Ohio University, 1980), 64.

22. Ulloa, "Informen los oficios del sueldo" (Report to the paymaster), Lima, Nov. 28, 1743, Latin America—Peru Mss., Manuscripts Department, Lilly Library, Indiana University.

Chapter IX The Dance of the Stars

1. On Graham and the zenith sector, see Allan Chapman, *Dividing the Circle: The Development of Critical Angular Measurement in Astronomy, 1500–1800* (New York: Ellis Horwood, 1990), 66–71, 87–89, 96–100, 144. The cost of Maupertuis' sector is from Terrall, *The Man Who Flattened the Earth*, 103.

2. Charles-Marie de La Condamine, *Journal du voyage fait à l'Equateur servant d'introduction historique à la Mesure des trois premiers degrés du Méridien* (Journal of the Voyage Made to the Equator, Serving as a Historical Introduction to the Measure of the First Three Degrees of the Meridian) (Paris: Imprimerie Royale, 1751), 102.

3. The known sources of astronomical error and their corrections were described in Jorge Juan y Santacilia and Antonio de Ulloa y de la Torre-Guiral, *Observaciones astronómicas y físicas hechas en los Reynos del Perú* (Astronomical and Physical Observations Made in the Kingdoms of Peru) (Madrid: Juan de Zúñiga, 1748), 270–286 (in which they described the corrections made by Godin, since he left no written records); Pierre Bouguer, *La Figure de la terre, déterminée par les observations de Messieurs Bouguer et de La Condamine, de l'Académie Royale des Sciences, envoyés par ordre du Roy au Pérou pour observer aux environs de l'équateur, avec une Relation abrégée de ce voyage qui contient la description du pays dans lequel les opérations ont été faites* (The Figure of the Earth, Determined by the Observations of the Messieurs Bouguer and La Condamine, of the Royal Academy of Sciences, Sent to Peru by Order of the King to Make Observations near the Equator, with an Abridged Relation of the Voyage Containing a Description of the Countries in Which the Observations Were Made) (Paris: Charles-Antoine Jombert, 1749), 172–226: and Charles-Marie de La Condamine, *Mesure des trois premiers degrés du méridien dans l'hémisphère austral, tirée des observations de MM. de l'Académie royale des sciences envoyés par le roi sous l'Équateur* (Measure of the First Three Degrees of the Meridian in the Southern Hemisphere, Drawn from the Observations of the Messieurs of the Royal Academy of Sciences, Sent by the King Below the Equator) (Paris: Imprimerie Royale, 1751), 105–120. These errors and corrections are analyzed in Antonio Lafuente

García and Antonio J. Delgado, *La geometrización de la tierra: Observaciones y resultados de la expedición geodésica hispanofrancesa al virreinato del Perú (1735–1744)* (The Geometrization of the Earth: Observations and Results of the Hispanic/French Geodesic Expedition to the Viceroyalty of Peru, 1735–1744) (Madrid: CSIC 1984), 173–247, and James R. Smith, *From Plane to Spheroid: Determining the Figure of the Earth from 3000 B.C. to the 18th Century Lapland and Peruvian Survey Expeditions* (Rancho Cordova, CA: Landmark Enterprises, 1986), 151–152.

4. Bouguer to unidentified recipient, Quito, January 14, 1742 and April 15, 1742, AARS Biographical Dossier Bouguer, unpaginated.

5. Diverse papers of Pierre Bouguer, AOP A.C. 2.8, unpaginated; La Condamine, *Journal du voyage*, 101.

6. Pierre Bouguer, "Relation abrégée du voyage fait au Pérou par Messieurs de l'Académie Royale des Sciences, pour mesurer les Degrés du Méridien aux environs de l'Equator, & en conclure la Figure de la Terre" (Abridged Relation of the Voyage Made to Peru by Gentlemen of the Royal Academy of Sciences, to Measure the Degrees of the Meridian Around the Equator, and to Deduce the Figure of the Earth), *HMARS 1744* (1748): 291–293.

7. La Condamine to Maupertuis, Quito, January 20, 1741 in *PVARS* 61 (1742): 2–9 (quote at 6); Pierre Bouguer, *Justification des Mémoires de l'Académie royale des sciences de 1744, et du livre de la Figure de la terre déterminée par les observations faites au Pérou* (Justification of Memoirs of the Royal Academy of Sciences of 1744, and of the Book The Figure of the Earth Determined by Observations Made in Peru) (Paris: Charles-Antoine Jombert, 1752), 41–42; Charles-Marie de La Condamine, *Supplément au Journal historique du voyage a l'équateur, et au livre de la Mesure des trois premiers degrés du meridien* (Supplement to the Historical Journal of the Voyage to the Equator, and of the Book of the Measure of the First Three Degrees of the Meridian), 2 vols. (Paris: Durand and Pissot, 1752–1754), 2:121–-122; Robert Finn Erickson, "The French Academy of Sciences Expedition to Spanish America, 1735–1744," PhD diss., University of Illinois, Champaign-Urbana, 1955, 140–141.

8. La Condamine to Maupertuis, Quito, January 20, 1741; La Condamine, *Journal du voyage*, 106–107.

9. La Condamine, *Journal du voyage*, 105–132. This first attempt at simultaneous observations is analyzed in Lafuente and Delgado, *La geometrización de la tierra*, 235–243.

10. Bouguer to unidentified recipient, Quito, January 14, 1742.

11. La Condamine, *Journal du voyage*, 123, 131.

12. La Condamine, *Journal du voyage*, 127–130.

13. La Condamine mentioned in his writings only the Yaruquí tablet, which is now at the Quito Observatory (*Journal du voyage*, 99, 109, 124, 162, 173). The Tarqui tablet was found many years later on the former estate of Pedro Sempértegui but has since disappeared; see Augustin Iglesias, "La Lápida de Tarqui" (The Tablet of Tarqui), *Revista del Centro de Estudios Historicos y Geograficos de Cuenca* (1928): vol. 13, 362–394, vol. 14, 25–51; and Miguel Díaz Cueva, *La Lápida de Tarqui* (The Tablet of Tarqui) (Quito: Casa de la Cultura Ecuatoriana Benjamín Carrión, 1988). The existence of the Cotchesqui tablet—which has not yet been located—was only recently revealed in a newly acquired letter from Bouguer (La Condamine to Bouguer, Cuenca, Jan. 24, 1743 and Bouguer to La Condamine, undated, AOP Ms 1113).

14. La Condamine's version of the story of the pyramids is in "Histoire des Pyramides de Quito" (History of the Pyramids of Quito), in *Journal du voyage*, 219–271; and "Copie authentique des pièces du procès soutenu devant l'audience royale de Quito par Charles-Marie de La Condamine contre D. Jorge Juan et D. Antonio de Ulloa . . . à propos de l'inscription à graver sur les pyramides de Yaruqui" (Authentic Copy of the Trial Statements Before the Audiencia of Quito by La Condamine Against Jorge Juan and Ulloa Apropos of the Inscription Engraved on the Pyramids of Yaruquí), BNF cotes Espagnol, 50. The Spanish side of the story is primarily documented in "Autos sobre las píramides de Yaruquí" (Case About the Pyramids of Yaruquí), AGI Quito 374, *DHAQ* 5:393–537.

Extended commentary on the controversy over the pyramids is given in Erickson, "The French Academy of Sciences Expedition," 320–326; Jorge Juan y Santacilia, Antonio de Ulloa y de la Torre-Guiral, and Luís Javier Ramos Gómez, *Epoca, génesis y texto de las 'Noticias secretas de América,' de Jorge Juan y Antonio de Ulloa, 1735–1745* (Epoch, Genesis and Text of the "Secret Notices of America" of Jorge Juan and Ulloa, 1735–1745), vol. 1, *El viaje a América* (Voyage to America) (Madrid: CSIC, 1985), 183–203; Neil Safier, *Measuring the New World: Enlightenment Science and South America* (Chicago: University of Chicago Press, 2008), 23–56; and Raûl Hernández Asensio, *El Matemático Impaciente; La Condamine, las Pirámides de Quito y la Ciencia Ilustrada, 1740–1751* (The Impatient Mathematician: La Condamine, the Pyramids of Quito and Enlightenment Science, 1740–1751) (Lima: Instituto Francés de Estudios Andinos, 2008).

15. La Condamine, *Supplément au Journal historique*, 2:194.

16. La Condamine, *Journal du voyage*, 133–137.

17. Jorge Juan, Ulloa and Ramos Gómez, *Las 'Noticias secretas de América'*, 1:214–219.

18. La Condamine, *Journal du voyage*, 147.

19. Bouguer to La Condamine, Quito, Feb. 1, 1742, BNF Nouvelles acquisitions françaises, 6197 fols. 20–21; La Condamine to Bouguer, Quito Dec. 3, 1742, AOP AC 2.7. Cipher in Godin to Bouguer, Quito March 28, 1742, AARS Biographical Dossier La Condamine, 50J.

20. Bouguer, *La Figure de la terre*, xxxiv; La Condamine, *Journal du voyage*, 142. Many studies have been done to identify El Niño events in the past; see for example, William H. Quinn and Victor T. Neal, "The Historical Record of El Niño Events," in *Climate Since A.D. 1500*, ed. Raymond S. Bradley and Philip D. Jones (New York: Routledge, 1992), 623–648.

21. La Condamine, *Journal du voyage*, 138.

22. Ibid., 146–147. The various inventories of collections belonging to the French Academy of Sciences show no evidence that the Jesuits' silver plaque exists today. (My thanks to Florence Greffe for this information.)

23. Ibid., 155.

24. Ibid., 157.

25. La Condamine to Bouguer, Cuenca, Jan. 24, 1743, AOP Ms 1113.

26. Pierre M. Conlon, "La Condamine the Inquisitive," in *Studies on Voltaire in the Eighteenth Century,* vol. 55, ed. Theodore Besterman (Geneva: Institut et Musée Voltaire, 1967), 361–393.

27. Bouguer, "Relation abrégée," 293; La Condamine, *Journal du voyage*, 122–123, 165.

28. Several notices of correspondence between the scientists and Jesuit colleagues, as far away as Arequipa, appear in Bouguer to La Condamine (undated), AOP Ms 1113.

29. Pierre Bouguer, *Traité du Navire, de sa Construction, et de ses Mouvemens* (Treatise of the Ship, Its Construction, and Its Movements) (Paris: Charles-Antoine Jombert, 1746), xxiv.

30. La Condamine, *Journal du voyage*, 165–172; La Gournerie, "Remarques historiques et critiques sur les observations faites au Pérou . . . par Bouguer" (Historical Remarks and Criticisms on the Observations Made in Peru . . . by Bouguer), in *Annales de l'observatoire de Paris* 14, no. 1 (Paris: Gauthier-Villar, 1877), D16, D25–D26.

31. La Condamine to Bouguer, Tarqui, Jan. 15, 1743, BNF Nouvelles acquisitions françaises 6197 fol. 31–32; quotation from La Condamine to Bouguer, Cuenca, Jan. 24, 1743, *AOP* ms1113.

32. La Condamine, *Journal du voyage*, 173.

33. Carlos Ortiz Arellano, *Pedro Vicente Maldonado: Forjador de la Patria Ecuatoriana, 1704–1748* (Pedro Vicente Maldonado: The Man Who Forged the Ecuadorian Fatherland) (Quito: Casa de la Cultura Ecuatoriana Benjamín Carrion, 2004), 207–209.

34. La Condamine, *Journal du voyage,* 175–177; La Condamine to Bouguer, Cuenca, Jan. 24, 1743, AOP Ms 1113.

35. The accepted length was from Bouguer's calculations; La Condamine's result differed slightly. See Bouguer, *La Figure de la terre*, 267–274; La Condamine, *Mesure des trois*

premiers degrés du méridien, 213–222. The modern World Geodetic System 1984 (WGS84) gives the length of the first degree of latitude at the equator as 110,574 meters or 362,776 feet, compared with the 362,899 feet found by Bouguer.

Chapter X The Impossible Return

1. Pierre Bouguer, *La Figure de la terre, déterminée par les observations de Messieurs Bouguer et de La Condamine, de l'Académie Royale des Sciences, envoyés par ordre du Roy au Pérou pour observer aux environs de l'équateur, avec une Relation abrégée de ce voyage qui contient la description du pays dans lequel les opérations ont été faites* (The Figure of the Earth, Determined by the Observations of the Messieurs Bouguer and La Condamine, of the Royal Academy of Sciences, Sent to Peru by Order of the King to Make Observations near the Equator, with an Abridged Relation of the Voyage Containing a Description of the Countries in Which the Observations Were Made) (Paris: Charles-Antoine Jombert, 1749), lxxix–xcvii (quote at xciii).

2. Ibid., 57; *PVARS*, March 24, 1744, 169; Robert Finn Erickson, "The French Academy of Sciences Expedition to Spanish America, 1735–1744," PhD diss., University of Illinois, Champaign-Urbana, 1955, 284.

3. "Mémoire sur feu M. [Daniel-René] Montaudouin" (Memorial to the Late M. [Daniel-René] Montaudouin), *Mercure de France* (October 1755): 36–44. On the place of the Bouguer family in training slave ship masters, see Robert Harms, *The Diligent: A Voyage Through the Worlds of the Slave Trade* (New York: Basic Books, 2002), 89–95.

4. Jean Mettas, *Répertoire des Expeditions Négrières Françaises au XVIIIe siècle* (Collection of French Slave Ship Expeditions of the 18th Century), 2 vols. (Paris: Société française d'histoire d'outre-mer, 1978), 1:276–277; Bouguer's certification of Pierre Fouré as pilot, June 3, 1744, Smithsonian Institution, Dibner Collection, National Museum of American History, Washington DC, MSS 155A RB; *Rôle de désarmement du Triton* (End-of-Voyage Roster of Triton), 1744, ADLA Séries anciennes B (Cours et juridictions), Admiralties of Nantes and Guérande (1613–1791), 120 J 378/B. On the turnaround of French slave ships, see Harms, *The Diligent*, 369–375.

5. Jean-Joseph Expilly, *Dictionnaire géographique, historique et politique des Gaules et de la France* (Geographical, Historical and Political Dictionary of the Gauls and of France), 6 vols. (Paris: Desaint & Saillant, 1764), 5:90–91.

6. Charles-Marie de La Condamine, *Journal du voyage fait à l'Équateur servant d'introduction historique à la Mesure des trois premiers degrés du Méridien* (Journal of the Voyage Made to the Equator, Serving as a Historical Introduction to the Measure of the First Three Degrees of the Meridian) (Paris: Imprimerie Royale, 1751), 122–123. Descriptions of previous voyages are in Anthony Smith, *Explorers of the Amazon* (New York: Viking, 1990).

7. La Condamine's accounts of his Amazon voyage are given in his "Relation abrégée d'un Voyage fait dans l'intérieur de l'Amérique Méridionale, depuis la Côte de la Mer du Sud, jusqu'aux Côtes du Brésil et de la Guyane en descendant la Rivière des Amazones" (Abridged Relation of a Voyage Made in the Interior of South America, from the Coast of the Southern Ocean, Until the Coasts of Brazil and Guyane while Descending the Amazon River), *HMARS 1745* (1749): 391–492 (published under the same title [Paris: Veuve Pissot, 1745]; second edition [Maastricht: Dufour et Roux, 1778]). A concurrent Spanish version came out as *Extracto del diario de observaciones hechas en el viaje de la Provincia de Quito al Pará por el Río de las Amazonas y del Pará a Cayena, Surinam y Amsterdam* (Extract from the Diary of Observations Made in the Voyage from the Province of Quito to Pará by Way of the Amazon River, and from Pará to Cayenne, Surinam and Amsterdam) (Amsterdam: Imprenta de Joan Catuffe, 1745). Critical analyses of La Condamine's voyage are in Anita McConnell, "La Condamine's Scientific Journey down the River Amazon, 1743–1744," *Annals of Science* 48 (1991): 1–19; and Neil Safier, *Measuring the New World: Enlightenment Science and South America* (Chicago: University of Chicago Press, 2008), 57–92.

8. La Condamine, *Journal du voyage*, 184.

9. Ibid., 187.

10. La Condamine, *Relation abrégée*, 46.

11. Maldonado's preparations for his voyage and sojourn in Europe are recounted in Ortiz Arellano, *Pedro Vicente Maldonado*, and Safier, *Measuring the New World*.

12. Jorge Pimentel Cintra and Janaina Carli de Freitas, "Sailing Down the Amazon River: La Condamine's Map," *Survey Review*, forthcoming.

13. La Condamine, *Extracto del diario*, 43.

14. Recent discoveries and ideas about pre-European Amazonia are described in Charles Mann, *1491: New Revelations of the Americas Before Columbus* (New York: Knopf, 2005).

15. La Condamine, *Journal du voyage*, 206.

16. Neptali Zúñiga, *La expedición científica de Francia del siglo XVIII en la Presidencia de Quito* (The Scientific Expedition of France in the 18th Century in the Audiencia of Quito) (Quito: Instituto Panamericano de Geografía e Historia, Sección Nacional del Ecuador, 1977/1986), 39.

17. Antonio Lafuente García and Antonio J. Delgado, *La geometrización de la tierra: Observaciones y resultados de la expedición geodésica hispanofrancesa al virreinato del Perú (1735–1744)* (The Geometrization of the Earth: Observations and Results of the Hispanic/French Geodesic Expedition to the Viceroyalty of Peru, 1735–1744) (Madrid: CSIC 1984), 247–257.

18. Jean and Isabel Godin des Odonais: Robert Whitaker, *The Mapmaker's Wife: A True Tale of Love, Murder and Survival in the Amazon* (New York: Perseus Books, 2004), 170. The return voyages of Jorge Juan and Ulloa are carefully documented in Arthur P. Whitaker, "Antonio de Ulloa, the *Délivrance*, and the Royal Society," *Hispanic American Historical Review* 46, no. 4 (1966): 357–370. Godin's inauguration at the University of San Marcos was announced in the *Gaceta de Lima* no. 8, Nov. 17, 1744. On the warship *San José el Peruano*, built in 1752 after the colonial navy was destroyed by the 1746 tsunami, see *Colección documental de la independencia del Perú* (Collection of Documents on the Independence of Peru), 27 vols. (Lima: Comisión Nacional del Sesquicentenario de la Independencia del Perú, 1971), 7, no. 1 *La marina, 1780–1822*, 5–125.

19. John Byron, *Loss of the Wager: The Narratives of John Bulkeley and the Hon. John Byron* (Woodbridge UK: Boydell Press, 2004), 225–226.

20. Election of Antonio de Ulloa to the Royal Society, nominated May 15, 1746, elected December 15, 1746, Royal Society archives EC/1746/13.

21. Quoted in Elisabeth Badinter, *Les passions intellectuelles*, vol. 1, *Désirs de gloire, 1735–1751* (Intellectual Passions: Desires of Glory, 1735–1751) (Paris: Fayard, 1999), 292.

22. Rose-Angelique Godin to Louis Godin and Pedro Vicente Maldonado, Paris, January 2, 1747 (summary in a notice of letters sold at auction in July 1991), AARS Biographical Dossier La Condamine, 50J.

23. Badinter, *Les passions intellectuelles*, 292 (noting that Godin's chair netted him 22,000 livres annually); Jorgé Ortiz Sotelo, "Los cosmógrafos mayores del Perú" (The Royal Astronomers of Peru), *Derroteros de la mar del sur* 7, no. 7 (1999): 135–146. The Marqués de Villagarcía, whose viceroyalty wholly coincided with the Geodesic Mission, left office in December 1745 and died on December 17, 1746, aboard ship off Cape Horn while returning to Spain.

24. Several riveting firsthand accounts of the 1746 earthquake and tsunami that wrecked Lima and Callao are found in the following sources: *A True and Particular Relation of the Dreadful Earthquake Which Happen'd at Lima, the Capital of Peru, and the Neighbouring Port of Callao, on the 28th of October, 1746* (London: T. Osborne, 1748); and Manuel de Odriozola, ed. *Terremotos: Coleccion de las relaciónes de los mas notables que ha sufrido esta capital y que la han arruinado* (Earthquakes: Collection of the Accounts of the Most Notable Ones Suffered by This Capital and Which Have Ruined It) (Lima: Aureila Alfaro, 1863), quote at 48.

25. Godin's role in the rebuilding of Lima, with the attendant tribulations of overcoming the resistance of the city's elite, is critically analyzed in Pablo Emilio Pérez-Mallaina Bueno, *Retrato de una ciudad en crisis: La sociedad limeña ante el movimiento sísmico de*

1746 (Portrait of a City in Crisis: The Society of Lima in the Face of the Seismic Movement of 1746) (Seville: CSIC, 2001); and Charles F. Walker, *Shaky Colonialism: The 1746 Earthquake-Tsunami in Lima, Peru, and Its Long Aftermath* (Durham NC: Duke University Press, 2008).

26. Joseph de Jussieu's wanderings in South America are detailed in Antoine-Laurent de Jussieu, "Jussieu," in *Encyclopédie Méthodique—Médicine* (Methodical Encyclopedia: Medicine), ed. Henri Agasse (Paris: Panckoucke, 1798), 7:765–772; and Gaston Lehir, "Joseph de Jussieu et son exploration en Amérique Méridionale (1735–1769)" (Joseph de Jussieu and His Exploration in South America, 1735–1769), master's thesis, University of Montreal, 1976. Patrick Drevet's haunting fictional account *Le Corps du Monde* (The Body of the World) (Paris: Seuil, 1997) reveals details of his voyages based on extensive archival research.

27. Antonio Lafuente García and Antonio Mazuecos, *Los Caballeros de Punto Fijo; Ciencia, politica y aventura en la expedición geodésica hispanofrancesa al virreinato del Perú en el siglo XVIII* (The Gentlemen of the Fixed Point: Science, Politics and Adventure in the Hispanic/French Geodesic Expedition to the Viceroyalty of Peru in the 18th Century) (Barcelona: Serbal/CSC, 1987), 124.

28. Antonio Lafuente García and Manuel Sellés, *El Observatorio de Cadíz (1753–1831)* (The Observatory of Cadiz, 1753–1831) (Madrid: Ministerio de Defensa, 1988), 148–151.

29. Lafuente and Mazuecos, *Los Caballeros de Punto Fijo*, 150–151.

30. Nicolas Louis de La Caille, *Journal Historique du Voyage fait au Cap de Bonne-Espérance* (Historic Journal of the Voyage Made to the Cape of Good Hope) (Paris: Libraire Guillyn, 1763), 121.

Chapter XI A World Revealed

1. On Clairaut leaving Bouguer out of his groundbreaking work *Théorie de la figure de la terre* (Theory of the Figure of the Earth) (Paris: Durand, 1743), see John Greenberg, *The Problem of the Earth's Shape from Newton to Clairaut: The Rise of Mathematical Science in Eighteenth Century Paris and the Fall of "Normal" Science* (Cambridge, UK: Cambridge University Press, 1995), 441.

2. Danielle Fauque, "The Introduction of the Octant in Eighteenth-Century France," in *Koersvast: Vijf eeuwen navigatie op zee* (On a Steady Course: Five Centuries of Navigation at Sea), ed. Leo Akveld (Zaltbommel, Netherlands: Uitgeverij Aprilis, 2005), 224–225. On the aftermath of the Maupertuis expedition and the acceptance of Newtonianism in France, see Mary Terrall, *The Man Who Flattened the Earth: Maupertuis and the Sciences in the Enlightenment* (Chicago: University of Chicago Press, 2002), 130–172; and John Bennett Shank, *The Newton Wars and the Beginning of the French Enlightenment* (Chicago: University of Chicago Press, 2008), 343–506.

3. Maupertuis, *La figure de la terre* (The Figure of the Earth) (Paris: Imprimerie Royale, 1738), 130; Clairaut, *Théorie de la figure de la terre*, 299; *PVARS* June 27, 1744, 333; Jean-Étienne Montucla and Jérôme Laland, *Histoire des mathematiques* (History of Mathematics), 4 vols. (Paris: Agasse, 1802), 4:159–160; Antonio Lafuente García and Antonio J. Delgado, *La geometrización de la tierra: Observaciones y resultados de la expedición geodésica hispanofrancesa al virreinato del Perú (1735–1744)* (The Geometrization of the Earth: Observations and Results of the Hispanic/French Geodesic Expedition to the Viceroyalty of Peru, 1735–1744) (Madrid: CSIC, 1984), 259–268.

4. Mary Terrall, "Heroic Narratives of Quest and Discovery," *Configurations* 6, no. 2 (1998): 223–242; Mary Terrall, "Mathematics in Narratives of Geodetic Expeditions," *Isis* 97 (2006): 683–699.

5. *PVARS* July 29, 1744, 418; *PVARS* Feb. 27, 1745, 40; Pierre Bouguer, "Relation abrégée du voyage fait au Pérou par Messieurs de l'Académie Royale des Sciences, pour mesurer les Degrés du Méridien aux environs de l'Equator, & en conclure la Figure de la Terre" (Abridged Relation of the Voyage Made to Peru by Gentlemen of the Royal Academy of Sciences, to Measure the Degrees of the Meridian Around the Equator, and to Deduce the Figure of the Earth), *HMARS 1744* (1748): 249–297, *HMARS 1746* (1751): 569–606.

6. Elisabeth Badinter, *Les passions intellectuelles*, vol. 1, *Désirs de gloire, 1735–1751* (Intellectual Passions: Desires of Glory, 1735–1751) (Paris: Fayard, 1999), 97.

7. *Mercure de France*, November 1744, 47–85 (quotes at 82–83); Anne-Marie Chouillet, "Rôle de la presse périodique de langue française dans la diffusion des informations concernant les missions en Laponie ou sous l'équateur" (Role of the French-Language Newspapers in the Diffusion of Information Concerning the Missions in Lapland or Under the Equator), in *La figure de la Terre du XVIIIe siècle à l'ère spatiale* (The Figure of the Earth from the 18th Century to the Space Age), ed. Henri Lacombe and Pierre Costable (Paris: Gauthier-Villers, 1988), 177.

8. Bouguer's pension: Grandjean de Fouchy, "Éloge de Clairaut" (Clairaut's Eulogy), *PVARS* Nov. 13, 1765, 383. Bouguer's *Traité du Navire, de sa Construction, et de ses Mouvemens* (Treatise of the Ship, Its Construction, and Its Movements) (Paris: Charles-Antoine Jombert, 1746) is discussed in Larrie D. Ferreiro, *Ships and Science: The Birth of Naval Architecture in the Scientific Revolution, 1600–1800* (Cambridge, MA: MIT Press, 2007). "Rue des Postes" appears on all of Bouguer's correspondence after 1744 (his home was further identified as "at the cul-de-sac near rue l'Estrapade"). Thanks to René Ghislaine for identifying rue Lhomond as the former rue des Postes, which can still be seen in the stonework of the buildings.

9. Pierre Bouguer, *Nouveau Traité de Navigation, contenant la Théorie et la Pratique du Pilotage* (New Treatise of Navigation, Containing the Theory and Practice of Piloting) (Paris: Guérin et Delatour, 1753), 377–378, 434–442 (quote at 435); John Robertson, *The Elements of Navigation*, 2 vols. (London: J. Nourse, 1764), 2:203–215.

10. On scientific expeditions with Spain and Portugal: James Pritchard, *Louis XV's Navy, 1748–1762: A Study of Organization and Administration* (Montreal: McGill-Queens University Press, 1987), 27. On the colonial repercussions of institutional science in France, see James E. McClellan III and Francois Regourd, "The Colonial Machine: French Science and Colonization in the Ancién Regime," *Osiris*, 2nd ser., 15 (2000): 31–50.

11. Olivier Bernier, *Louis the Beloved: The Life of Louis XV* (Garden City, NY: Doubleday, 1984), 125–133.

12. Charles-Marie de La Condamine, *Supplément au Journal historique du voyage a l'équateur, et au livre de la Mesure des trois premiers degrés du meridien* (Supplement to the Historical Journal of the Voyage to the Equator, and of the Book of the Measure of the First Three Degrees of the Meridian), 2 vols. (Paris: Durand and Pissot, 1752–1754), 2:209–210.

13. Jorge Cañizares-Esguerra, *How to Write the History of the New World: Histories, Epistemologies, and Identities in the Eighteenth-Century Atlantic World* (Palo Alto, CA: Stanford University Press, 2001), 1.

14. Charles-Marie de La Condamine, *Relation abrégée d'un Voyage fait dans l'intérieur de l'Amérique Méridionale, depuis la Côte de la Mer du Sud, jusqu'aux Côtes du Brésil et de la Guyane en descendant la Rivière des Amazones* (Abridged Relation of a Voyage Made in the Interior of South America, from the Coast of the Southern Ocean, Until the Coasts of Brazil and Guyane While Descending the Amazon River) (Paris: Veuve Pissot, 1745); La Condamine, *Lettre à Mme ***, sur l'émeute populaire excitée en la ville de Cuenca, au Pérou, le 29 d'août 1739, contre les académiciens des sciences envoyés pour la mesure de la terre* (Letter to Mrs. X, on the Popular Uprising Raised in the City of Cuenca, in Peru, 29 August 1739, Against the Scientific Academicians Sent to Measure the Earth) (Paris: Veuve Pissot, 1745).

15. The story of La Condamine and Maldonado's map is told in Neil Safier, *Measuring the New World: Enlightenment Science and South America* (Chicago: University of Chicago Press, 2008). On Graffigny's *Lettres d'une Peruvienne* (Paris: A Peine, 1747), and other popular works about Peru, see Deborah Poole, *Vision, Race and Modernity: A Visual Economy of the Andean Image World* (Princeton, NJ: Princeton University Press, 1997), 25–57.

16. La Condamine to Bouguer, Sept. 28, 1745, AARS Biographical Dossier La Condamine, 50J; La Condamine, *Journal du voyage*, 208.

17. Badinter, *Les passions intellectuelles*, 364–365.

18. The attack brochures were, in order of publication: Pierre Bouguer, *Justification des Mémoires de l'Académie royale des sciences de 1744, et du livre de la Figure de la terre déterminée par les observations faites au Pérou* (Justification of Memoirs of the Royal Academy of Sciences of 1744, and of the Book The Figure of the Earth Determined by Observations Made in Peru) (Paris: Charles-Antoine Jombert, 1752); La Condamine, *Supplément au Journal historique du voyage*, 2 vols., 1752–1754; Pierre Bouguer, *Lettre à M. *** dans laquelle on discute divers points d'astronomie pratique, et où l'on fait quelques remarques sur le supplément au Journal historique du voyage à l'équateur de M. de La C.* (Letter to Mr. *** in Which Is Discussed Diverse Points of Practical Astronomy, and Where Are Made Several Remarks on the Supplement of the Historical Journal of the Voyage to the Equator by Mr. de la Condamine) (Paris: Guérin et Delatour, 1754). The quarrel and the role of the press are analyzed in Yasmine Marcil, "La presse et le compte rendu de récits de voyage scientifique: Le cas de la querelle entre Bouguer et Las Condamine" (The Press and the Report of Accounts of a Scientific Voyage: The Case of the Quarrel Between Bouguer and La Condamine), *Science et Techniques en Perspective*, 2nd ser., 3, no. 2 (1999): 285–304.

19. Robert Finn Erickson, "The French Academy of Sciences Expedition to Spanish America, 1735–1744," PhD diss., University of Illinois, Champaign-Urbana, 1955, 314–318.

20. Marie Jean Antoine Nicolas de Caritat, Marquis de Condorcet, "Eloge de M. de La Condamine" (Eulogy of M. de La Condamine), *HMARS 1774* (1778): 108.

21. Jean-Baptiste Delambre, *Histoire de l'astronomie au XVIIIe siècle* (History of Astronomy in the 18th Century) (Paris: Bachelier, 1827), 367.

22. For a more detailed examination of Bouguer and his work, see Ferreiro, *Ships and Science*.

23. La Condamine's *Journal* was extensively excerpted in Antoine François Prévost, *Histoire générale des voyage* (General History of Voyages), 15 vols. (Paris: Didot, 1746–1759), 13:470–653, 14:24–54. His descriptions of the Amazon and the various flora and fauna formed the basis for numerous entries in Denis Diderot and Jean Le Rond D'Alembert, *Encyclopédie* (Encyclopedia), 17 vols. (Paris: Briason, 1751–1772).

24. Georges Louis Leclerc, Comte de Buffon, *Histoire naturelle, générale et particulière* (Natural History, in General and Specific), 15 vols. (Paris: Imprimerie Royale, 1749–1767), 9:104–105.

25. La Condamine, "Nouveau projet d'une mesure invariable, propre à servir de mesure commune à toutes les Nations" (New Project for a Fixed Measure to Serve as a Common Metric for All Nations)," *HMARS 1747* (1752): 489–514; and *Histoire de l'inoculation de la petite vérole* (History of Smallpox Inoculation) (Amsterdam: Société Typographique, 1773). For an examination of La Condamine's role in the acceptance of inoculation, see Andrea Rusnock, *Vital Accounts: Quantifying Health and Population in Eighteenth-Century England and France* (Cambridge, UK: Cambridge University Press, 2002), 71–95.

26. Alexis Piron, *Oeuvres complètes* (Complete Works) (Paris: L'imprimerie de la Société Typographique, 1777), 167–168.

27. La Condamine's illness and death were reported by Friedrich Melchior, Baron von Grimm in *Correspondance littéraire* (Literary Correspondence), 15 vols. (Paris: Furne, 1829–1831), vol. 5 (1766), 209–210, and vol. 8 (1774), 284–286. His symptoms were entirely consistent with a rare form of diabetes called MIDD, Maternally Inherited Diabetes mellitus and Deafness.

28. Godin, "Ecris sur le tremblement de terre qui s'est fait à Cadiz" (Note on the Earthquake at Cadiz), trans. Bouguer, *PVRS* vol. 35, 774, Dec. 16, 1755. This letter was written in Spanish and translated by Bouguer, showing that Godin was in fact now thinking in Spanish rather than in his native French. See also Paul-Louis Blanc, "The Tsunami in Cadiz on November 1, 1755: A Critical Analysis of Reports by Antonio de Ulloa and by Louis Godin," *Comptes Rendus Geoscience* 340, no. 4 (2008): 251–261.

29. Jean-Paul Grandjean de Fouchy, "Eloge de M. Godin" (Eulogy of Mr. Godin), *HMARS 1760* (1766): 192–193; Antonio Lafuente García and Manuel Sellés, *El Observatorio de Cadíz (1753–1831)* (The Observatory of Cadiz, 1753–1831) (Madrid: Ministerio de Defensa, 1988), 151–160.

30. The gazeteer was Jorge Juan y Santacilia and Antonio de Ulloa y de la Torre-Guiral, *Relación histórica del Viaje á la América Meridional* (Historical Relation of the Voyage to South America), 2 vols. (Madrid: Antonio Marín, 1748; facsimile, ed. José Merino Navarro and Miguel Rodriguez San Vicente, Fundación universitaria española, 1978); a recent abridged translation is *A Voyage to South America*, ed. Irving Leonard (New York: Knopf, 1964). The scientific work was Jorge Juan and Ulloa, *Observaciones astronómicas y físicas hechas en los Reynos del Perú* (Astronomical and Physical Observations Made in the Kingdoms of Peru) (Madrid: Juan de Zúñiga, 1748). The introduction to the 1978 edition of *Relación histórica* contains a detailed analysis of how the documents were produced, as does the first volume (351–397) of Jorge Juan y Santacilia, Antonio de Ulloa y de la Torre-Guiral, and Luís Javier Ramos Gómez, *Epoca, génesis y texto de las 'Noticias secretas de América,' de Jorge Juan y Antonio de Ulloa, 1735–1745* (Epoch, Genesis and Text of the "Secret Notices of America" of Jorge Juan and Ulloa, 1735–1745) (Madrid: CSIC, 1985).

31. In 1826 a copy of the *Discourse* was smuggled out of Spain and published in Britain under the title *Noticias Secretas de América* (Secret Information About America). It was republished and translated many times, the most detailed of which is Jorge Juan, Ulloa, and Ramos Gómez, *Las 'Noticias secretas de América'*. A recent abridged translation is *Discourse and Political Reflections on the Kingdoms of Peru*, ed. John J. TePaske (Norman: University of Oklahoma Press, 1978). Critical analyses of the works of Jorge Juan and Ulloa are found in Lewis Hanke, "Dos Palabras on Antonio de Ulloa and the *Noticias Secretas*," *Hispanic American Historical Review* 16, no. 4 (1936): 479–514; Luis Merino, "The Relation Between the *Noticias Secretas* and the *Viaje a la America Meridional*," *Americas* 23, no. 2 (1956): 111–125; and Kenneth J. Andrien, "The *Noticias secretas de América* and the Construction of a Governing Ideology for the Spanish American Empire," *Colonial Latin American Review* 7, no. 2 (1998): 175–192.

32. Hilda Krousel, *Don Antonio de Ulloa, First Spanish Governor to Louisiana* (Baton Rouge, LA: VAAPR, 1983).

33. Ulloa's career is analyzed in numerous papers presented during a 1995 symposium: Miguel Losada and Consuelo Losada, eds., *Actas de II Centenario de Don Antonio de Ulloa* (Proceedings of the Bicentennial of Antonio de Ulloa) (Seville: Escuela de estudios hispanoamericanos-CSIC, and Archivos general de Indias, 1995).

34. For Jorge Juan's career in the navy and his spy mission, see Ferreiro, *Ships and Science*. Jorge Juan's symptoms at the time of his death (paralysis, seizures, and coma) are consistent with primary amoebic meningoencephalitis caused by *Naegleria fowleri*, which are only found in warm, fresh waters like thermal springs. The disease was only identified in 1965, so Jorge Juan may have been the first recorded death from this rare parasite. See Rosa Die Maculet and Armando Alberola Romá, *La herencia de Jorge Juan: Muerte, disputas sucesorias y legado intellectual* (The Inheritance of Jorge Juan: Death, Disputes over his Succession and Intellectual Legacy) (Alicante: Fundación Jorge Juan, 2002), 69–77.

35. See Richard L. Garner, "Long-Term Silver Mining Trends in Spanish America: A Comparative Analysis of Peru and Mexico," in *An Expanding World: The European Impact on World History, 1450–1800*, vol. 19, *Mines of Silver and Gold in the Americas*, ed. Peter Bakewell (Aldershot, UK: Variorum Ashgate, 1997), 225–262.

36. Gaston Lehir, "Joseph de Jussieu et son exploration en Amérique Méridionale (1735–1769)" (Joseph de Jussieu and His Exploration in South America [1735–1769]), master's thesis, University of Montreal, 1976, 158–167; Robert Whitaker, *The Mapmaker's Wife: A True Tale of Love, Murder and Survival in the Amazon* (New York: Perseus Books, 2004), 288; Neil Safier, "Fruitless Botany: Joseph de Jussieu's South American Odyssey," in *Science and Empire in the Atlantic World*, ed. James Delbourgo and Nicholas Dew (New York: Routledge, 2008), 203–224.

37. Lehir, "Joseph de Jussieu," 194–200.

38. AARS Biographical Dossier of Verguin; Bernard Cros, "Verguin," *Neptunia* 250 (2008): 29–38.

39. Quote: Hugo to Joseph de Jussieu, Quito, Nov. 21, 1749, MNHN, Ms 179 no. 47. A letter from Hugo to La Condamine, Quito, Oct. 25, 1753, is summarized in a notice of

letters sold at auction in July 1991, AARS Biographical Dossier La Condamine, 50J. In the same notice is one letter mentioning an "Acte mortuaire de Mme. De Morainville" (Death Certificate of Mrs. Morainville).

Hugo's tileworks and family residence were located on the *calle* de Urcu Virgen, today known as *calle* Cotopaxi, near the corner of Manabí; see Fernando Jurado Noboa, *Calles, Casas y Gente del Centro Histórico de Quito* (Streets, Houses and People of Quito's Historic Center), 7 vols. (Quito: Fonsal, 2004–2009), 5:281–283.

Morainville constructed the towers of José Antonio Maldonado's church in Quinche in 1748, but by the beginning of the twentieth century they were demolished and new towers erected in their place. Morainville's surviving artworks are the Pyramids of Yaruquí and view of the baseline (watercolor on linen, 1742), Museum of the Banco Central del Ecuador, inventory no. 17-1-83; *San José* (oil, 1747), painted for Inmaculada Concepción Church in Cuenca, now in the Museum of the Monasterio de las Conceptas in Cuenca; and *San Ignacio* and *San Francisco de Borja* (oils, 1757), painted for the Jesuit Seminary in Quito, now in the La Compañia Jesuit church in Quito.

40. "Autos de Dn. Julio Salvador contra Dn. Antonio Ormasa, y Bernabé Chacón sobre cantidad de pesos" (Criminal Case of Julio Salvador Against Antonio Ormaza and Bernabé Chacón About the Amount of Money)," Quito, December 10, 1755, ANEC Minas Caja 2 expediente numéro 2; Germán Rodas Chaves, "J. de Morainville y el primer dibujo universal de la quina o cascarilla" (J. de Morainville and the First Universal Drawing of Cinchona or Cascarilla), *Bulletin de l'Institute Français d'Etudes Andines* 32, no. 3 (2003): 431–440.

41. Shortly before his death, La Condamine wrote a brief summary of what happened to the members of the mission: *Lettre de Monsieur de La Condamine à M. *** sur le sort des astronomes qui ont eu part aux dernières mesures de la terre, depuis 1735* (Letter of Mr. de La Condamine to Mr. X on the Fate of the Astronomers Who Took Part in the Latest Measures of the Earth, Since 1735) (Paris: n.p., 1773), rpt., Charles-Marie de La Condamine, *Relation abrégée d'un Voyage fait dans l'intérieur de l'Amérique Méridionale, depuis la Côte de la Mer du Sud, jusqu'aux Côtes du Brésil et de la Guyane en descendant la Rivière des Amazones* (Abridged Relation of a Voyage Made in the Interior of South America, from the Coast of the Southern Ocean, Until the Coasts of Brazil and Guyane while Descending the Amazon River) (Maastricht: Dufour et Roux, 1778). On Morainville's participation in the defense of Riobamba, see Eduardo Segundo Gonzalo Moreno Yánez, *Sublevaciones indígenas en la Audiencía de Quito: Desde comienzos del siglo XVIII hasta finales de la Colonia* (Indigenous Uprisings in the Audiencia of Quito: From the Beginnings in the 18th Century Until the Final Days of the Colony) (Quito: Ediciones de la Universidad Católica, 1985), 60–63, 90–93.

42. The story of Isabel Godin is wonderfully related in Robert Whitaker's *The Mapmaker's Wife*. Other recent accounts include Celia Wakefield, *Searching for Isabel Godin: An Ordeal on the Amazon—Tragedy and Survival* (Berkeley: Creative Arts, 1999); Carlos Ortiz Arellano, *Una Historia de Amor: Isabel Gramesón Godin, 1728–1792* (A Love Story: Isabel Gramesón Godin, 1728–1792) (Quito: Ediciones Abya-Yala, 2000); Anthony Smith, *The Lost Lady of the Amazon: The Story of Isabela Godin and her Epic Journey* (London: Constable, 2003). It also inspired a novel: Jorgé Velasco MacKenzie, *En nombre de un amor imaginario* (In the Name of an Imaginary Love) (Quito: Editorial El Conejo LIBRESA, 1996).

43. Nicolas Louis de La Caille, *Journal Historique du Voyage fait au Cap de Bonne-Espérance* (Historic Journal of the Voyage Made to the Cape of Good Hope) (Paris: Libraire Guillyn, 1763), 205–210.

Epilogue The Children of the Equator

1. "Mi delirio" was written in Loja on October 13, 1822. It was published posthumously in 1833 in Simón Bolívar, *Colección de documentos relativos a la vida pública del libertador de Columbia y del Perú* (Collection of Documents Relative to the Public Life of the Liberator of Colombia and Peru), 22 vols. (Caracas: Devisme, 1826–1833), 21:248.

2. Two recent anthologies discuss the relationship between science and empire: Londa

Schiebinger and Claudia Swan, eds., *Colonial Botany: Science, Commerce and Politics in the Early Modern World* (Philadelphia: University of Pennsylvania Press, 2005); and James Delbourgo and Nicholas Dew, eds. *Science and Empire in the Atlantic World* (New York: Routledge, 2008).

3. Gaston Lehir, "Joseph de Jussieu et son exploration en Amérique Méridionale (1735–1769)" (Joseph de Jussieu and His Exploration in South America, 1735–1769), master's thesis, University of Montreal, 1976, 108.

4. John Wilton Appel, *Francisco José de Caldas: A Scientist at Work in Nueva Granada* (Philadelphia: American Philosophical Society, 1994), 6–8, 52–56.

5. Humboldt, *Briefe aus Amerika 1799–1804* (Letters from America) (Berlin: Akademie Verlag, 1993), 200.

6. Juan de Velasco, *Historia del reino de Quito en la América meridional, 1789* (History of the Kingdom of Quito in South America, 1789), vol. 2 (Caracas: Biblioteca Ayacucho, 1981); Dionisio Alsedo y Herrera, *Aviso histórico, político, geográfico, con las noticias más particulares del Peru, Tierra-Firme, Chile, y nuevo reyno de Granada* (Historical, Political and Geographical Observations, with Particular Notices of Peru, Tierra-Firme, Chile and the New Kingdom of Granada) (Madrid: D.M. de Peralta, 1740). On the rebuilt sundial, see the beautifully illustrated history of the Jesuit church in Quito by María Antonieta Vásquez Hahn, *Luz a Través de los Muros* (Light Through the Walls) (Quito: Fonsal, 2005), 36–37.

7. For an insightful look at the influence of Enlightenment science and philosophy (including the impacts of the Geodesic Mission and the Patriotic Society) on the independence movement in Quito, see Ekkehart Keeding, *Surge la Nación: La Ilustración en la Audiencia de Quito (1725–1812)* (Emerging Nation: The Enlightenment in the Audiencia of Quito [1725–1812]) (Quito: Banco Central del Ecuador, 2005).

8. The adoption of "Ecuador" as the name of the country is detailed in Ana Buriano, "Ecuador, latitud 0: Una mirada al proceso de construcción de la nación" (Ecuador, Latitude Zero: A View of the Process of Building a Nation), in *Crear La Nación: Los Nombres De Los Países De América Latina* (Creating the Nation: Creating the Names of Latin American Countries), ed. José Carlos Chiaramonte et al. (Buenos Aires: Sudamericana, 2008), 172–202.

9. Charles Darwin and Francis Darwin, *The Life and Letters of Charles Darwin* (London: Francis Murray, 1887), 18.

10. For a critical history of geodesy, see Isaac Todhunter, *A History of the Mathematical Theories of Attraction and the Figure of the Earth*, 2 vols. (London: Macmillan, 1873).

11. The survey behind the metric system is described in Ken Alder, *The Measure of All Things: The Seven-Year Odyssey That Transformed the World* (London: Little, Brown, 2002).

12. Note that the Struve Arc was the first geodesic arc designated as a UNESCO World Heritage Site, in 2005.

13. Georges Perrier, *La mission française de l'équateur* (The French Equatorial Mission) (Paris: F. Levé, 1907); and Perrier, *La Figure de la terre, les grandes opérations géodésiques, l'ancienne et la nouvelle mesure de l'arc du méridien de Quito* (The Figure of the Earth, the Great Geodesic Operations, the Old and the New Measure of the Arc of the Meridian of Quito) (Paris: Delagrave, 1908).

14. Toise of Peru: Observatory of Paris, inventory no. 88. Morainville's watercolors: Museum of the Banco Central del Ecuador, inventory 17-1-83; MNHN Ms 111 and 1627. On Graham's pendulum clock, see Jean-Baptiste Joseph Dieudonné Boussingault, *Mémoires de Jean-Baptiste Boussingault*, 5 vols. (Paris: Chamerot et Renouard, 1892–1903), 3:113–114.

15. Caldas described his theft of the Tarqui tablet in his *Semanario de la Nueva Granada* (Seminar on New Granada) (Paris: Librería Castellana, 1849), 500–502. The return of the tablet to Cuenca is described in Miguel Díaz Cueva, *Lápida de Tarqui* [The Tablet of Tarqui] (Quito: Casa de la Cultura Ecuatoriana Benjamín Carrión, 1988), and José María Vargas, *Los Dominicos en El Ecuador* [The Dominicans in Ecuador] (Quito: Nacional Permanente de Conmemoraciones Cívicas, 1988), 88. The recovered tablet was

first affixed to the wall of the old Cuenca cathedral, until it was transferred to the Museo Municipal Remigio Crespo Toral in the early twentieth century. It was presumed lost for several decades, until it was rediscovered during an inventory in August 2012 (Diego Cáceres, "Se encontró placa de la Misión Geodésica [Geodesic Mission plaque found]", *El Tiempo* Cuenca, 29 August 2012).

16. Eliecer Enríquez Bermeo, *Quito a través de los siglos* (Quito Across the Centuries) (Quito: Imprenta Municipal, 1938), 128–129.

17. The story of the pyramids' gradual decay and reconstruction is told in Georges Perrier, "Histoire des pyramides de Quito" (History of the Pyramids of Quito), *Journal de la Société des Américanistes* 35, no. 1 (1943): 91–122; and Gabriel Judde, "Jean-Baptiste-Washington de Mendeville, premier consul frances en el Ecuador, y la restauración de las piramides de Caraburo y Oyambaro" (Jean-Baptiste-Washington de Mendeville, First French Consul in Ecuador, and the Restoration of the Caraburo and Oyambaro Pyramids), *Boletín de la Academia de Historia, Quito* 73, no. 155 (1990): 51–75.

18. Alexander von Humboldt, *Reise auf dem Río Magdalena, durch die Anden und Mexico* (Journey on the Magdalena River, Through the Andes and Mexico), ed. Margot Faak, 2 vols. (Berlin: Akademie-Verlag, 1986), 1:184–189; Boussingault, *Mémoires*, 5:232.

19. Edward Whymper, *Travels Amongst the Great Andes of the Equator* (New York: Scribner's, 1882), 291; Humberto Vera Herrera, *Historia y geografía del monumento a la línea ecuatorial: Intyñan: camino del sol* (History and Geography of the Equatorial Monument: Intyñan: Route of the Sun) (Quito: Casa del Album, 1972), 9–11.

20. Humberto Vera Herrera, *History and Geography of the Equatorial Monument* (Quito: Ediciones Equator, 1990). The monuments were located on the equator as determined by older World Geodetic System (WGS) coordinates; modern tourists with their handheld GPS, coded to WGS84 (or later) coordinates, will notice that the equatorial monument stands about eight hundred feet south of the line.

21. The trace of the original baseline crosses the runway just to the east of the passenger terminal. Acknowledgments to the staff at the Quito airport authority for providing me information on the new Quito International Airport.

Afterword

1. BIF Ms 2118, Papers of Jules Antoine René Maillard de La Gournerie on the Geodesic Mission; Robert Finn Erickson, "The French Academy of Sciences Expedition to Spanish America, 1735–1744," PhD diss., University of Illinois, Champaign-Urbana, 1955; personal communication from Robert Erickson to me, May 30, 2004.

2. Victor Wolfgang von Hagen, *South America Called Them: Explorations of the Great Naturalists: La Condamine, Humboldt, Darwin, Spruce* (New York: Knopf, 1945). Among Hagen's errors is his insistence that Maldonado met La Condamine on the coast of Peru and journeyed across the Esmeraldas jungle with him to Quito (30–42), when in fact they did not meet until well after the French had arrived in Quito. Hagen also invents an additional character in the expedition, Mabillon (11, 62), who he says went mad. In both cases Hagen completely misrepresented La Condamine's French-language accounts; in the first case, La Condamine had simply stated he took the same route as Maldonado's future road through Esmeraldas; in the second case, La Condamine had compared Jussieu's mental illness to that of the Benedictine scholar Jean Mabillon, who suffered a nervous breakdown and lost his power of speech for six years. Dr. Hagen's notes for his book have long since disappeared (personal communication with his daughter Adriana von Hagen, December 21, 2006).

3. Florence Trystram, *Le Procès des Étoiles* (The Trial of the Stars) (Paris: Seghers, 1979; Payot, 1997).

4. The conferences and exhibitions include: Nelson Gilberto Gómez Espinoza, ed., *250 años de la Primera Misión Geodésica* (250 Years from the First Geodesic Mission): *Cultura, revista del Banco Central de Ecuador* 8, no. 24 (1986); *Colloque International "La Condamine": La Condamine y la expedición de los académicos franceses al Ecuador* (Inter-

national La Condamine Colloquium: La Condamine and the Expedition of French Academicians in Ecuador) (Mexico City: Instituto Panamericano de Geografía y Historia; Nanterre: Université de Paris X, 1987); Ricardo Cerezo Martinez, ed., *La forma de la tierra: Medición del meridiano, 250 aniversario* (The Form of the Earth, Measure of the Meridian: 250th Anniversary) (Madrid: Museo Naval, 1987); Jacques Le Pottier and Marie-Dominique Heusse, eds., *La Condamine et la Mesure de la terre* (La Condamine and the Measure of the Earth) CD-ROM (Toulouse: Université des Sciences Sociales de Toulouse, 2001).

5. "Figure of the Earth," *Voyages of Discovery* TV series, Annabel Gillings, producer, Paul Rose, presenter, BBC, 2006.

6. Several letters and reports sold at auction in 1844 are listed in Ludovic Lalanne, *Dictionnaire de pièces autographes volées aux bibliothèques publiques de la France* (Dictionary of Signed Works Removed from French Public Libraries) (Paris: Panckoucke, 1851) 75, 177. AARS Biographical Dossier La Condamine, 50J, has a notice of letters sold at auction in Louviers, France, in July 1991, many of which went into private hands.

7. David A. Taylor, "Degree of Difficulty: Measuring the Earth Proved to be a Triumphant Fiasco," *Mercator's World* 4, no. 3 (1999): 18–25.

A Note on Language

1. "Geodesic" nowadays refers to a particular curve along the surface of the Earth. "Geodetic" refers to the overall three-dimensional picture. An excellent primer on the subject is given by James R. Smith, *Introduction to Geodesy: The History and Concepts of Modern Geodesy* (New York: Wiley, 1997). The observation that indigenous people call themselves Indians is from Charles C. Mann, *1491: New Revelations of the Americas Before Columbus* (New York: Knopf, 2005), xi.

Units of Measure and Currency

1. The first step was to convert the livre and peso into British pounds sterling, using tables found in John J. McCusker, *Money and Exchange in Europe and America, 1600–1775: A Handbook* (Chapel Hill: University of North Carolina Press, 1978), 96–105. The conversion rates in 1740 were £1 = 22.3 livres or 4.3 pesos. The next step was to use the Economic History Services website, http://eh.net/hmit/ (accessed May 2010), which has inflation-indexed calculators to convert past values to modern values; £1 in 1740 was worth £125 in terms of purchasing power in the year 2010. Finally, pounds were converted to dollars at the Organization of Economic Cooperation and Development 2010 purchasing-power exchange rate ($1.60 = £1).

INDEX